RAIN FOREST REGENERATION AND MANAGEMENT

MAN AND THE BIOSPHERE SERIES

MAN AND THE BIOSPHERE SERIES

Series Editor: J.N.R. Jeffers

VOLUME 6

RAIN FOREST REGENERATION AND MANAGEMENT

Edited by
A. Gómez-Pompa, T.C. Whitmore
and M. Hadley

PUBLISHED BY

PARIS

AND

The Parthenon Publishing Group
International Publishers in Science, Technology & Education

Published in 1991 by the United Nations Educational Scientific and Cultural Organization,
7 place de Fontenoy, 75700 Paris, France — Unesco ISBN 92-3-102647-X

and

The Parthenon Publishing Group Limited
Casterton Hall, Carnforth
Lancs LA6 2LA, UK — ISBN 1-85070-261-6

and

The Parthenon Publishing Group Inc.
120 Mill Road, Park Ridge
New Jersey 07656, USA — ISBN 0-929858-31-X

The designations employed and the presentation of the material in this publication do not imply the expression of any opinion whatsoever on the part of the publishers concerning the legal status of any country or territory, or of its authorities, or concerning the frontiers of any country or territory.
The authors are responsible for the choice and the presentation of the facts contained in this book and for the opinions expressed therein, which are not necessarily those of Unesco and do not commit the organization.

Composed by Ryburn Typesetting Ltd, Halifax, England
Printed and Bound in Great Britain by Butler and Tanner Ltd, Frome and London

British Library Cataloguing in Publication Data

Rain forest regeneration and management.
1. Tropical rain forests. Management
I. Gómez-Pompa, A. II. Whitmore, T.C. III. Hadley, M. (Malcolm) IV. Series
634.928

ISBN 1-85070-261-6

Library of Congress Cataloging-in-Publication Data

Rain forest regeneration and management / edited by A. Gómez-Pompa,
T.C. Whitmore, and M. Hadley
p. cm. — (Man and the biosphere series ; v. 6)
Includes bibliographical references.
ISBN 0-929858-31-X (Parthenon) : $65.00 (U.S.)
1. Rain forests — Management. 2. Forest management —Tropics
3. Forest reproduction — Tropics. 4. Rain forest ecology. I. Gómez-Pompa, A.
(Arturo) II. Whitmore, T.C. (Timothy Charles) III. Hadley, M. (Malcolm) IV. Series:
MAB (Series) : 6.
SD247.R35 1990
634.9'0913—dc20 90-6711
CIP

PREFACE

Unesco's Man and the Biosphere Programme

Improving scientific understanding of natural and social processes relating to man's interactions with his environment, providing information useful to decision-making on resource use, promoting the conservation of genetic diversity as an integral part of land management, enjoining the efforts of scientists, policymakers and local people in problem-solving ventures, mobilizing resources for field activities, strengthening of regional co-operative frameworks. These are some of the generic characteristics of Unesco's Man and the Biosphere Programme.

Unesco has a long history of concern with environmental matters, dating back to the fledgeling days of the organization. Its first Director General was biologist Julian Huxley, and among the earliest accomplishments was a collaborative venture with the French Government which led to the creation in 1948 of the International Union for the Conservation of Nature and Natural Resources. About the same time, the Arid Zone Research Programme was launched, and throughout the 1950s and 1960s this programme promoted an integrated approach to natural resources management in the arid and semi-arid regions of the world. There followed a number of other environmental science programmes in such fields as hydrology, marine sciences, earth sciences and the natural heritage, and these continue to provide a solid focus for Unesco's concern with the human environment and its natural resources.

The Man and the Biosphere (MAB) Programme was launched by Unesco in the early 1970s. It is a nationally based, international programme of research, training, demonstration and information diffusion. The overall aim is to contribute to efforts for providing the scientific basis and trained personnel needed to deal with problems of rational utilization and conservation of resources and resource systems, and problems of human settlements. MAB emphasizes research for solving problems: it thus involves research by interdisciplinary teams on the interactions between ecological and social systems; field training; and applying a systems approach to understanding the relationships between the natural and human components of development and environmental management.

MAB is a decentralized programme with field projects and training activities in all regions of the world. These are carried out by scientists and technicians from universities, academies of sciences, national research laboratories and other research and development institutions, under the auspices of more than a hundred MAB National Committees. Activities are undertaken in co-operation with a range of international governmental and non-governmental organizations.

Further information on the MAB Programme is contained in *A Practical Guide to MAB*, *Man Belongs to the Earth*, a biennial report, a twice-yearly newsletter *InfoMAB*, MAB technical notes, and various other publications. All are available from the MAB Secretariat in Paris.

Man and the Biosphere Book Series

The Man and the Biosphere Book Series has been launched with the aim of communicating some of the results generated by the MAB Programme to a wider audience than the existing Unesco series of technical notes and state-of-knowledge reports. The series is aimed primarily at upper level university students, scientists and resource managers, who are not necessarily specialists in ecology. The books will not normally be suitable for undergraduate text books but rather will provide additional resource material in the form of case studies based on primary data collection and written by the researchers involved; global and regional syntheses of comparative research conducted in several sites or countries; and state-of-the-art assessments of knowledge, or methodological approaches based on scientific meetings, commissioned reports or panels of experts.

The series will span a range of environmental and natural resource issues. Currently available, in press, or in preparation are reviews on such topics as control of eutrophication in lakes and reservoirs, sustainable development and environmental management in small islands, reproductive ecology of tropical forest plants, the role of land/inland water ecotones in landscape management and restoration, ecological research and management in alpine regions, structure and function of a nutrient-stressed Amazonian ecosystem, assessment and control of non-point source pollution, research for improved land-use in arid northern Kenya.

The Editor-in-Chief of the series is John Jeffers, until recently Director of the Institute of Terrestrial Ecology, in the United Kingdom, who has been associated with MAB since its inception. He is supported by an Editorial Advisory Board of internationally-renowned scientists from different regions of the world and from different disciplinary backgrounds: E.G. Bonkoungou (Burkina Faso), Gonzalo Halffter (Mexico), Otto Lange (Federal Republic of Germany), Li Wenhau (China), Gilbert Long (France), Ian Noble (Australia), P.S. Ramakrishnan (India), Vladamir Sokolov (USSR) and Anne Whyte (Canada). Bernd von Droste and Malcolm Hadley of Unesco's Division of Ecological Sciences are *ex officio* members of the board.

A publishing rhythm of three to four books per year is envisaged. Books in the series will be published initially in English, but special arrangements will be sought with different publishers for other language versions of individual volumes.

Rain Forest Regeneration and Management

The aim of this book is to explore the implications to management of present scientific knowledge on rain forest regeneration. In addition to providing an overview of scientific information on rain forest regeneration, subsidiary aims are to identify gaps in information and understanding, in respect to both scientific hypotheses and the needs of management, and to explore directions for future collaborative research and action. The intention is not to present an encyclopedic or comprehensive literature review. Rather, the concern is with a review of selected technical issues and ecological processes within the context of management. The motivation is to help bridge the gap between the sciences associated with the wet tropics and on-the-ground management.

The book is based on thematic reviews, complemented by case studies. Synthesis reviews deal with such topics as sylvigenesis and architectural diversity, regeneration dynamics at various spatial scales, physiology of fast-growing species, reproductive biology and genetics, fruit and seedling ecology, nutrient cycling, current management programmes. Case studies deal with research and management experience in particular locations and regions. A dual challenge to the authors of case studies is to inform a wider audience of the experience gathered in a particular project or technical field, but also to suggest what might be the wider practical applications of a given case study for rain forest management.

The intended audience spans research scientists, resource managers and persons involved in teaching and education who are interested in land-use planning and resource management issues in the forested lands of the humid and subhumid tropics. The concern is with natural rain forest ecosystems (both "intact" and secondary) but not with plantations.

The book is based on invited contributions presented and discussed at an international workshop held at Guri (Venezuela) from 24–28 November 1986. The workshop was hosted by the state hydroelectric company CVG Electrificación del Caroní, CA (EDELCA), and was organized as a joint venture of the Man and Biosphere (MAB) Programme of Unesco, the Decade of the Tropics of the International Union of Biological Sciences (IUBS), the United Nations Environment Programme (UNEP) and the Instituto Venezolano de Investigaciones Científicas (IVIC). Additional technical and financial support was provided by the Food and Agriculture Organization (FAO), the World Resources Institute (WRI) and the Commonwealth Science Council (CSC). Participants included some 20 research workers and forest managers from Venezuela together with some 40 invited specialists from 20 other countries and international organizations. A summary report on the workshop was published in 1988 by IUBS, as Special Issue 18 of Biology International.

Plans for the Guri workshop had been elaborated by a small working party organized at Harvard Forest in July 1985. This working party was, in turn, convened in the light of a recommendation by the International Coordinating Council for the Man and Biosphere (MAB) Programme, that rain forest regeneration might provide a suitable focus for future comparative work on tropical ecology within the framework of MAB.

In terms of scope, links between science and technology, and between research and resource-use, came within the ambit of the reviews and case studies. The broader economic and societal context of rain forest regeneration and management was not within the direct terms of reference of the workshop. Not that issues like land tenure, economic evaluation of forestry schemes, international trade, etc., are not important. The inverse is true – their importance is overriding. But such an objective and focus would have called for a completely different type of workshop agenda and participants' list. Though the focus was on ecological and biological concerns, social and economic issues were inevitably raised during discussions, and this is reflected in particular case studies.

During the Guri workshop, drafts of the thematic reviews were examined in small working groups, and were subsequently revised by their authors in the light of discussions at Guri and (in some cases) of the comments of additional peer reviewers. Some case studies presented at the workshop do not appear in the present volume, since they have already been published substantively elsewhere. On the other hand, the case studies published here include a couple of contributions not presented at Guri, but commissioned for inclusion in the book. The case studies, as a whole, have only been lightly revised and edited.

Unesco extends its warm thanks to all those who have contributed to the preparation and review of the volume. Special thanks are due to the three volume editors – Arturo Gómez-Pompa, Tim Whitmore and Malcolm Hadley – for seeing the longish period of preparation through to publication.

CONTENTS

Section 2. *Topical reviews*

Section 4. *In guise of a conclusion*

LIST OF CONTRIBUTORS

S. del Amo R.
Gestion de Ecosistemas
A.C. Apartado Postal 19-182 C.P.
03910 México, D.F.
México

A. Anderson
Museu Goeldi
Caixa Postal 399
66.000 Belem
Brazil

S. Appanah
Forest Research Institute Malaysia
Kepong
Selangor
52100 Kuala Lumpur
Malaysia

K.S. Bawa
Department of Biology
University of Massachusetts
Boston
MA 02125
USA

F.A. Bazzaz
Department of Organismic and Evolutionary Biology
Harvard University
Cambridge
MA 02138
USA

E.F. Bruenig
Institute for World Forestry and Ecology
Leuschenstrasse 91
D-2050 Hamburg 80
FRG

F.W. Burley
World Resources Institute
1735 New York Avenue N.W.
Washington, D.C. 20006
USA

R.P. Capote
Institute of Ecology and Systematics
Cuban Academy of Sciences
Carretera de Varone km 3 1/2
Apartado Postal 8010
Código Postal 10800
Havana 8
Cuba

F.H.J. Crome
CSIRO
Division of Wildlife and Ecology
Tropical Forest Research Centre
P.O. Box 780
Atherton
Queensland 4883
Australia

J. van Dijk
Department of Forestry
Wageningen Agricultural University
P.O. Box 342
6700 AH Wageningen
The Netherlands

J.C.L. Dubois
Project IICA-Tropicos
Caixa Postal 33110 (Leblon)
22442 Rio de Janeiro
Brazil

A. Gómez-Pompa
Department of Botany and Plant Sciences
University of California
Riverside
CA 92521
USA

S. Gorzula
Division de Cuencas e Hidrologia
C.V.G. Electrificacion del Caroni C.A.
Apartado 62413
Caracas
Venezuela

N.R. de Graaf
Department of Forestry
Wageningen Agricultural University
P.O. Box 342
6700 AH Wageningen
The Netherlands

C.V.S. Gunatilleke
Department of Botany
University of Peradeniya
Sri Lanka

I.A.U.N. Gunatilleke
Department of Botany
University of Peradeniya
Sri Lanka

M. Hadley
Division of Ecological Sciences
Unesco
7 place de Fontenoy
75700 Paris
France

R.A. Herrera
Institute of Ecology and Systematics
Cuban Academy of Sciences
Carretera de Varona km 3 1/2
Apartado Postal 8010
Código Postal 10800
Havana 8
Cuba

P. von Hildebrand
Fundación Puerto Rastrojo
Apartado Aereo 241438
Bogotá
Colombia

D.H. Janzen
Department of Biology
University of Pennsylvania
Philadelphia
PA 19104
USA

C.F. Jordan
Institute of Ecology
University of Georgia
Athens
GA 30602
USA

R.J. Kalliola
Department of Biology
University of Turku
SF-20500 Turku
Finland

K. Kartawinata
Unesco Regional Office for Science and Technology for South East Asia
Jalan Thamrin 14
P.O. Box 273/JKT
Jakarta 10002
Indonesia

S.L. Krugman
Timber Management Research
U.S. Department of Agriculture
Washington, D.C. 20013
USA

D. Lamb
Botany Department
University of Queensland
Brisbane
Australia

H.F. Maître
Division d'Inventaire et Aménagements
Centre Technique Forestier Tropical (CTFT)
45 bis avenue de la Belle Gabrielle
92130 Nogent sur Marne
France

G. Maury-Lechon
CNRS
45 bis avenue de la Belle Gabrielle
94736 Nogent sur Marne
France

L. Menéndez
Institute of Ecology and Systematics
Cuban Academy of Sciences
Carretera de Varona km 3 1/2
Apartado Postal 8010
Código Postal 10800
Havana 8
Cuba

L. Castro Morales
Divisiőn de Cuencas e Hidrolgía
C.V.G. Electrificacion del Caroní C.A.
Apartado 62413
Caracas 1060
Venezuela

C.T.S. Nair
Kerala Forest Research Institute
Peechi 680 653
Kerala
India

D.U.U. Okali
Department of Forest Resources Management
University of Ibadan
Ibadan
Nigeria

R.A.A. Oldeman
Chair of Silviculture and Forest Ecology
Wageningen Agricultural University
P.O. Box 342
6700 AH Wageningen
The Netherlands

H.D. Onyeachusim
Forestry Research Institute of Nigeria
P.M.B. 5054
Ibadan
Nigeria

J.R. Palmer
Oxford Forestry Institute
University of Oxford
South Parks Road
Oxford
OX1 3RB
UK

C.M. Peters
Institute of Economic Botany
New York Botanical Garden
Bronx
NY 10458
USA

P.S. Ramakrishnan
School of Environmental Sciences
Jawaharlal Nehru University
New Delhi 110067
India

S. Riswan
Herbarium Bogoriense
Puslitbang Biologi-LIPI
Jalan Raya Juanda 22–24
Bogor 16122
Indonesia

M.E. Rodríguez
Institute of Ecology and Systematics
Cuban Academy of Sciences
Carretera de Varona km 3 1/2
Apartado Postal 8010
Código Postal 10800
Havana 8
Cuba

J.G. Saldarriaga
The Tropenbos–Colombia Programme
Apartado Aéro 42209
Bogotá, D.E.
Colombia

Salleh Mohd. Nor
Forest Research Institute Malaysia
Kepong
Selangor
52100 Kuala Lumpur
Malaysia

J.S. Salo
Department of Biology
University of Turku
SF-20500 Turku
Finland

S. Saulei
Papua New Guinea National Forest Research Institute
P.O. Box 314
Lae
Papua New Guinea

R.C. Schmidt
Forest Resources Division
Forestry Department
FAO
Rome
Italy

C. Uhl
Department of Biology
The Pennsylvania State University
University Park
PA 16802
USA

C. Vázquez-Yanes
Centro de Ecología
Universidad Nacional Autónoma de México
México, D.F.
México

T. Walschburger
Fundación Puerto Rastojo
Apartado Aereo 241438
Bogotá
Colombia

T.C. Whitmore
Department of Geography
University of Cambridge
Downing Place
Cambridge
CB2 3EN
UK

N.D. de Zoysa
Department of Botany
University of Peradeniya
Sri Lanka

Section 1

Setting the scene: introduction

CHAPTER 1

THE MANAGEMENT OF NATURAL TROPICAL FORESTS

A. Gómez-Pompa and F.W. Burley

INTRODUCTION

Conserving biological diversity in the tropics has become an issue of increasing priority and urgency in recent years. However, the options available to address this issue, and to slow or arrest the decline in tropical ecosystems and biological diversity, are limited and discouraging. More than 11 million hectares (ha) of mature tropical forests are converted into agriculture, pasture lands or other uses every year (Lanly 1982, WRI 1985). Less than 10% of the land being deforested is replanted each year. Although the amount of tropical forest land coming under protection or conservation management is growing, the future of many of these areas is in doubt due to rapidly increasing pressures of development and exploitation.

To reverse this trend, major changes in land management and forest policy will be necessary (Dourojeanni 1984). The success in reversing deforestation will depend on political leadership and appropriate policy changes by developing country governments and by the international development assistance community. It is well accepted that, if protected areas and management of natural forests are included as important parts of the global solution to deforestation, land-use planning and zoning will be needed on a scale vastly greater than we see throughout the tropics today. All of this, of course, will require innovation and real political leadership.

One of the most important subjects of discussion and controversy in the tropical sciences has been the conservation and management of the tropical forests (Budowski 1984, Mergen 1981, Unesco 1978, National Research Council 1980, Srivastava *et al.* 1982, Hallsworth 1982). The issues have been addressed by foresters, developers, agronomists, ecologists and conservationists, and a stream of publications has been devoted to the subject. Yet the results to date have been very poor, far too little tropical forest is now under sustained yield management, and each day most tropical forest areas that are cut have no implementable plan for their regeneration or replacement.

The last 3 decades have been rich in research on tropical biology, including research on forest ecosystems. Yet this increase in knowledge of forests and how they work has not resulted in vastly improved management or in much greater areas of tropical forest coming under some form of management. The primary contribution of all this research has been to corroborate the complexity and diversity of tropical ecosystems and our pervasive ignorance about them. One problem has been our inability to set real priorities for research – a sad situation given that the human factor and financial resources to address these issues are scarce.

DEFORESTATION IN THE TROPICS

A large potential area for forest management in the tropics is the land area that increases daily through logging without any subsequent measures for regeneration. This area, during the next 5 years, could be well over 60 million ha (WRI 1985). Even assuming that 10% of this area could be rehabilitated into a managed forest, the impact on the biological resources (ecosystems and species) will be enormous. Unfortunately, few attempts are being made to ensure that a significant amount of this converted land is reforested or put into plantations.

Only about 600 000 ha of industrial plantations are being established in developing countries each year world-wide, less than 10% of the natural forest being converted. The area of natural forest coming under management is still very limited, for example, and one calculation puts this area at 37 900 ha in 1982 (Lanly 1982). The figures are lowest in Latin America. In contrast to these figures, the World Resources Institute (1986) reports that there were 522 000 ha of closed forests under some kind of management plan for all of tropical America, though the bulk of that surface area is under coniferous forests. The great differences in various calculations reflect, in part, different definitions and degrees of management.

SOME QUESTIONS

Many questions come immediately to mind when one considers the issues of natural forest management in the tropics. Can existing natural forests supply the lumber and fuelwood needs of the tropical countries? Why are single species tree plantations taking the lead in reafforestation projects?

Do we need more basic ecological research in order to manage natural forests? What type of ecological research is most needed by forest managers? Assuming the world needs to regenerate natural tropical forests, do we know enough to do this? If a developing country chooses to include management of tropical forests in its development plans, can we suggest basic policy steps and management guidelines to be followed? What information is necessary to start a large-scale natural forest management or forest regeneration project? These are some of the questions which were discussed at the Guri Workshop, and are addressed in the summary report of that workshop (Hadley 1988) and in this book.

DIFFERENT PERCEPTIONS OF TROPICAL FOREST MANAGEMENT

The term "management" means different things to different people, as Bawa and Krugman point out in their contribution to this volume. We may envisage a forest management spectrum with strict forest protection at one end and nearly complete forest conversion at the other end. But for convenience and simplicity, we may distinguish three main trends or "management perspectives" along this spectrum:

(1) The first trend entails the nearly complete conversion of the original forest ecosystem into another system that better suits the material and economic needs of those who decide to "manage" the system.

(2) The second trend includes the management of the natural forests through the extraction of some of their products, without severely disturbing the system. This is forest manipulation and partial conservation through an efficient natural or induced regeneration of the forest ecosystem that is only lightly disturbed.

(3) The third trend includes the total or strict preservation of representative samples of tropical forest for conserving biological diversity, especially to conserve the genetic variation of many tree species, some of which are likely to prove useful to us and to be exploited.

These three perspectives along the forest management spectrum are defended for often quite different management objectives. The first trend, essentially a replacement approach, is the prevailing one in most of the tropics today. It is simply the conversion of the species-rich tropical forests into species-poor ecosystems, including rangelands and other agro-ecosystems. This conversion can be done very quickly and efficiently by the use of modern machinery and chemicals. In the short-term, it is the most financially rewarding system, and for that reason it is the dominant trend throughout the world. The most common replacement ecosystems are herbaceous agro-ecosystems such as grasslands and annual crops. More rare is the conversion into other types of forest or agro-forestry systems. Proponents of this method are looking first for short-term economic return. The result is a strong tendency toward single species agricultural systems, such as sugar-cane, coffee, rubber (*Hevea*), *Gmelina*, *Pinus*, *Eucalyptus*, etc.

The second trend, in the middle of the forest management spectrum, is the management of so-called natural forests through the careful and selective removal of some of their products (wood, wildlife, fruit, etc.). This is done with the intention of allowing some type of natural or induced tree regeneration that may permit the long-term renewability of the system so that it is sustainable.

The third trend is the increasing allocation of forest areas to preserve large tracts of natural ecosystems. The primary objective here is usually to preserve ecosystem processes, species diversity, and genetic variation within species. All these are necessary to maintain future options and to keep available for human use and exploitation the widest array of biological resources possible. Some of the benefits of this type of management are often called "ecosystem services", e.g. watershed protection.

Today these three management perspectives are too often seen as opposing views and basically incompatible management options. The possibilities of linking different management objectives, and so to plan land-use more broadly, are often overlooked. In traditional economic terms, at least in the near future for the less developed tropical countries, the short-cycle agricultural systems of many traditional crops are better than long-cycle systems such as forests. The consequence of this apparent reality of economics is that forests are being simplified or eliminated everywhere.

THE HUMAN FACTOR

Many rural people live in the areas where development projects have failed either socially or economically, or both. They subsist in increasing poverty, on lands that increasingly exhibit patterns of deflected vegetation succession, invasion by weeds and animal pests, forest degradation toward savannas, and other symptoms of environmental degradation.

Modern society has failed to provide viable alternatives for the great majority of the rural population in the tropics. Substitution of their ancient and often successful methods of production has been proposed with little success, provoking mass emigrations to urban areas or to other forests, including those that our modern society is trying unsuccessfully to protect. The cases of poaching, deforestation and invasion of reserved forests and protected areas of various kinds by poor farmers who have been "expelled" from "developed" regions are well-known and increasingly a major land management problem.

Possibly the most important alternative that our modern industrial society has, is the possibility that the remaining natural tropical forests, along with the large areas of degraded land, can provide adequate and sustainable means of living for the local people, in addition to products for industrial society. This alternative necessarily requires broad-scale and innovative management schemes for large areas of forest that have been relatively undisturbed in recent years, as well as reafforestation of other areas, to meet the demands of the local and regional market.

Assuming that society recognizes its responsibility towards this degradation of tropical lands, how can biological diversity be maintained, and where necessary partially restored, in such a situation? How can expanding areas of secondary vegetation and degraded soils be managed to be more productive, to the benefit of local people? What role could forest patches, including, for example, the core areas in biosphere reserves, play in the restoration process? Since most primary forests have disappeared, what sort of forest can we re-establish that could be stable and productive and ensure conservation of biological diversity? These are among the difficult questions that will continue to be on the agenda of development planners for years to come.

Very little tropical forest is pristine in the sense of never having been inhabited by man. Most forest areas are, or have been, inhabited (Webb 1982), and large areas such as the Peten in Guatemala, Mexico and Belize have been exploited for centuries. In this latter case, what today looks like a vast area of pristine tropical forest is actually a mosaic of forest patches that have "recovered" from small-scale agriculture and forest clearance over many years in times past. Most tropical forests have been submitted to numerous cycles of management and abandonment by human societies since remote times.

We need to understand this fact clearly, if only to give us a perspective on the seemingly devastating rates of deforestation today. Forest management must be considered in the broadest sense of land-use planning, including: clearing of forests for shifting cultivation, habitation, or hunting; the voluntary and intentional protection of forests or individual species; the extraction, at various levels of intensity, of a wide variety of forest products, not only trees; and the enrichment of the forest, with species from other places.

Management also needs to be considered in terms of the inhabitants of these regions – traditional societies that belong to a great diversity of ethnic groups. These societies have lived in and from the forests for eons and have coexisted with them. They have evolved through time and produced an enormous, still untapped, knowledge about the ecosystems in which they evolved. Some have had great successes and consequently have expanded geographically and/or culturally, while others remain stable or deteriorated over time.

Unfortunately, many of these traditional societies have been ignored by the dominant modern societies in the march of development. Many of them are nearing extinction today, after flourishing in times past. Their knowledge of the trees and the forests remains largely an untapped resource likely to be lost before it can be studied or preserved.

SILVICULTURAL SYSTEMS, REGENERATION AND FOREST MANAGEMENT

The term "regeneration" is used here as a practical concept that includes not only natural secondary successions but also types of forest manipulation that can lead intentionally to new and more productive stages for forest growth. This definition is a practical one because it includes many types of manipulation of the forests, with respect to tree density, standing volumes at various stages, and species composition.

Whatever method we choose for managing or regenerating a forest, knowledge of the causes, mechanisms and factors that drive the processes of species change, population change, and replacement through time – or ecological succession – allows us to be far more efficient with different manipulative management schemes.

One of the primary purposes of the contributions that follow is to provide those professionals in charge of tropical forest land management with some insights on the structure and function of tropical forest ecosystems and the ecology of tropical trees that may be important in forest management (FAO 1985). Without such a baseline of ecological knowledge, not only are our management options more limited, but we shall be less effective with those few management schemes that are applied.

One of the best descriptions of tropical forest management systems is by Baur (1968). Although there have been more recent publications on management (FAO 1985), few, if any, important technical advances or systems have been proposed. A useful experience in different regions is found in the chapter by Schmidt. Here we will present only a practical classification of the systems available, focusing on the ecological processes involved in the regeneration of tropical forests. For convenience, four main groups of silvicultural systems may be identified: natural regeneration systems; clearing systems; replacement systems; and restoration systems.

The natural regeneration systems

This first group includes the simplest systems of selecting only a few trees at a time for removal, allowing natural regeneration to fill in the gaps created, and maintaining standing volumes of all tree species. It is assumed that the regeneration of the desired species will occur naturally without further manipulation or enrichment. Additional logging may often occur 20–30 years later to extract the smaller regenerating trees. Such systems are often termed "selective cutting", or the "selection system" under the broader category of "polycyclic systems". Under this group also falls the system of "regeneration by selective felling of individual mother trees".

The natural regeneration systems are the ones most commonly used by the local people and are being employed more and more by commercial foresters and forestry departments. The removal of isolated, often widely-spaced, individual trees resembles natural tree-fall, and regeneration occurs normally as in undisturbed forests. From the strictly commercial viewpoint, natural regeneration systems are often unpopular because of the assumed high costs of extraction (Wadsworth 1981).

After removing or killing a large tree from the forest, a light gap is created and at least five different groups of species will begin to take over:

(1) First are the "functional" species that were already on the site and have survived the drastic mechanical and environmental changes in light, temperature and nutrients, either as standing individuals, as stumps, or as underground propagules (Martínez-Ramos 1985).

(2) Second are the latent species in the seed-bank of the topsoil but not growing as seedlings or adult trees in the immediate area. They are triggered by the change in light intensity and/or temperature, and in other mechanical factors (Whitmore 1983).

(3) Third are those species represented usually by dormant individuals - ones that normally are found on the floor of the forests in the seedling or juvenile stages only, with little or no real growth toward adult size (Gómez-Pompa and del Amo 1985). With the opening of the canopy, the growth of these species is stimulated by the new micro-environment.

(4) Fourth are those species coming from rootstocks from the neighbouring standing forest (Peñalosa 1985).

(5) Fifth are the newcomers or invaders that arrive during the time that the light gap remains open.

These five groups of species interact and compete for the available resources. Collectively, they will form the new regeneration forest, with different sets of species dominating in different stages of the succession. It is important to remember that each species may have quite different ecological requirements. The success of any species, and of individual trees, depends on growth rates, architectural construction that facilitates its competition for light, nutrient requirements relative to competitors and nutrient availability at the site, and other interactions with numerous plant and animal species.

The filling of the gap by economically desirable species is not ensured, and many tropical tree invaders are not at this time considered desirable from the forest production standpoint. Some of the "most successful" natural silvicultural systems have been so judged because of the natural presence of seedlings or juveniles of the desirable species. Other topographically similar sites, perhaps nearby, may not contain this right mix of species.

This natural regeneration process can, of course, be "assisted" or "directed" to increase representation of any particular species. Plants can be brought to the gap at the right time and at the right age. Competitors may be systematically eliminated by weeding and thinning. Theoretically, this direction may seem easy, but it is often difficult and expensive in practice. To increase representation successfully, much needs to be known about the ecology and specific requirements of the desirable species in order to provide the correct treatments at the right time.

This type of harvesting with somewhat predictable regeneration afterwards is the most commonly used method of "management". It may really be considered a mining operation if only one extraction is planned and the value of the forest is determined by what can be harvested at that first cutting. This management is a one-time affair, little investment is put back into the site, and little stumpage value remains after logging. The economic rewards are great because very low long-term investment is needed. Much recent ecological research applies very well to this type of forest management.

Several chapters of this book address the problem encountered in this type of forest management. Whitmore discusses gap-phase dynamics and the processes related to it and its implications for tropical forest management. Janzen and Vázquez-Yanes discuss the importance of seeds and seedlings in forest dynamics and the need to know the requirements of seeds and seedlings to ensure regeneration of the desirable species. Bazzaz stresses the importance of micro-environmental factors and their key role in understanding the functioning of the seeds and seedlings in the regeneration process.

One potentially widespread application of this type of management is in national parks and other conservation areas, where some type of forest exploitation could benefit local inhabitants and where biological diversity can be maintained. The protected status of parks and reserves can help to ensure that the remaining standing forests will be disturbed minimally over time.

These silvicultural systems are widely used by traditional cultures that live in the tropical forest areas of the world. Their hunting and gathering activities (including some lumbering) are most often done in mature forests. There is evidence that some cultures also select, protect and even plant some useful trees, increasing the value of their forests (Gómez-Pompa 1987).

The clearing systems

A second group of silvicultural systems involves the extraction of some (or all) commercially valuable trees and also the elimination of unwanted individuals through cutting, girdling or poisoning operations. Regeneration, either natural or artificial, is then allowed to occur. The most important feature of these systems is the presence of seedlings or juveniles of the important commercial species in addition to medium or large size trees of those desirable species.

Among the best known systems of this group is the "Malayan Uniform System" which involves the clearing of the old trees in the forest in a single commercial operation, followed immediately by systematic poisoning of unwanted species to release natural regeneration (Wyatt-Smith 1963). This system has been successful in the dipterocarp forests of South-east Asia; these forests have several dominant species of the single family Dipterocarpaceae (National Academy of Sciences 1982).

Several other related silvicultural systems exist that are variations of the Malayan Uniform System and are based on the percentage of valuable trees left, timing of cutting, etc. One of the best known is the "Shelterwood System", a delayed silvicultural system better adapted to non-dipterocarp forests (Catinot 1965) and used extensively in Africa (Lowe 1978). A notable modification of this is being attempted in the Palcazu Valley of Peru under the name of "strip shelter belt" (Hartshorn 1983). It is based on extracting trees in long, narrow clear-cut areas that are rotated through the primary forest in such a way that the undisturbed forest will be the source of propagules for the regeneration of the strips on 30-year cycles.

Theoretically, these systems include the protection of valuable species and the elimination of unwanted ones in order to promote the enrichment of the future forest with propagules coming only from useful or commercial species. In addition to the mature trees, the seedlings and juveniles of most undesirable trees are eliminated, leaving the regeneration trees to be only those selected for originally. The extraction of some commercially valuable individual trees helps finance the initial operation.

The dynamics of regeneration in this type of management are similar to those of natural regeneration. The important difference is the decrease in numbers of new seedlings of undesirable tree species and the larger size of the light gaps created. The intent is to promote dominance of seedlings of the desirable species. This does not always happen however, because seeds from many undesirable tree species are found in the soil seed-bank and will germinate if proper conditions exist. Some additional cleaning or weeding will have to be done. An additional complication is that some desirable species may not flower for several years, and seedlings of these species will not appear immediately as potential regenerators. This difficulty is discussed by Bawa and Krugman and also by Janzen and Vázquez-Yanes in this volume. Also, some species may produce seeds that will require some environmental factor not present.

The replacement systems

A third set of systems includes those by which the entire forest is cut down and replaced immediately, or within a few years, by tree plantations. Often replacement is with introduced species of *Eucalyptus* or *Pinus*, for example. Tree plantations *per se* are not the subject of this volume.

However, this group also includes shifting agriculture systems, which should be regarded as silvicultural systems in many ways because the end result is a tree crop for multiple uses. A good example is the system in which a forest ecosystem is substituted by a temporary agro-ecosystem, later allowing a natural or semi-natural succession to produce a secondary forest ecosystem. This method, either intentional or unintentional, is found on millions of hectares of land in the tropics today.

The ecological processes operating in this mode of regeneration are extremely variable and complex. Regeneration will depend on the seed-bank, the remaining standing trees, the species remaining as stumps, and the seeds coming from nearby or remote forest areas or introduced by man. Shifting cultivation in many tropical areas is practised as a "coppice with standards" silvicultural system, because many useful trees are left behind in the slash process as standing trees or as living stumps. In general, this method has not been looked upon as a separate silvicultural system and evaluated for its reafforestation potential.

This method has great potential for the future of the tropical areas. Much research is needed to evaluate the efficiency of these traditional systems and to find ways to improve the process. However, very little is known about the role of protected standing trees that remain from the fallen forests as sources of seeds. What are the possibilities of survival of different species? How does the individual adapt to a completely different environment? Hundreds of questions can be asked that have great importance for the regeneration of the future forests in these sites.

The management of these abandoned areas by the introduction of selected species of trees is an old and proven technique. In fact, one of the best known methods of tropical silviculture is the "taungya" system, a combination of a shifting agriculture with a forest plantation. It consists of planting native or exotic tree species mixed with annual crops; after 2 or 3 years, the field is abandoned, the planted trees are protected against other undesirable competitors, and the field develops into a forest plantation (Wiersum 1982). Many agro-forestry systems fall into this group, as for example the planting of cacao under native forest cover, or planting cacao or coffee under useful shade trees (CATIE 1979).

A major disadvantage of this system is the danger of losing species of animals and plants from the forest ecosystem – a major decrease in biological diversity – if large tracts of forests are not also conserved. Another difficulty is that this system may be linked exclusively with the local economy of the farmer and may not necessarily be a part of the larger national, industrial economy. For this reason, the whole system is often discouraged or disparaged by governments and financial institutions.

Another important method is the agro-forestry system in which a plantation of useful trees and agricultural crops are mixed with native forest species. It is difficult to define as a silvicultural system because the methods are diverse, but the final results may be similar. Some examples, in addition to coffee and cacao systems, are the forest gardens of tropical Asia (Wiersum 1982), the "kitchen" gardens of the Maya, the huastec te'lom (Alcorn 1983), the pet kot Maya (Gómez-Pompa *et al.* 1987), the palm gardens of several ethnic groups of Brazil (Anderson in this volume), and the "talun-kebun" system (Soemarwoto 1984).

The restoration systems

These are the least known methods and, increasingly, the most needed. They include management of heavily disturbed and frequently abandoned forested areas as well as areas that have been denuded for some time. This is not a well-defined group of silvicultural systems – the common element is the degraded land base. Today, millions of damaged tropical areas are abandoned and unproductive. Because the forests have disappeared, these areas remain in "arrested succession" and are of little use. The *Imperata* grasslands of Asia, the guinea savannas of Africa (Kio 1983) and the sabanoid vegetation of many areas in Latin America are well-known types of degraded land.

Restoration systems, also known as "assisted natural regeneration systems", include a great variety of techniques for planting desirable species in the logged forests and in the light gaps produced by falling trees, logging tracks, etc. One interesting variation is now being tested in Mexico (del Amo 1985, this volume) by introducing the seedlings of desirable species in the secondary stages of succession. There are examples of restoration management in different parts of the world tropics. In most cases, the work is based on local experience, without any substantial ecological or biological information about the forests or the species. The results are often encouraging, but of small scale.

Local, empirical knowledge should never be underestimated. In the case of many traditional tropical cultures, this knowledge is rich, and results based on experience are good, as in the case of taungya systems. In contrast, there are many examples of failures based on wrong empirical information applied by unskilled professionals and forest farmers. Two well-known Mexican examples illustrate the difference: the Lacandon forest management systems (Nations and Nigh 1980) and the Uxpanapa forest management plan (Gómez-Pompa 1979). The first is a successful traditional system that gave us the biologically rich Lacandon rain forests. The second is a wrongly designed forest management project that is considered by some to be the best documented ecological disaster of the tropics of Mexico.

However, all the silvicultural systems described above have value. Each system, if done well, can accomplish the goals of management and conservation of the resources that are so needed. A major problem has been the lack of long-term commitment that is needed to make any management system work. Budowski (1981, 1984) points this out very clearly and mentions the rarity of development projects that follow the specific recommendation of any particular system.

It is clear that ecological knowledge is greatly needed as a basis for enlightened silviculture and to understand the ecological processes of forest regeneration. We need continually to ask ourselves what kind of knowledge is needed to manage tropical forests better? What data do we lack? Which are the most important problems in tropical forest regeneration that need ecological research? What forest management experiences could be suggested from the available ecological knowledge?

Even though there is a great need for more long-term research on the available silvicultural systems, we already have many good systems that can be applied in most situations in the tropics. Good ecological studies and monitoring of the ecological processes in different silvicultural systems should be encouraged in any long-term forest management plan. The role of nutrients in the long-term success of any tropical management effort is an important issue to be taken into consideration, as discussed by Jordan in this book.

KNOWLEDGE OF THE SPECIES

It is clear that, if we review all the general methods of regeneration, the basic unit that is common to all of them is the need to select the species that will be protected, planted and encouraged. We need to know the life cycles, ecological requirements in all stages of development, competitive abilities, predators, diseases, etc. There is no substitution for this type of fundamental knowledge of the species being exploited (see chapter by Oldeman and van Dijk).

This lack of knowledge of the requirements of individual species is the greatest gap in our knowledge. We know almost nothing of the autecology of most of the world's valuable tree species. We need to select sets of valuable species for priority in research in each region. The genetics of individual species, as stressed by Bawa and Krugman, need to be better known. A good selection of the population to be used for reafforestation or enrichment has incredible value – but this selection of the best genotypes cannot be made without considerable testing and research beforehand. The same cost and effort are needed to introduce a well selected variety as are needed to introduce a bad one – a sobering point to remember.

We must understand that we are dealing with plants in a tropical environment, where success is very unpredictable. One pest or one disease can destroy years of effort. The problems of tropical forestry are the same as those of tropical agriculture. The best way to ensure success with low capital investment is to use as much of the great species and genetic diversity as possible. This fact may be of even greater importance to tropical forest management than to agriculture.

POLICY IMPLICATIONS

One may well ask why natural forest management is not practised more widely in the tropics. The reasons are not fundamentally technical or silvicultural – the research that has been done in the three major geographical regions and the pilot projects that have been completed leave little doubt that managing tropical forests for substantial economic benefits is feasible, often on a scale larger than might be expected. The reasons why this type of forest management is not more widespread are primarily political, administrative, and economic. Short-term economic demands force administrators into short-term projects from which they expect substantial financial return. This failing is aggravated by a lack of forest planning in general, and a lack of political commitment to slowing or halting deforestation and the decline of forest resources. The result is that the resource is mined rather than managed.

We must recognize that, although there is great potential for bringing large areas of tropical forest under some form of sustained yield management, and in the long-run this objective must be achieved, it will not happen very rapidly. In the near-term, therefore, to meet enormous and growing demands for wood and other forest products, more plantations are needed, larger areas need to be reforested, and already degraded lands must be brought back into various forms of production, including fuelwood plantations, agro-forestry, and forests for lumber.

For any nation to address its loss of forests and ensure that its forest estate yields economic benefits well into the future, three fundamental prerequisites must be met: first, a forest law or basic forestry policy must be written which clearly states the objective of long-term, sustainable management of the forest estate; second, forest regulations or management guidelines and procedures must be written and followed; and third, sufficient financial and human resources must be allocated to the effort to do the job. Several countries have the first and second prerequisites in place, but they have not yet allocated sufficient resources to their forestry departments to carry out the policy and regulations.

Until forestry and forest management are considered an integral part of land-use and natural resource planning, we will continue to see the inexorable depletion of forest capital and the inevitable deterioration of the remaining tropical forests. We can no longer afford to treat forestry – and forestry departments – in a strictly sectoral fashion, in isolation from the rest of agricultural planning and natural resource development in general.

Finally, we need to get our cost accounting straight. Until nations begin to recognize the real value of natural resources as capital to be managed, they will continue to deplete their own resources at wasteful rates. Inappropriate forest revenue systems that

(1) leave enormous economic rents to concessionaires,

(2) provide concessionaires little reason to practise sustainable long-term forestry, and,

(3) encourage highly selective harvesting of tropical forests with undue wastage of remaining trees, are among the forestry policies that lead to rapid depletion (Repetto 1987, 1988).

NON-GOVERNMENTAL ORGANIZATIONS IN FOREST MANAGEMENT

Deforestation cannot be reversed and sustainable patterns of forest land-use established without the active participation of millions of small farmers and landless people who daily depend on forests and trees for survival. The non-governmental organizations (NGOs) have a crucial role to play in fostering this grassroots participation, and in ensuring that development assistance funds reach the local levels where they can foster such basic programmes as tree planting, agro-forestry projects, and forest management.

Hundreds of NGOs world-wide are already involved in forestry activities in the developing countries. They range from small local cooperatives or village development groups, to national organizations or coalitions of NGOs, to large international NGOs based in developed countries. They vary greatly in objectives, capabilities and technical expertise, scale and mode of operation, and funding. They often have a very strong commitment to helping the rural poor and an intimate knowledge of local conditions and community relationships. In some cases (India and Kenya, for example), they are well-situated to carry out re-afforestation projects, if only the government and development assistance agencies were to make better use of them.

REFERENCES

Alcorn, J.B. (1983). El Te'lom Huasteco: Presente, pasado y futuro de un sistema de silvicultura indígena. *Biotica*, **8** (3), 315–25

Amo, R. S. del. (1985). Algunos aspectos de la influencia de la luz sobre el crecimiento de estados juveniles de especies primarias. In Gómez-Pompa, A. and Amo, R.S. del, (eds.) *Investigaciones Sobre la Regeneración de Selvas Altas en Veracruz, México*. Volumen II. pp. 79–91. (INIREB: Xalapa and Editorial Alhambra: México)

Baur, G.N. (1968). *The Ecological Basis of Rain Forest Management*. (Forestry Commission of New South Wales: Sydney)

Budowski, G. (1981). The place of agro-forestry in managing tropical forests. In Mergen, F. (ed.) *Tropical Forests: Utilization and Conservation*, pp. 182–94. (Yale School of Forestry and Environmental Studies: New Haven)

Budowski, G. (1984). The role of tropical forestry in conservation and rural development. *The Environmentalist*, **4** (7), 68–76

CATIE (Centro Agronómico Tropical de Investigación y Enseñanza). (1979). *Proceedings of Workshop on Agro-forestry Systems in Latin America*. Turrialba, Costa Rica. 26–30 March 1979. (CATIE: Turrialba)

Catinot, R. 1965. Sylviculture tropicale en forêt dense africaine. *Bois et Forêts Tropicaux*, **100**, 9–18

Dourojeanni, M.C. (1984). *Manejo de Bosques Naturales en el Trópico Americano: Situación y Perspectivas*. (International Forestry Congress: Quebec)

FAO. (1985). *Intensive Multiple-use Forest Management in the Tropics. Analysis of Case Studies from India, Latin America and the Caribbean*. FAO Forestry Paper No. 55. (FAO: Rome)

Gómez-Pompa, A. (1979). Antecedentes de las investigaciones botánico-ecologícas en la región del Rio Uxpanapa, Veracruz, México. *Biotica*, **4** (3), 127–33

Gómez-Pompa, A. (1987). On Maya silviculture. *Mexican Studies/Estudios Mexicanos*, **3** (1), 1–17

Gómez-Pompa, A. and Amo, R.S. del (eds.) (1985). *Investigaciones Sobre las Selvas Altas en Veracruz, México*. Volumen II. (INIREB: Xalapa and Editorial Alhambra: México)

Gómez-Pompa, A., Flores, S. and Sosa, V. (1987). The "Pet Kot": a man-made tropical forest of the Maya. *Interciencia*, **12** (1), 10–15

Hadley, M. (ed.) 1988. *Rain Forest Regeneration and Management*. Report of a workshop. Guri,

Venezuela, 24–28 November 1986. Biology International – Special Issue **18** (IUBS: Paris)
Hallsworth, E.G. (ed.) (1982). *Socio-Economic Effects and Constraints in Tropical Forest Management* (Wiley: Chichester)
Hartshorn, G.S. (1983). *Sustained Yield Management of Natural Forests: A Synopsis of the Palcazu Development Project in the Peruvian Amazon.* (Tropical Science Center: San José, Costa Rica)
Kio, P.R.O. (1983). Management potentials of the tropical high forest with special references to Nigeria. In Sutton, S.L., Whitmore, T.C. and Chadwick, A.C. (eds.) *Tropical Rain Forest: Ecology and Management*, pp. 445–55. Special Publications Series of the British Ecological Society Number 2. (Blackwell Scientific Publications: Oxford)
Lanly, J.P. (1982). *Tropical Forest Resources.* FAO Forestry Paper No. 30. (FAO: Rome)
Lowe, R.G. (1978). Experience with the shelterwood system of regeneration in natural forest in Nigeria. *Forest Ecology and Management*, **1**, 193–212
Martínez-Ramos, M. (1985). Claros, ciclos vitales de los árboles tropicales y regeneración natural de las selvas altas perennifolias. In Gómez-Pompa, A. and Amo, R.S. del (eds.) *Investigaciones Sobre la Regeneración de Selvas Altas en Veracruz, México*, pp. 191–239. Volumen II. (INIREB: Xalapa and Editorial Alhambra: México)
Mergen, F. (ed.) (1981). *Tropical Forests: Utilization and Conservation.* (Yale School of Forestry and Environmental Studies: New Haven)
National Academy of Sciences. (1982). *Ecological Aspects of Development in the Humid Tropics.* (National Academy Press: Washington D.C.)
Nations, J.D. and Nigh, R.B. (1980). The evolutionary potential of Lacandon Maya sustained-yield tropical forest agriculture. *Journal of Anthropological Research*, **36** (1), 1–29
National Research Council. (1980). *Research Priorities in Tropical Biology.* (National Academy of Sciences: Washington, D.C.)
Peñalosa, J. (1985). Dinámica de crecimiento de lianas. In Gómez-Pompa, A. and Amo R. S. del (eds.) *Investigaciones Sobre la Regeneración de Selvas Altas en Veracruz, México,* pp. 147–69 Volumen II. (INIREB: Xalapa and Editorial Alhambra, México)
Repetto, R. (1987). Creating incentives for sustainable forest development. *Ambio*, **16** (2–3), 94–9
Repetto, R. (1988). *The Forest for the Trees? Government Policies and the Misuse of Forest Resources.* (World Resources Institute: Washington, D.C.)
Soemarwoto, O. (1984). The talun-kebun system, a modified shifting cultivation, in West Java. *The Environmentalist*, **4** (7), 96–8
Srivastava, P.B.L., Abdul Manap Ahmad, Kamis Awang, Ashari Muktar, Razali Abudul Kader, Freezaillah Che' Yom and Lee Su See (eds.) (1982). *Tropical Forest. Source of Energy through Optimization and Diversification.* (Penerbit Universiti Pertanian Malaysia, Serdang)
Unesco. (1978). *Tropical Forest Ecosystems.* A state-of-knowledge report prepared by Unesco, UNEP and FAO. Natural Resources Research Series 14. (Unesco: Paris)
Wadsworth, F. (1981). Management of forest lands in the humid tropics under sound ecological principles. In Mergen, F. (ed.) *Tropical Forests. Utilization and Conservation*, pp. 168–80. (Yale School of Forestry and Environmental Studies: New Haven)
Webb, L.J. (1982). The human face in forest management. In Hallsworth, E.G. (ed.) *Socio-economic Effects and Constraints in Tropical Forest Management,* pp. 143–57. (Wiley: Chichester)
Whitmore, T.C. (1983). Secondary succesion from seed in tropical rain forests. *Forestry Abstracts*, **44**, 767–79
Wiersum, K.F. (1982). Tree gardening and taungya on Java: examples of agroforestry techniques in the humid tropics. *Agroforestry Systems,* **1**, 53–70
WRI (World Resources Institute). (1985). *Tropical Forests: A Call for Action.* Report of a Task Force of WRI, World Bank and the United Nations Development Programme. (World Resources Institute: Washington, D.C.)
WRI (World Resources Institute). (1986). *An Assessment of the Resource Base that Supports the Global Economy.* (World Resources Institute: Washington, D.C.)
Wyatt-Smith, J. (1963). *Manual of Malayan Silviculture for Inland Forests.* Vol. I and II. Malayan Forest Records, 23

Section 2

Topical reviews

CHAPTER 2

DIAGNOSIS OF THE TEMPERAMENT OF TROPICAL RAIN FOREST TREES

R.A.A. Oldeman and J. van Dijk

INTRODUCTION

In both silviculture and forest ecology, the behaviour of trees has always been considered of the utmost importance, because forests are defined by their trees. A traditional notion in European silviculture therefore is the "temperament" of a tree species. This term is used in such a self-evident way (e.g. Cochet 1959, Lanier 1986) that it is not strictly defined in the literature. From the way it is employed, the temperament of a tree may be defined as the set of growth-and-development reactions shown by a tree towards its environment during its life cycle. For each tree species, or even provenance, this requires a closer definition both of the reactions and of the environmental factors to which the tree reacts. Light is the one most used, and wood increment has been the main growth criterion. Classical examples of specific tree temperaments then are "light-demanding" versus "shade-tolerant" trees, "pioneers" versus "climax species", or tree species growing well on poor versus rich soils.

The temperament is usually pinpointed for a tree species by combining its environmental preferences, for germination and for the mature tree. For instance *Cecropia* species (Moraceae) are generally known as light-demanding pioneers, because they require open conditions with much light both for germination and in their mature phase. In classical silviculture geared to wood production, forest stands are thought to develop from a "seedling phase", through a "sapling phase" and a "pole phase" into a "mature phase", when trees can be harvested. Even in this simplified context, the neglect of the sapling and pole phase constitutes a clear gap in the understanding of tree temperaments.

In natural regeneration silviculture in tropical rain forests (e.g. de Graaf

1986, Jonkers 1987, Queensland Department of Forestry 1983), with its much more complicated forest development patterns, this gap is extremely troublesome. The environment of growing trees in rain forests often changes rapidly and profoundly during their life cycle.

SYSTEMS ANALYSIS

The subject of tropical rain forest tree temperaments cannot be dealt with in simple terms of " a tree" and "its environment". The reaction of a tree may be total and involve the whole organism or be partial and take place in a complex of organs, like a leaf-bearing branch, or at the level of the organs themselves. On the other hand, the environment of the tree does not only consist of abiotic patterns, determined by climatic and soil factors. These factors are filtered by the surrounding vegetation, composed of a patchwork of young, building, mature and decaying forest patches (e.g. see Whitmore, this volume). And within a particular patch, the incoming nutrients and energy are filtered again by many interacting organisms before they reach the tree under consideration.

This situation can be ordered systematically in terms of systems analysis. The main concept of this type of analysis, which implicitly underlies most scientific research, is that a system may be explained in terms of interactions among its subsystems. For instance, the interactions among leaf- and flower-bearing shoots or axes explain the tree crown (cf. Hallé *et al.* 1978). The hierarchy of systems and subsystems which are in turn explained by their own subsystems, cannot be treated extensively in this contribution, and the reader is referred for a more complete discussion of concepts and definitions to Oldeman (1974*a*, 1974*b*, 1979, 1983, 1990 in press) and to Hallé *et al.* (1978).

The immediate environment of a tree can be referred to as an eco-unit. This is one patch in the forest patchwork, or mosaic of patches, mentioned above. A forest eco-unit (regeneration unit) can be defined (Oldeman 1990, in press) as "every surface on which, at one moment in time, a vegetation development has begun, of which the architecture, ecophysiological functioning and species composition are ordained by one set of trees until the end" (Fig. 2.1).

This definition not only includes "patches", "gaps" or "chablis" of modest sizes, e.g., between 0.01 and 0.1 ha (cf. Shugart 1984), but extends to all deforested surfaces where forest starts to develop, e.g. also clear-cuts, cleared and abandoned lands, or intermediate-sized patches (cf. Budowski 1965). The development of patches is thought to occur in a step-wise fashion (e.g. see Whitmore, this volume), as has long been recognized in silviculture (cf. K. Gayer, ca 1888, ex Van Schermbeek 1898). The

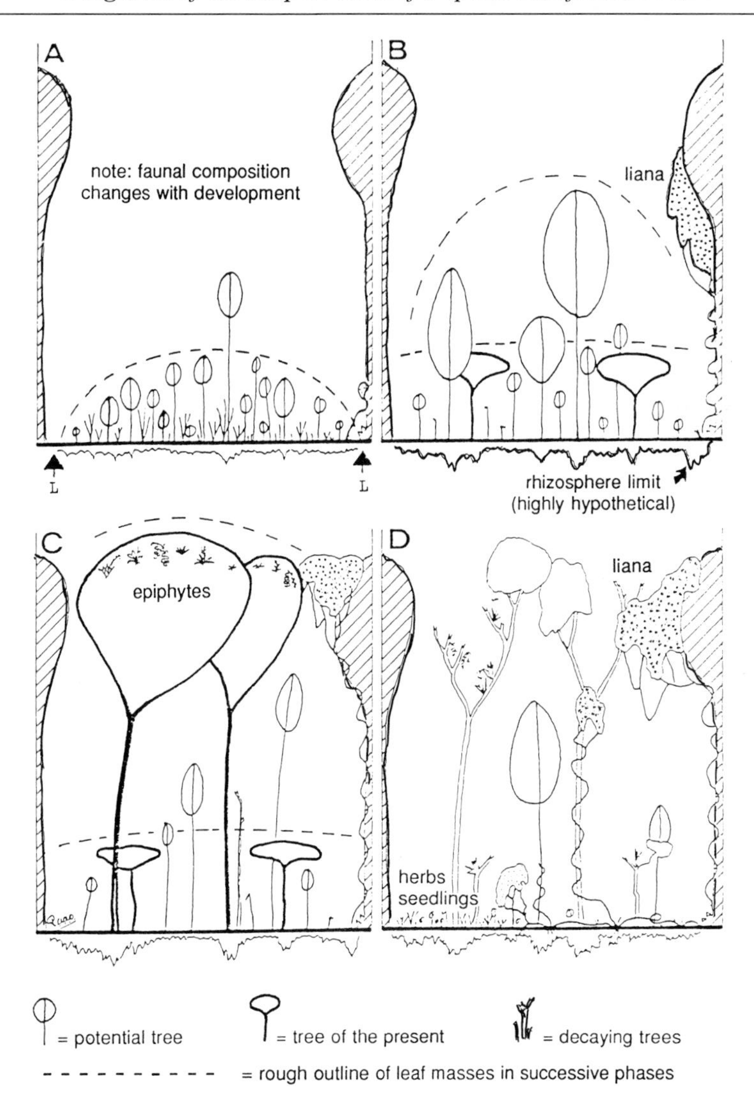

= potential tree   = tree of the present   = decaying trees

- - - - - - - - - - = rough outline of leaf masses in successive phases

Fig. 2.1 Simplified analysis of eco-unit development, the eco-unit being the smallest coherent forest ecosystem determining the immediate environment of trees during their whole life cycle, from sapling (potential tree) to decay. Simplifications of this diagram: linear limits (L) instead of a transition zone; unrealistic unchanging surrounding forest (hatched), unrealistic low number of component organisms. A – Innovation phase ("gap phase", Whitmore, this volume); B – Aggradation phase ("building phase", Whitmore, this volume); C – Biostatic phase ("mature phase", Whitmore, this volume); D – Degradation phase (no Whitmore equivalent). Note that trees of the present, incapable of further crown expansion but stable survivors, determine layering in the biostatic phase, the younger phases not being completely layered as yet, the degradation phase no longer being layered. Eco-units occur with widely differing sizes ("gap size" at origin; cf. Fig. 2.9) and with widely varying architectures, offering many different and changing niches for growth of trees with different temperaments. From Oldeman (1990, in press)

insertion of architectural criteria in the definition is to account for different shapes of an eco-unit, e.g. funnel-shaped, cylindrical or bowl-shaped, and the differences in forest development induced by this shape. Architectural analysis begins by the mapping of the outer geometrical limits of an eco-unit and then the mapping of the geometrical disposition of internal compartments with a certain function. This allows the introduction of ecophysiological production (see Ramakrishnan, this volume) by each compartment, e.g. dominant trees (trees of the present), suppressed (potential) trees, seedlings or insects. The mode of production is determined in turn by adapted organisms which implement it, i.e. the species building an eco-unit in interaction with each other.

Although such a discrete model is only one way of interpreting forest ecosystems – continuous modelling based on gradients rather than mappable limits being another (cf. Kessell 1979, Bongers and Popma 1988) – the eco-unit approach has been chosen because it permits refinement of tree biotope analysis without rejection of the notion of gradients. Indeed, if borders between eco-units are considered to be transition zones with steep climatic, biotic and soil gradients, both approaches can ultimately be harmonized.

In the following paragraphs, eco-units are mainly categorized as small, medium-sized or large, and according to their development phase, i.e. successively as young or open, dynamic or building, mature (biostatic), and degrading. All references to quantities and qualities of incoming light or nutrients have to be seen in this context, as well as matters concerning biological competition between trees. The eco-unit is considered as the system, in a certain state, and the trees are seen as the sub-systems determining its overall function.

TREES, THEIR SUBSYSTEMS AND DIAGNOSIS OF THEIR TEMPERAMENT

The function of a tree in its forest eco-unit has to be considered in two ways. On the one hand, the tree participates in building the eco-unit and so contributes to the survival of the eco-unit. On the other hand, the tree reacts to all biotic and abiotic inputs coming from its immediate biotope. Can this behaviour be seen directly, and can a diagnosis of a tree's temperament be based upon visible and measurable symptoms?

Here again, no simple answer is possible. Trees are too complex. Different behaviour can become apparent at the level of the whole organism, of parts of the crown, of branches with their lateral organs, and of these organs. These levels are not independent of each other. As a frame of reference, their interdependence can be sketched as follows:

Organs, in interaction, build shoots. The nature of the organs, particularly leaves (photosynthesis) and flowers and fruits (reproductive strategy), is a determinant of tree temperament. But their arrangement on the shoot also has to be considered;

Shoots, in interaction, build trunks-cum-crowns. This occurs according to a limited number of constructional modes, the architectural models of Hallé and Oldeman (1970; also see Hallé *et al.* 1978). Diagnostic criteria here can be found both in shoot and crown architecture;

Crowns, in interaction, build the canopies of large trees. The process of building extra crowns in the same tree, each according to the architectural model of the tree, has been called reiteration by Oldeman (1974*a*, also see Hallé *et al.* 1978). A more programmed way to build additional crowns has been described under the name metamorphosis by Edelin (1984). Diagnostic criteria indicating the tree temperament can be expected to be found both in crown architecture and in tree canopy architecture with a particular arrangement of reiterated "crownlets".

It is to be understood from the outset that organs, shoots, branching patterns and reiteration patterns are all "adaptive tools", leading to a certain tree temperament. Between each of these features and each of the possible environments, there are no simple, straightforward and clear-cut relationships. In the overall biological efficiency of a tree in a certain environment, the combination of the features at different levels of organization is decisive. For leguminous trees, this has been analyzed by Oldeman (1988).

Unfortunately, this combination of features has not been studied thoroughly for any tree species, and certainly not over its whole life cycle. Insights can, however, be gleaned from piecing together published specialist information, complemented by unpublished data from Oldeman's field work files in French Guyana (1965–1974). The emerging image remains hypothetical but is falsifiable by observation and experimentation. No firmly proven tables or keys to tropical tree temperaments can be provided as yet for the forest manager.

A first review of published studies provides an overview of findings at the different system levels mentioned above. Thus, several attempts have been made to relate architectural models (Hallé and Oldeman 1970) and reiteration patterns (Oldeman 1974*a*, Hallé *et al.* 1978) to environments, as adaptive growth structures. Hallé *et al.* (1978) attempted to link them to K- and r-selected strategies, and Fournier (1979) tried to correlate models and specific biotopes. Only in exceptional cases could such correspondances be demonstrated clearly. For instance, Attims' model is particularly frequent in trees of inundated forests, like mangroves, and rare in other environments, while Holttum's model clearly is geared to an r-strategy. Ashton (1978) and Givnish (1984) examined the problem from other angles, and found no conclusive evidence for clear-cut, immediate adaptive significance in the tree models described by Hallé and Oldeman (1970).

There is more convincing evidence for adaptive features at the hierarchical level of shoot architecture. Unbranched, monocaul trees, showing Corner's model, are well adapted to the forest undergrowth (Fournier 1979), and so is Cook's model which is morphologically different but physiognomically convergent (Hallé *et al.* 1978). The ecophysiologically derived hypothesis that multistemmed plants will be low in stature and single-stemmed plants will be tall (Givnish 1984) seems to hold true for temperate forest understoreys with their multistemmed shrubs, but not in the tropical rain forests where such shrubs are conspicuously rare. This difference cannot be explained as yet. In any case, monopodial, sparsely branched growth habits of many fast-growing tropical pioneer trees (see Whitmore, this volume) seem to enhance rapid height growth.

Continuous or intermittent growth of leader shoots in trees is well correlated with their successional status (Ashton 1978). Continuous growth is adaptive in uniform environments and rhythmic (intermittent) growth is adaptive in habitats with pronounced fluctuations in environmental conditions (Hallé *et al.* 1978, Fournier 1979). However, there is a complete range of intermediates (Fournier 1979, Lipman 1981).

That orthotropic axes are well adapted to sunny environments, and plagiotropic axes to shade is generally accepted (Hallé *et al.* 1978, Fournier 1979, Givnish 1984). Givnish has argued that: orthotropic axes favour height growth; erect axes can support more leaf mass per unit twig mass; plagiotropic branches or branch systems minimize self-shading, this being favourable for low-light photosynthesis near the compensation point; self-shading of leaves on orthotropic axes reduces transpiration losses in high-radiation environments.

The hypothesis that adaptation is easier to pinpoint at the level of shoot architecture than in whole tree architecture is strengthened by tree species displaying different architectural models in different environments (Kahn 1975, Hallé 1978). Such cases are rare in the tropics, however, and concern architectural models that resemble each other in organization. They differ most often by one structural differentiation only. For instance, a tree showing Leeuwenberg's model in the sun may replace it by Scarrone's model in the shade, substituting the terminally flowering orthotropic epicotyledonary axis by an indeterminate orthotropic shoot which functions as a perduring leader shoot. Again in trees conforming to Rauh's model, branches may be clearly orthotropic in the open and show some plagiotropic tendencies in the shade.

Fisher (1984, p. 580) aptly summarizes the conclusion in stating: "It appears that the architectural models of Hallé *et al.* (1978), are important for deterministic development, but a model *per se* need not be of adaptive significance. Crown shape, including branch and leaf geometry, is produced by the iteration of small scale elements and by the process of reiteration. Thus, seemingly small changes in the elemental parameters (branch angle, branch length) may have major effects on overall crown shape. What seems of major biological significance is the capacity for architectural change, i.e., plasticity. By this mechanism, a tree can respond opportunistically to changes in its environment".

Both tree architecture and shoot architecture therefore yield only partial answers in the quest for visible criteria for the diagnosis of tree temperament. Therefore, there is need to shift attention to the next lower level, that of the organs, where architectural features are considered like the means of support, supply and spatial arrangement of the organs. Photosynthetic organs that have been accorded special consideration include leaves and other photosynthetic units like phyllomorphs (cf. Hallé and Oldeman 1970) and short shoots (cf. Oldeman 1989). The key issue is the spatial arrangement of photosynthetic tissues, which is variously optimized so as to maximize assimilation in a tree under different environmental conditions.

Let us assume that, under given environmental circumstances, the leaf arrangement shown in Figure 2.2*a* is optimal. Several architectural models may now be apt to serve as "leaf carriers" providing the stems, branches and petioles needed to support the leaves and connect them with the roots (Fig. 2.2*b*). Physical and ecophysiological factors linked to the sap stream (Oldeman 1974*a*, 1990 in press) make certain architectural models more "economical" than others. This reduces the number of models once more (Fig. 2.2*c*). In this way, a leaf arrangement model can, in principle, predict the probability of one or a few well-defined leaf arrangements, each carried by a limited number of architectural models occurring under a given set of environmental conditions (also see the next section of this contribution). Such a model combines criteria at the three levels of organs, shoots and the architectural model, and hence is not strictly architectural.

Except for standard growth dynamics, conforming to tree models, this image is static. Such stereotypical growth patterns are an answer to rather constant climatic and soil conditions (Hallé and Oldeman 1970). But, during the life cycle of a tree, these conditions may change, sometimes in a drastic way. A tree should be adjusted optimally to a rather constant environment, or else have the capability to adjust rapidly to changing conditions. For abrupt increase of radiation or nutrient inputs, for instance, fast and "cheap" leaf formation might be implemented by either unbranched or rhythmically growing models (Fig. 2.2*d*). Reiteration of the model is a

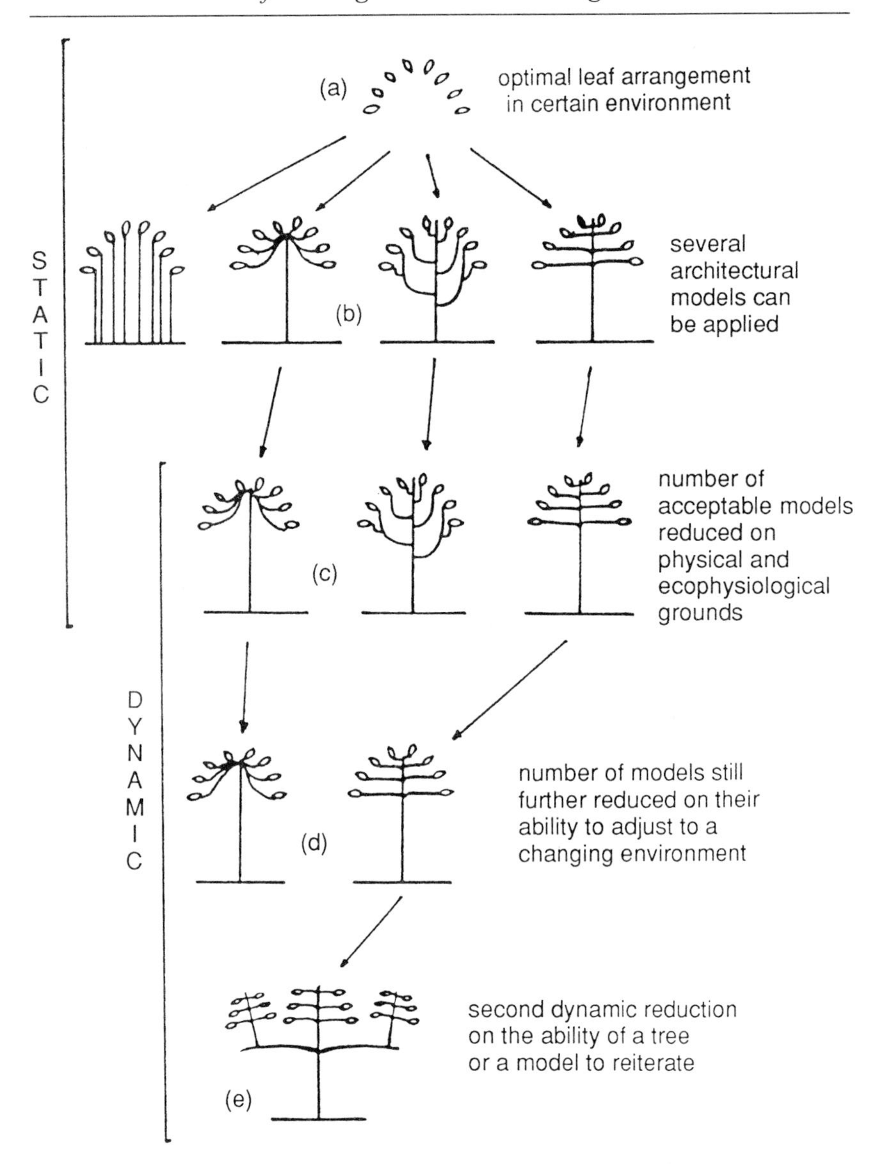

Fig. 2.2 Stepped selection of tree architecture as a support for adapted leaf arrangement. (a) Leaf canopy as needed, without support. (b) First selection of architectural tree models able to carry such a leaf arrangement. (c) to (e) Following steps of selection, models and trees without sufficient reiteration capacity being successively eliminated by the dynamic qualities of the immediate environment

particularly opportunistic mechanism (Hallé *et al.* 1978) and therefore the acme of flexibility is selected in the species that can reiterate rapidly and profusely (Fig. 2.2*e*). This selection mechanism is not to be seen apart from the eco-unit determining the quality and quantity of radiation and nutrient inputs, as well as the selection pressure by competitors adapted to comparable environments.

The question might properly be posed as to what one can expect to gain by organizing the research needed to check this leaf arrangement model and its validity in the field, for selected tree species.

(1) Such a model should predict which tree is best adapted to a given sequence of environmental rain forest conditions encountered in its particular ecotope (see Oldeman 1974*b*, 1979, Hallé *et al.* 1978). However, not all tree species occupying the given ecotope will show a complete set of adaptive characters. Trees displaying the complete set will only be more frequent than less well-adapted species. Being optimally adapted is no guarantee that a tree will complete its life cycle and reproduce well enough to maintain its proportion of genes in the population: it only enhances its chances of doing so.

(2) The model should account for apparent but temporary low adaptiveness. Suppressed juvenile trees of canopy species may show little or no growth and seem ill-adapted when compared with other species that grow and reproduce in the same densely shaded undergrowth. However, such juveniles often respond quickly to the opening of the canopy, the quick response "over-ruling" the apparent imperfections of the suppressed state. In comparison with light-requiring seedlings and saplings of pioneer trees, survival ability under low light conditions for a sometimes considerable time is adaptive. Adaptiveness is a comparative notion.

(3) The model should allow comparison between tree performances. Intrinsic photosynthetic leaf properties may differ considerably. Hence, leaf arrangements that are apparently identical may be functionally different, adapting their trees to different environments (Ramakrishnan, this volume). However, if the photosynthetic performance is compared in two different leaf arrangements, both with a first type of leaves, their relative performance will not change if both are equipped with the same leaves again, but of a second type. In this sense, form reflects function.

(4) The model cannot be expected to cover all aspects of tree temperament. The leaf-arrangement model does not take root architecture into account. A measure of root architecture would certainly add to the performance of the model, but existing classifications of root shapes (e.g. Jeník 1978) are not formulated in a

way that allows them to be linked to above-ground architecture in a single architectural or functional model. The same is true for trunk models, even if centred around cambial functions.

It has been demonstrated above that no simple symptoms for tree temperament diagnosis are provided by the 23 architectural tree models (Hallé and Oldeman 1970). An improvement would be obtained by more detailed "architectural diagrams" (Edelin 1977), which add characters up to the specific level and so involve shoot and organ architecture in a more detailed way. This degree of precision is needed in autoecological studies, but it is costly and time-consuming to achieve in the species-rich tropical rain forest. Therefore, in the following paragraphs, particular architectural features are combined with other crown characteristics. The combination reflects the tree's environmental requirements and its flexibility in a changing environment. This attempt is again limited to visible, diagnostic features.

CROWN AND LEAF CHARACTERISTICS

Crown characteristics

Literature on crown characteristics is scarce. In the tropics, tree measurements have generally been restricted to the usual silvicultural parameters of height, diameter or girth at breast height, and clear bole height (crown depth). Moreover, these measures have nearly always been confined to trees > 10 cm diameter at breast height, leaving aside saplings and poles. Data on crown characteristics of young trees are scarcer still. As a consequence, the subject remains a largely hypothetical one (though the following section includes our own field data, in order to start checking the theoretical viewpoints proposed here).

The following discussion focusses successively on dicotyledons, monocotyledons and gymnosperms with emphasis on seedlings, saplings and poles. Mature tree crowns and reiteration will not be included.

Dicotyledons

Horn (1971) distinguished two basic leaf arrangements, monolayers and multilayers: in monolayers, leaves are densely packed in one single layer, while multilayers dispose their diffused leaves into many layers. Horn hypothesizes that multilayers are more productive in the sun, whereas monolayers produce more in the shade. Multilayers contain a far greater leaf surface than monolayers and, in open places, have so many leaves that, if spaced accurately, they are able to exceed monolayers in photosynthesizing performance at their light saturation rates. In the shade, the leaves in monolayers can photosynthesize above their compensation points, whereas many of the lower leaves in multilayers will function below that point. Horn therefore predicts that early successional trees will tend towards multilayered leaf arrangements, and monolayers will occur in juvenile trees of late successional trees in the undergrowth.

Horn's predictions seem to apply quite well in temperate forests, but there are some obvious aberrations in tropical forests – with their great variety of life forms. The first are sparsely branched, large-leaved, nearly monolayered pioneer trees, such as *Cecropia* spp., *Musanga cecropiodes* or *Macaranga gigantea*, from the neotropics, tropical Africa and tropical Asia, respectively. They seem to be adapted exclusively to rapid height growth without "wasting" biomass to build multilayered leaf arrangements. Large leaves on "cheap" petioles serve as throw-away photosynthetic units and the trees can put most photosynthates into the growth of a single shoot (Coombe and Hadfield 1962, Bazzaz and Pickett 1980, Givnish 1984).

Another aberration from Horn's predictions is the "pauci-layer". This pagoda tree, in which leaves are disposed in a limited number of distinct layers is exemplified in genera such as the pantropical *Terminalia* or the palaeotropical *Alstonia.* These "pagoda" trees (Corner 1952) are very flexible, can be found among both pioneers and emergents (Pickett 1983) and are distinctive for later seral succession (Ashton 1978, Unesco 1978). Their branches show plagiotropy by apposition (Hallé and Oldeman 1970), are adaptive in the shade, in combination with a single orthotropic leader shoot, which grows rapidly in height (Ashton 1978). Several architectural models may be shown by pagoda trees (Fisher 1978, Hallé *et al.* 1978), but Aubréville's model, in which the leaves are most evenly spaced within each branch tier (Hallé *et al.* 1978), seems to be the most frequent and successful one (Kahn 1975).

A third aberration from Horn's predictions is represented by suppressed juveniles of canopy trees, waiting for a canopy gap in which to grow, and which have tall and narrow crowns (Hallé *et al.* 1978, Bruenig 1983) with a high leaf area index. Hence, these trees are essentially multilayers.

To extend the validity of Horn's temperate model to the tropics, Givnish (1984) substituted the "ecological compensation point of a leaf" for Horn's "photosynthetic compensation point". Givnish added six cost items in terms of energy. The leaf arrangement of a tree represents a certain balance amongst these cost items, in summary:

(1) nightly leaf respiration,

(2) cost of leaf construction,

(3) cost of root, xylem and phloem building and maintenance,

(4) cost of tissues supporting leaves,

(5) losses due to herbivory or diseases,

(6) cost of nutrient extraction and displacement from shaded leaves towards young, well-lit leaves. We add a seventh item, i.e.

(7) cost of strategic investment.

This cost is illustrated by a juvenile tree building and maintaining a multilayered leaf arrangement in the shaded undergrowth, which only becomes an energetic advantage instead of a drawback when the canopy opens and it helps the tree to outgrow its competitors.

Monocotyledons

Palms are by far the most prominent tropical monocots. Many are found in dynamic forests, in which new eco-units are frequently opened up by wind, unstable soils or flooding (Givnish 1978, de Granville 1978, 1984, Oldeman 1990, in press). Palms contain many efficient "gap colonists" with single erect axes and large compound leaves as throw-away photosynthesizers. Givnish (1978) hypothesizes that "feather palms" with long, pinnate fronds are colonists of small eco-units ("gaps") and "fan palms" with shorter palmate fronds tend to colonize large new eco-units.

He argues that the compact, (hemi-) spherical crowns of fan palms are vulnerable to overtopping and shading that menace them in small eco-units. Fan palms moreover have a high degree of self-shading, minimizing transpiration losses in sunny or dry habits. In contrast, the umbrella-shaped crowns of feather palms (Hallé *et al.* 1978) are, according to Givnish, much less vulnerable to overtopping and shading, while long pinnate fronds minimize self-shading. De Granville (1978) adds the function of funnel-shaped palm crowns in the understorey of tropical American rain forests as "litter collectors", concentrating litter masses at the foot of the palm. This function also fits in small eco-units, which are exposed to litter falling from nearby trees.

Givnish's hypothesis is supported by amphistomaty, characteristic of light-requiring plants (Parkhurst 1980, ex Givnish 1984) and being frequent in fan palms, whereas most feather palms are hypostomatous (Tomlinson 1961, ex Givnish 1984). Other, non-monocotyledonous trees such as *Araliaceae* species and tree ferns show leaf arrangements similar to feather palms and are colonists of rather small gaps at higher altitudes in the tropics (Givnish 1978, 1984). De Granville (1978) drew attention to palm habitats being more frequently filled by tree ferns with increasing altitude. The pioneer tree ferns of the Antilles and the high Andes seem to be an exception to this rule-of-thumb.

Quite a few short-stemmed or stemless palms are characteristic of the rain forest understorey in tropical America and Asia. They most often have funnel-shaped crowns, but flattened disks also occur (Hallé *et al.* 1978, de

Granville 1978). Some understorey palms, like the Asian *Eugeissonia tristis* or American *Astrocaryum* species form dense, monolayered thickets virtually blocking the germination and establishment of other plants (Ashton 1978, Dransfield 1978, de Granville 1978, Kahn 1986).

Gymnosperms

Many conifers conform to their archictectural model throughout their life cycle, showing little or no reiteration (Hallé *et al.* 1978; but see Edelin 1977). Their strongly monopodial growth habit results in characteristic narrow and elongate, cylindrical or conical crowns (Whitmore 1975). Most conifers are heliophilous and act either like small eco-unit dwellers (e.g. *Agathis* and *Araucaria* species), or large eco-unit colonizers (e.g. *Pinus* species; Whitmore 1975, Denslow 1980). In terms of available light, this almost certainly means that many newly formed, small, but narrow eco-units in high forest are lethal environments for tropical gymnosperms. Probably, these latter tropical conifers are all multilayers, as was found by Horn (1971) for all temperate coniferous gap species, such as *Pinus* and *Sequoia* species.

Leaf characteristics

At a more detailed level of the organs being carried by the crowns, Givnish (1984) has provided an excellent review of leaf characteristics and their ecological significance, the most relevant conclusions of which are summarized here.

Leaf size

Many authors have observed decreases in leaf size towards sunny, dry or nutrient-poor sites. Indeed, the mathematical model of Givnish and Vermeij (1976), which calculates differences between photosynthetic gains and transpiration costs, predicts that the smallest leaves will be found in xeric, sunny environments. In moist, shaded rain forest understoreys, leaf size is not very adaptive and so a wide range of leaf sizes may be expected there (Givnish and Vermeij 1976, Chiariello 1984). Tropical forest trees may indeed often have large juvenile leaves and much smaller adult ones (e.g. *Shorea curtisii*, Ashton 1978), as has been mentioned in rain forest literature since the beginning of the century (e.g. Schimper 1903). The large, simple leaves of many pioneer tree species may be an energetically cheap solution for adapting trees to rapid height growth.

Compound leaves

Givnish (1978) argues that deciduous compound leaves reduce costs in warm, arid habitats, the shedding of highest-order branches or large rachises minimizing transpiration losses. Their ecological signifiance in early successional trees has been discussed above.

Leaf inclination
Trees with peripheral leaves or leaflets steeply inclined with respect to the horizontal plane are noticeable in sunny, dry or nutrient-poor habitats. Such inclination reduces heat load and consequently also transpiration of the outer leaves. It may also increase the potential leaf area index of a crown, light penetrating more deeply into it, which is quite important for multilayer arrangements. Indeed, Ashton (1978) found that trees with leaves oriented more perpendicularly to incident light tend to have relatively low leaf area indices.

The leaf characters mentioned above can be seen immediately and are therefore diagnostic. Other characteristics fall beyond the scope of this contribution. Leaf translocation and leaf turnover rates, for instance, are of prime importance in nutrient turnover and conservation (cf. Ramakrishnan this volume), but they have to be measured and cannot be seen directly in the field, except by linking them to visible architectural symptoms, as has been done by Givnish. Other intrinsic but invisible properties of leaves are, for instance, leaf production and longevity (Bazzaz and Pickett 1980, Bazzaz 1984); deciduousness (Givnish 1978, 1984, Medina 1983, Carabias-Lillo and Guevara Sada 1985); palatability (Hartshorn 1978, 1980, Hladik 1978); autotoxicity; reflectivity (Givnish 1984); thickness and consistency; indument; and still more hidden physiological properties (e.g. Bazzaz and Pickett 1980, Mooney *et al.* 1984).

LIFE HISTORY TYPES AND REGENERATION STRATEGIES

As noted in the introduction, foresters have for several centuries recognized different life histories of forest trees, distinguishing at least three "temperaments" in light-demanding (heliophilous), half-tolerant and shade-tolerant (sciadophilous) trees (for definitions, see Mayer 1980; for a critical review, Lanier 1986).

> Trees living and reproducing in the dense shade of the forest understorey (not distinguished in classical forestry);
>
> Trees living and reproducing in the understorey but benefiting from light gaps; (would be very tolerant trees in forestry; "Schattbaumarten", see Mayer 1980);
>
> Trees with juvenile establishment in the shaded understorey and

needing a gap to grow up and mature (classical shade-tolerant trees in forestry; "Schattbaumarten", see Mayer 1980);

Trees that become established and grow up in gaps (light-demanding trees in forestry; "Lichtbaumarten", see Mayer 1980).

Trees that tolerate medium to light shade when establishing themselves and that need gaps to grow up and mature (half-tolerant; "Halbschattbaumarten", see Mayer 1980).

And recently, a temperament that had dropped out of forestry manuals with the decreased interest in coppice-with-standards, a traditional European system of farm silviculture, was rediscovered in the tropics. The word "late-tolerant" was coined to name this temperament (Oldeman 1990, in press).

Trees that germinate, that become established and mature in the light of gaps and that survive a long time in the shade of later-overtopping larger trees.

Kahn (1982) described *Microdesmis puberula* and *Myrianthus liberica* in Côte d'Ivoire. They display this late-tolerant temperament. These species can form a mature understorey (structural set, Hallé *et al.* 1978) below a canopy of overtopping pioneer trees. Some *Inga* species in French Guyana may do the same (Poncy 1985, Lescure 1986). In Mexico and Central America, *Piper amalago* and other *Piper* species exemplify this temperament (Gómez-Pompa *pers. comm.*, 1986). In the case of *M. puberula*, *M. liberica* and *Inga* spp., these small trees have been followed through successive steps of their establishment and growth. They have certainly not originated from tolerant seedlings.

During the last decade, such life history types have been linked to eco-unit size in the forest ecosystem (e.g. Hartshorn 1978, 1980, Oldeman 1978, 1983, 1987, Whitmore 1978, Denslow 1980, Brokaw 1985, Vooren 1986). Tolerant trees do best in small eco-units, created by the fall of a single tree. Intermediate eco-units and border zones of eco-units would yield good early environments for semi-tolerant trees. And light-demanding or late-tolerant species need gaps of about 200 m^2 or more. Denslow (1980) therefore distinguished small-gap specialists (shade-tolerant) and large-gap specialists (light-demanding, late-tolerant).

Life history types is a notion associated with the vegetative development of individual trees. In terms of species, i.e. populations, it is also important to consider reproductive strategies. Following Bazzaz and Pickett (1980) and Pickett (1983), contrasting "gambler" and "struggler" strategies can be distinguished. Species with a gambler strategy produce numerous seedlings which are unable to survive in the shaded understorey of old eco-units and which need light in young, larger eco-units where they may grow rapidly. Gamblers expend massive efforts in reproduction and thus the chances are high that at least one juvenile tree per effort will become established in a suitable young eco-unit.

Species with a struggler strategy produce few, very persistent seedlings, which struggle but survive and show some growth in the densely shaded understorey. Some can even complete their life cycles there. Reproductive efforts of strugglers are effective at low intensities, while the life expectancy of each juvenile is higher.

The struggler-versus-gambler model of Bazzaz and Pickett (1980) is an updated and modified version of van Steenis' (1958) nomad-versus-dryad concept. The new version is a step forward in that it links the van Steenis view to other biological population models in such a way that it can be expressed numerically. This issue can be placed easily in the context of the discussion on r- and K-selection, a notion which is akin to the concepts mentioned here, (cf. Begon *et al.* 1986, their chapter 14).

It is useful to visualize the result of the "gambler" and "struggler" strategies in terms of eco-unit development. In Figure 2.3, these abstract considerations are translated into an image of the kind of patterns in a tree population that may be observed in the field, when observing different developmental phases of an eco-unit. Scale-drawings of real situations, depicting such observed developments, have been at the base of the rediscovery of late-tolerant species by Kahn (1982) and Lescure (1986). It is useful to realize that the gambler strategy works best in large young eco-units, where not only the germination environment is exposed to high light intensities, but these conditions persist throughout the life of the tree. In contrast, the struggler strategy is most suited to the successive opening of small to very small eco-units, or even temporary damage to the canopy of a persisting eco-unit, because it allows a resumption of height growth with each successive modest increase in light.

A more comprehensive classification of juvenile tree temperaments may be reached by combining their vegetative life history, according to criteria found at different levels and discussed above, and their reproductive strategy. An analogous effort for all Mexican plant types in the Veracruz rain forest has been visualized schematically by Gómez-Pompa and Vázquez-Yanes (1985). In the following list of tropical forest tree temperaments, the reproductive strategy has been used to define extremes, i.e. hard strugglers and hard gamblers. Between these extremes, intermediate temperaments depend very much upon crown and leaf characteristics.

(1) Hard strugglers. Offspring are produced infrequently in low numbers.

Crown and leaf characteristics enable completion of the whole life cycle in dense shade within older forest eco-units with a closed canopy. Shade tolerance remains high during the life cycle. Increased light levels may be beneficial but are not indispensable, whereas abrupt exposure to high light intensity is lethal.

(2) Strugglers. Offspring are produced infrequently in low numbers. Crown and leaf characteristics are flexible, permitting a development from shade-tolerant seedlings to later development stages that are less tolerant, whether they grow up into the upper canopy or not. Abrupt exposure to full sunlight is often detrimental.

(3a) Gambling strugglers. Offspring are produced frequently in high numbers. This is a characteristic of gambling strategies though crown and leaf characteristics show much of the struggler temperament. From the shade-tolerant seedling – which, however, demands a light intensity steadily increasing with height from the ground – gambling strugglers grow very fast. Later they can be overtopped by larger trees and complete their life cycle as strugglers under the canopy.

(3b) Struggling gamblers. Offspring are produced frequently in high numbers. Crown and leaf characteristics are flexible, but, from the shade-tolerant seedling, juveniles soon pass into a phase of light-demanding development which is irreversible. Hence, they can only complete their life cycle in the upper canopy of a mature eco-unit (see Fig. 2.1); if overtopped, they die.

(4a) Gamblers. Offspring are produced frequently in high numbers. Crown and leaf characteristics are so flexible that the trees, which arise from seedlings and juveniles that tolerate no shade at all, can later survive under a certain amount of shade. The difference with gambling strugglers is not only in the light tolerance of the seedling, but also in crown and leaf characteristics that are not especially adapted to live in the shade, but flexible enough to accommodate some shade.

(4b) Hard gamblers. Offspring are produced frequently in vast numbers. Given their leaf and crown characteristics, from the seedling on, these trees can only survive and grow in full light throughout their life cycle. This condition is to be found mainly in large-sized young eco-units with little remaining vegetation, and maintained only for those individual trees that grow ahead of the dense population as a whole (Fig. 2.3).

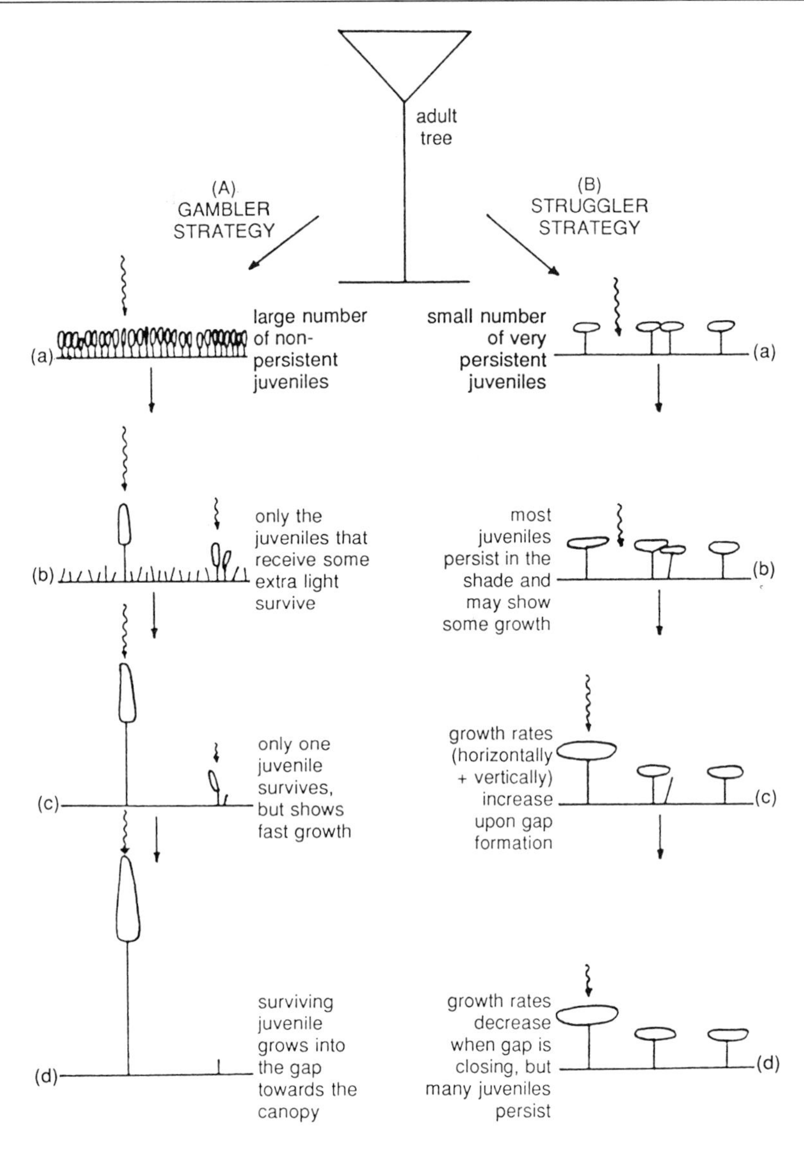

Fig. 2.3 Visualization of two extreme reproduction strategies, geared to be successful in different eco-units (Figs. 2.1 and 2.9). A – Hard gamblers saturate large and sunny young eco-units (cf. Fig. 2.9, to the right of the middle) with diaspores, survival and further growth demanding light and depending on crown characteristics (Table 2.5). B – Hard strugglers produce few diaspores in closed, older eco-units (Fig. 2.1C), survival and further growth tolerating quite heavy shade, being vulnerable to abrupt increases of light, and depending on crown characteristics (Table 2.5). Undulating arrows: incident solar radiation, length indicating intensity. (a) to (d) show population development during development of the eco-units, the hard gamblers becoming most often overtopped so that few survivors are left, and the hard strugglers maintaining a small but rather constant population as long as no major disturbance destroys the "mature" eco-unit in which they live

Light is the environmental factor appearing in each description of a temperament. As already noted, foresters have long recognized the major importance of this factor in the definition of temperaments, and Whitmore (1975, this volume) gives thorough descriptions of the two extreme temperaments that were distinguished. Light-demanding trees or pioneers, colonizing large open terrains, have long been known for their combination of special crown and leaf characteristics, their fast growth and often short life span, and their abundant seed production.

But is there only one category of non-pioneers? Given the many variations on the theme of eco-unit development as schematically shown in Figure 2.1 – characterized by different sizes and shapes, different development and growth rates, and different species compositions (including that of species other than trees) – those trees which are not pioneers may encounter all kinds of sequences of light and shade during their lifetime. If the pioneer displays an extreme temperament, the other extreme is not the non-pioneer but rather the anti-pioneer, for which intensive light is lethal. Pioneers are characterized by the temperament described here as "hard gamblers" and anti-pioneers as "hard strugglers". The other four temperaments mentioned above are intermediates. Their descriptions with the recurrent reference to flexibility already point to the fact that they do not concern discrete categories, but are landmarks in what may well be a continuum of environmental conditions, some of which are more frequent and more easily observed in the forest than others.

Figure 2.4 shows the ecological setting for these six tree temperament landmarks. They are clearly linked to the size of the eco-unit concerned (Oldeman 1983), because large eco-units are high-radiation environments from the outset and small eco-units show intermediate to low light intensities at ground level at the start of their development. But these strategies are also linked to eco-unit development and architecture (Fig. 2.1A to D). This is the time dimension in Figure 2.4, from young below to mature above. In different eco-unit development sequences, different shifts in temperament of a tree species enhance survival. The importance of the seedling and juvenile stages, as the time interval during which such shifts occur, is clear, and the neglect of this period in most classical silvicultural temperament studies can therefore be recognized for the handicap it is.

The vertical axis in Figure 2.4 also has a spatial connotation. Because temperaments are defined by crown and leaf characteristics and by reproductive strategies, the lower reaches of the figure concern the undergrowth, and the upper parts indicate the potential or the real upper crown canopy level of the eco-unit concerned. Because there is a gradual increase in light intensity from forest floor to forest canopy, also in eco-units, struggling for light is less intense higher up in an eco-unit than lower down. The hard strugglers therefore are indeed struggling more than struggling gamblers and gambling strugglers (Fig. 2.4).

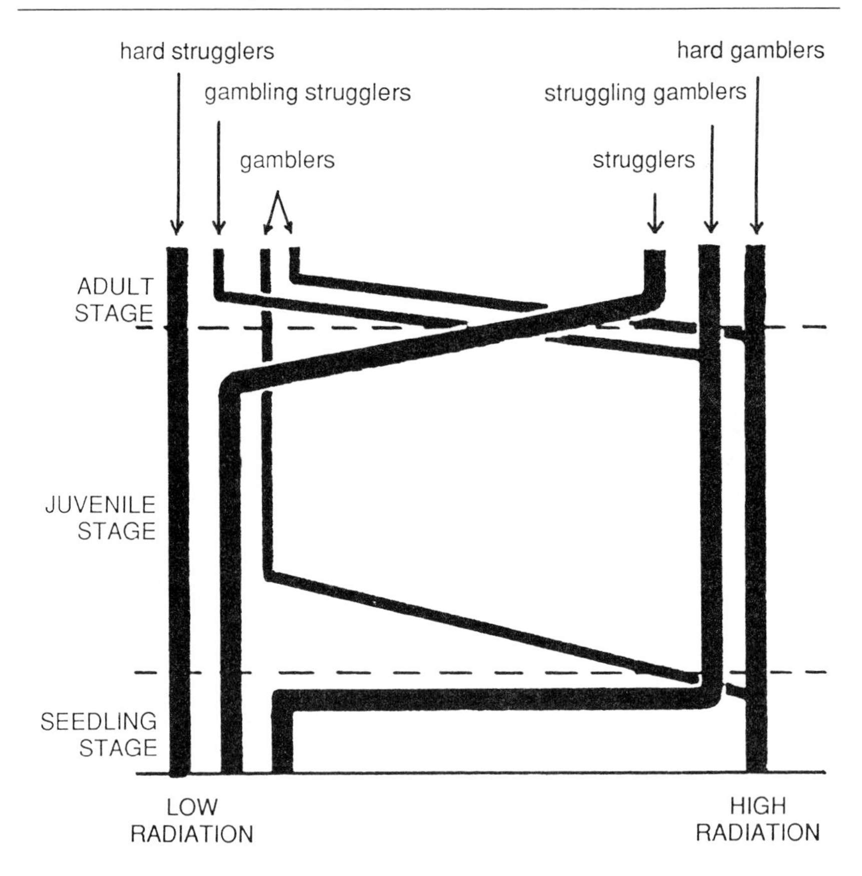

Fig. 2.4 Visualization of six tree temperaments. Each temperament may either remain constant from seedling to adult tree (hard strugglers, hard gamblers), or be flexible and change its shade tolerance at some stage of the life cycle (others). The vertical dimension of this diagram evaluates height above ground level, which will be different for each tree species. It also refers to processes, but the tempo and rhythm of these processes differ for each tree species, and thus this is also a relative scaling. The strugglers closely correspond to the classical "shade-bearers" or "climax species", the hard gamblers to classical "light-demanding" or "pioneer" species. The relative frequency of a temperament is indicated by the thickness of the bar

The figure would be compatible with a continuum of temperaments, on which the frequent ones have been indicated schematically. This frequency is determined by the most regularly occurring geometrical and dynamic patterns in eco-unit development, determining the immediate environment of the tree and its ecological properties. The continuum, therefore, is not homogeneous, in which case it would be sufficient to characterize the extreme strategies in the sense of van Steenis (1958) as "nomads" and "dryads". Neither is its heterogeneity determined exclusively by eco-unit (gap) size as a geometrical parameter, so that the simple division in "small and large gap specialists" (Denslow 1980) also is an oversimplification. Temperaments of tree species have to contain reactions to factors determined by both geometry and dynamics of forest development, and therefore the frequent strategies in Figure 2.4 are closely linked to eco-unit development (Fig. 2.1).

Some data from the literature confirm the danger of oversimplification in exclusive categories such as "pioneers" versus "non-pioneers", "nomads" versus "dryads" or "light-requiring" versus "shade-tolerant" tree species. Vázquez-Yanes and Orozco-Segovia (1985) drew attention to the necessary increase in the red/far-red ratio of the spectrum, if maximum light intensity is to trigger pioneer germination. Florence (1981) found that pioneer seeds in Gabon germinate in light intensities below full exposure. For instance, *Musanga cecropioides* could germinate in eco-unit border zones with light intensities between 40% and 80% of full sunlight. This fact has been established experimentally for *Cecropia surinamensis* seeds from Suriname by Holthuijzen and Boerboom (1982). The seedling phase alone, therefore, takes place in a very complex environment, be it shaded or not, and studies that combine the quantitative and qualitative aspects of radiation in this environment seem to be very rare.

Figure 2.4 is admittedly not comprehensive, because some rarer strategies are not shown, e.g. the "sciaphilous nomads" (Oldeman 1974*a*, Hallé *et al.* 1978) which grow in "ecological chimneys" with a slow and small intensive vertical light gradient. However, it does indicate the most frequently encountered tree species strategies – recognizable in the juvenile stage by visible characters of crown and branch architecture and leaf arrangement.

Hard strugglers (1) are found in older eco-units under closed canopy (Fig. 2.1B and C). Their crowns are monolayered and more broad than deep. Leaves are borne on plagiotropic branches or, in monocaul trees, in a whorl just below the stem apex. Growth may be mainly horizontal, but also vertical when gap formation slightly increases light availability (cf. Hartshorn 1980).

Strugglers, gambling strugglers and struggling gamblers (2 and 3) are taken together here, because studies are as yet lacking on representative species which are precise enough to order the visible characteristics of each with sufficient confidence. These temperaments often occur in large, economically important trees. For example, many South-east Asian Dipterocarpaceae may be regarded as being shade-tolerant or even needing shade during part of their life cycle. In this family of large trees, shade-bearers and light-demanding species can be recognized readily (cf. Whitmore 1975, his references). Among the strugglers are slow-growing, shade-tolerant "heavy hardwoods", e.g. the balau group of *Shorea* and species of *Vatica*, *Hopea* or *Neobalanocarpus heimii*. Struggling gamblers are to be found among the light-demanding "light hardwoods" which require shade only during the first life phases and later grow very fast in high radiation environments, such as the red meranti group in *Shorea*, or species of *Parashorea* and *Dipterocarpus*. The need for shade before canopy closure is reflected in the observation that high soil temperatures are lethal for the quite specific mycorrhizal fungi that are associated with these dipterocarps in Kalimantan (Smits *et al.* 1987). This is yet another indication that temperaments are no simple categories linked to light requirements or gap sizes, but result from complex strategies adapted to eco-unit development.

The architecture of all canopy trees favours rapid height growth, allowing them to reach the canopy quickly. Monopodial growth is therefore beneficial, shade tolerance being visible by plagiotropic branch orientation, pointing to a struggler metabolism. Orthotropic axes rather favour gamblers. The nec-plus-ultra in architectural flexibility is attained in the Leguminosae (Oldeman 1988), where, in a single species, axes and branched complexes are capable of either struggler or gambler behaviour, depending on the ambient environment. Leguminosae with sympodial growth models dominate the high canopy tree flora of the tropical American and African rain forests. These tree species most often have plagiotropic mixed axes, being capable of either horizontal or vertical extension.

Leaf arrangement may tend to be monolayered in strugglers and multilayered in gamblers. Families like the Dipterocarpaceae have clear-cut architectures in this sense, whereas Leguminosae represent the other extreme of being able to build the two arrangements, both in the same species and in different development periods of the same tree.

Gamblers and hard gamblers (4) are adapted to continuously high radiation levels and do not support shade. Their survival depends upon outgrowing their competitors. This is made possible by the well-known characteristic architecture of pioneer trees, growing in very large eco-units. Their single leader shoot is usually sparsely or not branched and bears large leaves in a single, subapical layer. Such monolayers may not be most productive in terms of photosynthesis, multilayers being more productive in smaller gaps (see above), but they certainly are most competitive in placing their entire crowns above their competitors (Ashton 1978).

However, tree components of large eco-units do exist that build multilayered leaf arrangements. Examples are *Macaranga hypoleuca* from tropical Asia (Ashton 1978), *Goupia glabra* and *Ceiba pentandra* in tropical America (see section below) and some tropical conifers (Whitmore 1975). It may be relevant that, among such pioneers, there are long-lived and very large trees, as is the case in *G. glabra* and *C. pentandra* (see also the paragraphs on emergents in Hallé *et al.* 1978). According to Oldeman (1990, in press), they may be termed long-lived pioneers and play a predominant role in the fragmentation of large eco-units into small eco-units.

A FIRST STEP TOWARDS FIELD CHECKS

As stated above, the proposed hypotheses cannot be validated as yet from literature. Therefore, use was made of an unpublished file by Oldeman, containing data on over a hundred tree and shrub species from French Guyana in the form of drawings, slides, photographs and notes. This information mostly concerned juvenile plants, and has been gathered between 1965 and 1974. We processed the information on those species for which the data were sufficient for a first check of their strategy and environment.

Tables 2.1 to 2.4 summarize important architectural features for each species, to which is attached one of the temperaments above. The first column gives the architectural model (Hallé and Oldeman 1970), the second the maximum attainable height, if known. In the third column, the mode of height growth is given as monopodial or sympodial, and the fourth column provides information on branch orientation (either plagiotropic or orthotropic). In the last column, branching is characterized as rhythmical (intermittent) or continuous, this last term includes diffuse branching with no discernible pattern of intermittence.

The files only contained a few miscellaneous notes on the eco-units in which the trees were observed. Therefore, the tables do not give any

Table 2.1 Hard strugglers and strugglers from French Guyana. These are small, monolayer trees from the understorey of the tropical rain forest, with a simple architecture. From Oldeman (unpub.)

Species	*Arch.*[1] *model*	*Height*[2] *max.(m)*	*Height*[3] *growth*	*Branch*[4] *orient*	*Branching sequence*
Carica cf. spec.nov.	CORNE	2	mono	.....	.....
Casearia bracteifa	ROUX	2	mono	plagio	diffuse
Cedrela cf. *barbata*	CHAMB	–	sym	.....	.....
Geonoma stricta	TOMLI	1	mono	.....	.....
Guarea richardiana	CORNE	2	mono	.....	.....
Pavonia cf. *flavispina*	CHAMB	0.5	sym	.....	.....
Picrolemma cf. *pseudocoffea*	CHAMB	(1.5)	sym	.....	.....
Quiina oyapocensis var. *bicolor*	CHAMB	–	sym	.....	.....
Ryania speciosa	COOK	(7)	mono	plagio	diffuse
Talisia mollis	CHAMB	(6)	sym	.....	.....

LEGEND – = not known; ..... = undefined or not applicable

Column (1) – AUBRE = Aubréville's model; CHAMB = Chamberlain's model; COOK = Cook's model; CORNE = Corner's model; FAGER = Fagerlind's model; KORIB = Koriba's model; LEEUW = Leeuwenberg's model; MANGE = Mangenot's model; MASSA = Massart's model; NOZER = Nozeran's model; PETIT = Petit's model; PREVO = Prévost's model; RAUH = Rauh's model; ROUX = Roux's model; TOMLI = Tomlinson's model; TROLL = Troll's model.

Column (2) – numbers in parentheses are not the maximum heights observed for the species, but the heights of the actual trees in the file (Oldeman, unpub.).

Column (3) – mono = monopodial; sym = sympodial

Column (4) – plagio = plagiotropic; ortho = orthotropic

Table 2.2 Hard strugglers and strugglers from French Guyana. These are trees between 6 and 18 m high, from the lower reaches of the tropical rain forest. Their crown characteristics are a little more complex than in small strugglers and leaves are diffuse throughout the crown. Legend: See Table 2.1. From Oldeman (unpub.)

Species	*Arch.*[1] *model*	*Height*[2] *max.* (m)	*Height*[3] *growth*	*Branch*[4] *orient*	*Branching sequence*
Anaxgorea acuminata	TROLL	6	sym	plagio	.....
Byrsonima verbascifolia	FAGER	–	mono	plagio	inter
Couepia cf. *versicolor*	TROLL	(4)	sym	plagio	.....
Dugutia cf. *obovata*	ROUX	12	mono	plagio	diffuse
Guatteria cf. *ouregou*	ROUX	–	mono	plagio	diffuse
Himatanthus articulatus	KORIB	12	sym	ortho	.....
Hirtella velutina	TROLL	–	sym	plagio	.....
Licania aff. *ovalifolia*	TROLL	–	sym	plagio	.....
Miconia plukenetii	LEEUW	12	sym	ortho	.....
Mouriri frankavillana	MANGE	12+	sym	plagio	.....
Perebea guianensis	ROUX	15	mono	plagio	diffuse
Quararibea turbinata	FAGER	17	mono	plagio	inter
Tachigalia spp.	PETIT	6	mono	plagio	diffuse

Table 2.3 Struggling gamblers, gambling strugglers and a few higher strugglers from French Guyana. These are trees able to form an upper forest canopy, which grow up in small eco-units having resulted from small gaps. Their crown characteristics are more complex and leaves occur in pauci-layers or pagoda-like arrrangements. Legend: See Table 2.1. From Oldeman (unpub.)

Species	*Arch.*[1] *model*	*Height*[2] *max.* (m)	*Height*[3] *growth*	*Branch*[4] *orient*	*Branching sequence*
Aniba sp.	RAUH	–	mono	ortho	inter
Aparisthmium cordatum	KORIB	25	sym	ortho	.....
Aspidosperma sp.	MASSA	'large'	mono	plagio	inter
Cordia spp.	PREVO	–	sym	ortho	.....
Diospyros sp.	MASSA	25+	mono	plagio	inter
Eriotheca sp.	RAUH	'large'	mono	ortho	inter
Iryanthera hostmanii	MASSA	25	mono	plagio	inter
Manilkara bidentata	AUBRE	'large'	mono	plagio	inter
Ocotea splendens	MASSA	–	mono	plagio	inter
Pourouma minor	AUBRE	15+	mono	plagio	inter
Tetragastris altissima	RAUH	'large'	mono	ortho	inter
Theobroma speciosum	NOZER	30	sym	plagio	.....

reference to a temperament per tree species. In each successive table, a group of species has been placed in a certain height category. Among these trees, different temperaments are present, and these will be discussed below. The result of this discussion is illustrated by Figures 2.5 to 2.8. The image that emerges is coherent and supports the existence of a series of temperaments, from hard gamblers to hard strugglers, with intermediates.

Table 2.4 Hard gamblers and gamblers from French Guyana. These are trees that form a low or a high forest canopy in large clearings and deforested areas and often form huge eco-units, which may live only a short time. Their crown characteristics are quite diverse (also see figures and Table 2.5). Legend: see Table 2.1. From Oldeman (unpub.)

Species	*Arch.*[1] *model*	*Height*[2] *max.*(m)	*Height*[3] *growth*	*Branch*[4] *orient*	*Branching sequence*
Annona paludosa	TROLL	–	mono	plagio	.....
Apeiba burchelli	TROLL	15	mono	plagio	.....
Ceiba pentandra	MASSA	50+	mono	plagio	inter
Didymopanax morototoni	LEEUW	20+	mono sym	ortho	.....
Goupia glabra	ROUX	50+	mono	plagio	diffuse
Jacaranda copaia	CHAMB LEEUW	15+	sym	ortho	.....
Toulicia guianensis	LEEUW	30	mono sym	ortho	.....

Hard strugglers

Hard strugglers all occur in the understorey, with a maximum height of around 15 m. A group of small, monolayered trees mostly much under 10 m (Table 2.1) is distinguished from a group of larger trees with more complex crown characteristics that may reach a height of 10 to 15 m (Table 2.2).

In the first group, there are 10 examples (Table 2.1). Most display a monocaul habit, built sympodially (Chamberlain's model) or monopodially (Corner's model), with a few large, simple or compound leaves at the top. *Geonoma stricta* (Palmae, Tomlinson's model) can be considered as a cluster of such monocaul trees. The two branched species remain functionally close to the monocauls. *Ryania speciosa* var. *bicolor*, with its phyllomorphic branches, conforms to Cook's model, whereas *Casearia bracteifera* shows Roux's model, but with branches tending to the phyllomorphic state. Both are Flacourtiaceae. Because of this architecture, *C. bracteifera* has a broad, shallow, monolayered, shade-adapted habit. In all 10 species, tree architecture is very simple (Fig. 2.5). Their photosynthetic organs, mostly large leaves, or small ones arranged in phyllomorphs, form a single subapical layer, minimizing mutual shading within the crown.

The second group contains 13 examples (Table 2.2), and shows greater crown heterogeneity. The only common feature is the plagiotropic orientation of branches or branch complexes (plagiotropy by apposition). This is an expression of shade-tolerance. However, the degree of shade-tolerance varies considerably, ranging from *Himathantus articulatus* (which is found in young secondary thickets near Cayenne) to *Miconia plukenetii* or *Quararibea turbinata*, both of which are found only in mature forest understoreys.

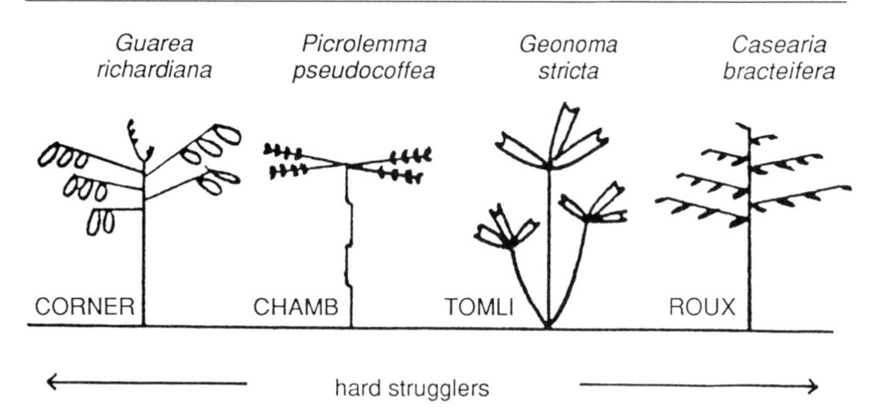

Fig. 2.5 Four examples of hard strugglers from the rain forest in French Guyana, with their typical "simple and poor" architecture and horizontal leaf arrangement; also see Table 2.5. From Oldeman (unpub.)

The species displaying Troll's, Roux's and Petit's models have elongated, diffusely branched crowns with sparsely foliated plagiotropic branches or branch complexes, minimizing mutual shading. *Mouriri francavillana* alternates between Mangenot's model and Massart's model (Oldeman 1974*a*) and so displays great flexibility in leaf arrangement by either spacing the branches apart (Massart), or placing them closely over each other (Mangenot) in an elongate crown, indicating that this species may grow to a considerable height. Distinct superposed branch tiers, plagiotropic by apposition, are found in *Byrsonima* cf. *verbascifolia* (not the savanna genotype) and *Quararibea turbinata*, both conforming to Fagerlind's model. Mutual shading by branches may be great. *Himatanthus articulatus* may also have such a pagoda-like crown architecture, but in more open habitats the modules of branches displaying plagiotropy by apposition form orthotropic shoots. This is an indication of the flexibility of Koriba's model, shown by *H. articulatus*. *Miconia plukenetii* has the same model, but its subapical, rather large leaf clusters on orthotropic axes are true monolayers, like those in the first group.

Strugglers, gambling strugglers and struggling gamblers

Twelve examples of strugglers, gambling strugglers and struggling gamblers can be distinguished, mostly capable of living in the upper canopy (Table 2.3). Their adult stages are light-demanding, but the juveniles show different light requirements. Architecturally, monopodial height growth and intermittent branching are common features.

In Massart's and Aubréville's models, with their plagiotropic branches or branch complexes, this structure results in pagoda trees (see above). In Rauh's model, with orthotropic branches, the growth habit resembles a candelabrum, with essentially monolayered subapical leaf clusters. Examples like *Aniba* sp., *Tetragastris altissima* and especially *Eriotheca* sp. with its large palmate leaves, quite closely resemble pioneer trees like *Cecropia*, *Musanga* or *Macaranga* species (see earlier). This resemblance indicates that the examples cited have gambler strategies and are among the least shade-demanding species in their juvenile state.

More shade-tolerant examples have plagiotropic branches or branch complexes, with marked differences in foliage density denoting different tolerance ratings. *Pourouma minor* and *Manilkara bidentata* (both Aubréville's model) show a very dense and even foliage on plagiotropic branch complexes. *P. minor*, a tree 5 m in height, may have as many as four superposed branch tiers. Monopodial branches in Massart's model have a lower leaf density and therefore less mutual shading of superposed tiers. The number of branch tiers, number of branches per tier, and degree of branch ramification all co-determine the leaf arrangement and density within one crown, linked to tolerance. *Iryanthera hostmannii*, with five branches per plagiotropic tier of Massart's model, is probably less shade-tolerant than *Aspidosperma* sp. with three.

In short, the examples considered range from the struggler to the struggling gambler temperaments (Fig. 2.6). *Aniba* sp., *Tetragastris altissima* and *Eriotheca* sp. are the least tolerant species, *P. minor* and *Manilkara* sp. are somewhat more tolerant. *Iryanthera* sp. is quite tolerant and *Aspidosperma* sp. is most tolerant, although this is not the case for all species of this genus.

Among the species with sympodial height growth, *Cordia* species (Boraginaceae, Prévost's model) have a marked pagoda habit, although their plagiotropic branch modules bear leaves in lower densities than pagoda trees in Aubréville's model. Therefore, their shade tolerance may be comparable to that of trees conforming to Massart's model, converging leaf arrangements being carried on crowns showing the two kinds of plagiotropy. In *Theobroma speciosum* (Nozeran's model), the plagiotropic branch tiers are confined to the upper parts of the sympodial trunk, giving the tree a monolayered crown and shade tolerance. This relative of the cocoa tree indeed lives in deep shade below the forest canopy. *Aparisthmium cordatum* (Koriba's model) has a rather loose, diffusely foliated crown with potentially little mutual shading of leaves and is indeed shade tolerant.

Like the hard strugglers, the examples described here show that different architectural models can lead to convergent crown characteristics. Crown architecture therefore appears once more as one crown characteristic

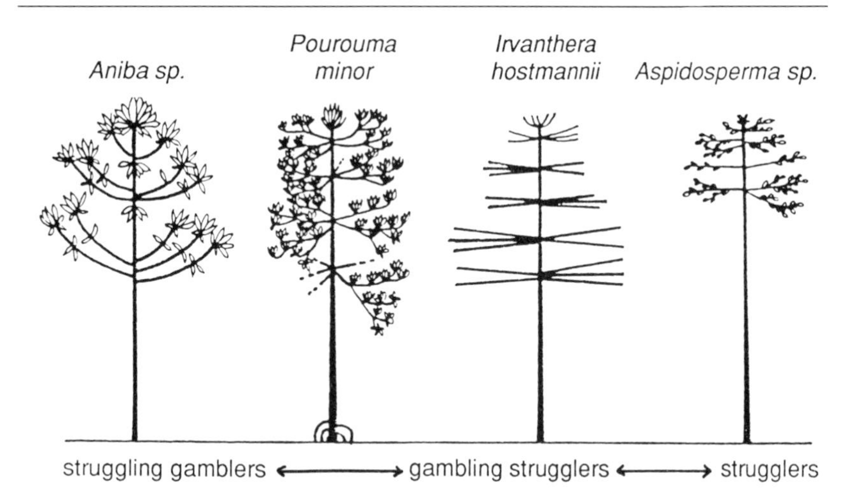

Fig. 2.6 Intermediate temperaments, from strugglers to struggling gamblers. Examples from the rain forest of French Guyana, where they all occur in small to rather small eco-units. Other species of the genera *Aspidosperma* and *Aniba* may behave differently from those observed. Compare with crown characteristics given in Table 2.5. From Oldeman (unpub.)

among others, all being drawn upon in the diagnosis of temperament. Leaf size is also an important feature, Figure 2.6 being consistent with the leaf considerations given earlier.

Gamblers and hard gamblers

Gamblers and hard gamblers are represented by seven examples (Table 2.4), and grow in large eco-units which they colonize. Two groups may again be recognized. The most frequent strategy among these examples is fast height growth at the cost of horizontal crown spreading. *Didymopanax morototoni* (Leeuwenberg's model, Fig. 2.7) has an epicotyledonary axis with prolonged growth up to a height of 10 to 15 m. In this phase, large compound leaves on long petioles are borne in a spiral at the top of the solitary axis (Fig. 2.7A, B). After this monocaul phase, ended by terminal flowering, orthotropic relay axes may prolong height growth by extending vertically. In the mature tree, leaves are very densely packed at the top of terminal modules in a large umbrella (Fig. 2.7C, D). *Jacaranda copaia* follows the same strategy, but its monocaul phase is built sympodially according to Chamberlain's model. *Toulicia guianensis* is similar to *J. copaia*, but data are scarcer.

Height growth concentrated on one orthotropic axis also characterizes the long-lived pioneers mentioned above, namely *Ceiba pentandra* (Massart's model with secondary plagiotropy) and *Goupia glabra* (Roux's

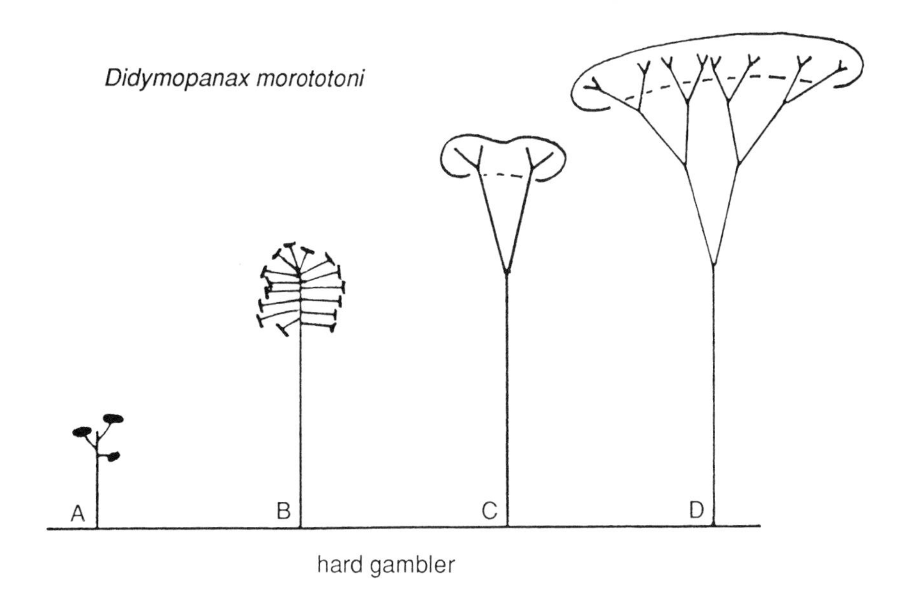

Fig. 2.7 A hard gambler: *Didymopanax morototoni* (Araliaceae) from French Guyana (Oldeman, unpub.). This species occurs in large to very large eco-units ("early succession") and initially show fast, monopodial height growth (A, B) before entering a period of crown spreading, here according to Leeuwenberg's model. For other crown characteristics see Table 2.5

model). They attain heights of over 50 m and persist a long time as emergent survivors of large pioneer eco-units (cf. Hallé *et al.* 1978), while passing their juvenile stage exactly like the common short-lived pioneers. As a matter of fact, juvenile *Goupia glabra* trees in a large, young eco-unit show a striking and profound likeness with the short-lived pioneer tree *Trema occidentalis*, with which it is often confused. Their strong monopodial growth is coupled to rhythmic (*C. pentandra*) or continuous branching (*G. glabra*). The juveniles of these trees are rather sparsely branched, but gradually the trees shift towards more elongate and more densely foliated crowns showing a tendency to multilayering.

The second group of examples follows a completely different strategy (Troll's model, Fig. 2.8). *Annona paludosa*, a pioneer from the tree savannas around Cayenne, and *Apeiba burchellii*, a rain forest pioneer, build densely foliated, spreading, phyllomorphic crowns. They already outshade competitors when juvenile, by occupying the crown space horizontally (Fig. 2.8A). Height growth is accomplished in these species by secondary erection of the basal part of the central plagiotropic axis (Fig. 2.8B, C), and, in other species, by sympodial addition of phyllomorphs of the same kind. This last growth pattern is found in many Leguminosae (see above) and is very flexible.

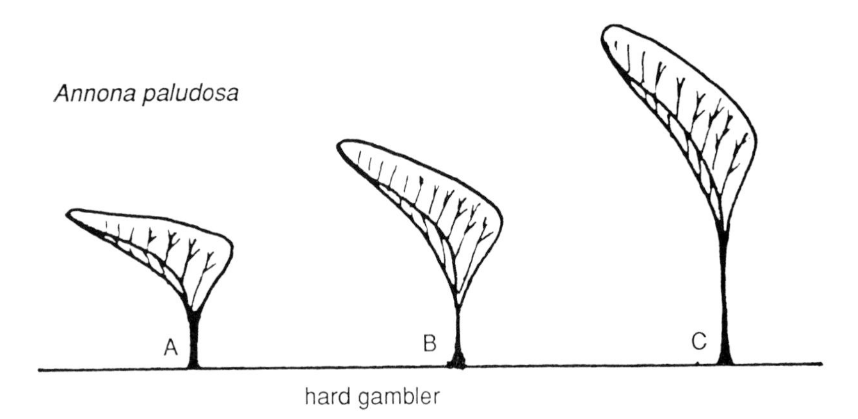

Fig. 2.8 A hard gambler: *Annona paludosa* (Annonaceae) from French Guyana (Oldeman, unpub.). This species occurs in eco-units on poor soils where the forest regenerates very slowly. Other hard gamblers following this architectural principle of phyllomorphic crowns occur in less extreme, large forest eco-units and produce sympodial sequences of such phyllomorphic elements (e.g. *Apeiba burchelli*). Also see Table 2.5

DISCUSSION

On the basis of the preceding data and interpretations, a tree temperament classification that is testable in the field can be tentatively established (Table 2.5). Various crown forms, leaf arrangements and architectural features are classified in the table according to their hypothetical ecological functions, in terms of species strategy, successional status, origin of eco-unit (kind of gap), light requirements and shade tolerance.

Columns A to D (Table 2.5) show hard gamblers and gamblers, and their leaf arrangements. These range from monocaulous or sparsely branched, monolayered trees like *Cecropia* species and others (see above) in column A, to the elongate, multilayered crowns of long-lived pioneers like *Ceiba pentandra* and others in column D. In between, leaves are densely packed – as in fan palms (column Ba) or *Didymopanax*-like umbrellas (column Bb) where mutual leaf shading is enhanced – or by contrast are spread in overtopping phyllomorphs outshading competitors (column C).

Columns E and F mainly concern the intermediate temperaments of struggling gamblers and gambling strugglers. Feather palms (column D) are abundant in small eco-units, frequently appearing in forest mosaics with high dynamics. Other specialists of these environments are pagoda trees or paucilayers, shown in column F as other "small-gap specialists". The density of the foliage within each crown will determine shade tolerance, according to the criteria enumerated above, i.e. number of branch tiers, distance between them, organization of each branch or branch complex and its foliation. The higher the density, the more mutual shading there is amongst leaves, and more light is required.

Column G shows strugglers with elongate but diffusely branched crowns. Again, their degree of tolerance is determined by foliage density according to the criteria mentioned. Column H contains simply constructed, monolayered hard strugglers completing their life cycles in the shade of the undergrowth near the forest floor.

Table 2.6 is a condensation of the preceding tables, and recapitulates the most important diagnostic characters for establishing tree temperaments by visual criteria. Observation of these criteria, which are physiognomical, architectural and morphological, may serve to establish a diagnosis of the status of a forested surface. The preponderance of hard gamblers and gamblers, for instance, indicates large-scale deforestation in the past, having created large young eco-units. On the other hand, the presence of many strugglers points to a forest having been subject to small-scale disturbances of low intensity (e.g. Pickett and White 1985), and such a forest mainly contains small, mature eco-units. In an inverse way, knowledge of tree temperaments is useful in the design of silvicultural systems for certain management aims.

If the diagnosis of tree temperament can be made in a juvenile stage, as has been advocated here, it may contribute to the ecological diagnosis of the eco-unit composition of a forest mosaic ("forest diagnosis"). It may also provide silvicultural insights for forest managers, in facilitating their intervention early in the development of managed stands, before developments have begun that conflict with the management aims.

THEORY AND PRACTICE: RECOMMENDATIONS FOR FURTHER RESEARCH

As shown above, in many silvicultural and ecological concepts and theories, two contrasting temperaments are described, sometimes with intermediates (Mayer 1980). For the tropics, we may list the following pairs although the list is certainly not exhaustive.

- heliophiles (light-demanding) versus sciadophiles (shade-bearers);
- pioneers versus primary species (cf. Whitmore this volume), which is
- analogous to nomads versus dryads (van Steenis 1958) or scar species versus forest species (Mangenot 1958);
- multilayers versus monolayers (Horn 1971);
- r-strategies versus K-strategies (e.g. see Hallé *et al.* 1978, for trees);
- weed trees versus long-lived trees (Corner 1952);
- gamblers versus strugglers (Bazzaz and Pickett 1980).

Table 2.5 Tree temperament classification

CROWN PHYSIOGNOMY (schematic)	(a) (b) A	(a) (b) B	C	D
Temperament	hard gambler	hard gambler	hard gambler	gambler
Leaf arrangement	mnolayer	spherical to hemispherical	phyllomorphic (s.l.)	multilayer
Successional status	early successional	early successional	early successional	later successional
Origin of eco-unit	large gap	large gap	large gap	medium-sized gap
Light requirement and shade tolerance	light-demanding shade-intolerant	light-demanding shade-intolerant	light-demanding shade-intolerant	light-demanding some tolerance
Common architectural models	Corner, Holttum (a) Rauh, Scarrone (b)	Corner (a) Leeuwenberg (b)	Troll	Rauh, Massart Roux, Attims
Branch architecture	(a) unbranched orthotropic leader, most often monopodial, few large leaves at the apex; (b) orthotropic leader with few orthotropic branches and large leaves	(a) single, unbranched orthotropic leader with large palmate leaves; (b) tridimensional sympodium with large, compound leaves densely packed on outer shoot ends	main axis mixed; proximal part erect, distal part horizontal with many densely foliated and strongly ramified branches; height growth by further erection of main axis	single orthotropic leader with orthotropic or plagiotropic branches or branch complexes
Examples	(a) first life stage of many pioneers, e.g. *Cecropia*, *Jacaranda*, *Didymopanax*; (b) *Musanga*, *Cecropia*, *Macaranga gigantea*	(a) 'fan palms'; (b) *Didymopanax morototoni*, *Jacaranda copaia*, *Toulicia guianensis*	*Annona paludosa*, *Apeiba burchellii* (latter sympodial)	Gymnosperms; *Trema*, *Macaranga hypoleuca*, *Goupia glabra*, *Ceiba pentandra*

(continued)

Table 2.5 continued

CROWN PHYSIOGNOMY (schematic)	E	F	G	H
Temperament	Struggling gamblers, gambling strugglers	struggling gamblers, gambling strugglers, strugglers	strugglers, hard strugglers	hard strugglers
Leaf arrangement	'umbrella'	'pagoda', paucilayer	elongate, 'diffuse'	monolayer
Successional status	later successional	later to late successional	late successional	late successional
Origin of eco-unit	intermediate/small gap	small gap	small gap or no gap	no gap
Light requirement and shade tolerance	shade-tolerant to intolerant favoured by light	shade tolerant to intolerant favoured by light	shade tolerant, favoured by moderate light levels	very shade tolerant, favoured by moderate light-levels
Common architectural models	Corner, Tomlinson	Aubréville, Massart	Troll, Roux and others	Chamberlain, Corner, Roux, Cook
Branch architecture	single, unbranched orthotropic leader with large pinnate compound leaves	single orthotropic leader with distinct tiers of plagiotropic branch complexes; shade tolerance determined by number of tiers, distance between tiers and density of foliation within one tier	great architectural flexibility, but crowns with diffuse branching , plagiotropic branches and sparse foliation	'poor' and simple crown architecture, size very small, short orthotropic leader with either few large leaves at the top, or few sparsely foliated (phyllomorphic) branches near the top
Examples	'feather palms'	species of *Terminalia*, *Manilkara*, *Pourouma*, *Iryanthera*, *Aspidosperma*	species of *Anaxagorea*, *Duguetia*, *Hirtella*, *Perebea*	species of *Pavonia*, *Guarea*, *Casearia*

Table 2.6 The important diagnostic characters for establishing tree temperaments

Temperament	*hard struggler*	*(hard) strugglers*	*gambling strugglers struggling gamblers*	*gamblers, hard gamblers*
Orientation	horizontal	horizontal to vertical	vertical	vertical, sometimes sympodia of horizontal elements
Branch distribution	unbranched or subapical	diffuse or subapical	distinct tiers	see Table 2.5
Common architectural models	Chamberlain, Corner, Roux	Troll, Roux	Aubréville, Massart	Rauh, Troll, Leeuwenberg (but see Table 2.5)
Height growth	monopodial or sympodial	sympodial or monopodial	mostly monopodial	monopodial or sympodial
Branch orientation	plagiotropic	plagiotropic	platiotropic or orthotropic	orthotropic or plagiotropic
Branching	diffuse, if any	diffuse	mostly intermittent (rhythmic)	diffuse or intermittent (rhythmic)
Higher-order branching	sparse	sparse	abundant	sparse to abundant (age)
Leaf arrangement	very little leaf surface in single apical layer	little leaf surface sparsely distributed through the crown	leaves more or less densely packed in several distinct layers	see Table 2.5
Leaf size	large if unbranched, small if branched	small	small	often large

Shugart (1984) required more refined temperaments for his Kiambram model in species-rich Australian forests, and therefore resorted to a matrix of four, the pairs being trees that cause or do not cause large gaps when falling and trees that require or do not require large gaps to germinate. He gave no criteria for the diagnosis of these temperaments.

Dualistic classifications are only elegant in appearance. On the one hand, they neglect site factors other than light (e.g. soil, water and biotic factors). On the other hand, they neglect the dynamics of all factors, including light, which cause many tree species to have to face several successive crown environments in growing up. However, it has to be agreed that light availability modulates the intrinsic photosynthetic performance, and so influences all metabolic processes in a tree. Hence light is indeed an essential factor, which justifiably is considered first and foremost. This is what has been done in Tables 2.5 and 2.6, in which all criteria refer directly or indirectly to photosynthesis, including reproduction.

Because adaptation to actual conditions and adjustment to changing conditions are often conflicting, many trees do one after the other instead of reaching a compromise. If changes in their environment are predictable, which in fact they are as far as they depend on eco-unit development (Fig. 2.1), flexible temperaments (Fig. 2.4) can also be predicted. This is most apparent in juveniles, because mature trees, in mature eco-units, face environmental change of a macroclimatic and therefore more stochastic than structural nature (e.g. Vooren and Offermans 1985).

In tropical rain forest mosaics, the diversity in size and architecture of eco-units is much greater than in temperate forests. This explains both the incompleteness of temperate-based theories like that of Horn (1971) in the tropics, and a higher flexibility in the temperament of many tree species than was previously recognized. For instance, species like *Musanga cecropioides* (Florence 1981, in Gabon) or *Cecropia* sp. pl. (Holthuijzen and Boerboom 1982, Suriname) were considered as completely light-demanding, until it was proven that they can germinate in conditions of between 40% and 100% of full sunlight. Using the criteria of Table 2.5, they can therefore be placed more precisely between the extremes of hard gamblers (pioneers) and hard strugglers (very shade-bearing species), on condition that enough information on their behaviour is available. This information should cover the whole life cycle of a tree species, in view of late adjustments in temperament (Fig. 2.4).

The intrinsic leaf properties remain a determinant of tree temperament which is difficult to assess visually. However, this does not render worthless the other criteria. It introduces a vagueness in classifications like the one in Table 2.5, which could be corrected by taking into account the silvicultural growth rates of the species examined, given that cambial activity is a parameter for the whole metabolic performance of a tree which is determined in the first place by photosynthesis (cf. Oldeman 1974*a*, 1978).

The growth rates of trees are most often a central criterion in silviculture, because forest management most often aims at wood production. Innovative operational systems of high quality, implemented by motivated people, are too often neglected as decisive factors in strategic planning of enterprises, as has been proven by a McKinsey and Cy team (Peters and Waterman 1982). It is our claim that one of the key factors for improving the silvicultural quality of forest management enterprises in the tropics is the knowledge of tree temperaments, because this is essential in promoting the growth of the trees to be harvested. The same is true in the case of other management aims requiring the growth of healthy trees, e.g. in protection forestry and nature conservation management.

Choosing tree species with the wrong temperament to implement a silvicultural system is an error in both ecological and economic terms. Keeping trees alive and growing in an environment to which they are not optimally adapted costs work and money, in the form of many silvicultural interventions. Generally, such increased cost is not warranted by increased income. This is an empirical truth, supported by the history of forest management.

Figure 2.9 illustrates the role of tree temperaments in different regeneration systems. Selective cutting (only a few trees being harvested per hectare), is contrasted to all-timber cutting (e.g. if all sizes over 25 cm diameter in all or most tree species are cut for pulp production). Selective cutting favours forest regeneration involving many strugglers and gambling strugglers (Table 2.5). All-timber cutting, leaving behind huge young eco-units, will favour hard gamblers, gamblers and to a lesser degree, struggling gamblers. In time, after tens of years, more struggling temperaments will become more frequent in the latter case, when the pioneer forests (gamblers) fall apart.

The temperaments present in populations of juvenile trees can be assessed by examining the criteria mentioned above. Lists of the precise temperament for each useful tree species do not exist as yet. These would be needed for the precise design of high performance silvicultural systems. Preliminary assessment of the temperaments among the juvenile trees at least is a method to estimate the growth potential of groups of wanted species present, even if they may hold surprises in later life (Fig. 2.4).

The smaller the young eco-unit (Fig. 2.1), the higher the proportion of strugglers among the juveniles; the larger the young eco-unit, the more frequent are hard gamblers and gamblers (cf. Fig. 2.9). This simple principle may serve to establish a list of silvicultural systems and required tree temperaments as shown in Table 2.7.

Later silvicultural operations, like tending and liberation thinning, may create the conditions needed to favour certain temperaments among the

Table 2.7 Tree temperaments best suited to silvicultural systems employing different sized regeneration units (eco-units)

Size of regeneration unit (= Eco-unit)	*Temperaments required (Fig 2.4; Table 2.5)*
Large eco-units	
Clear-cutting and plantation	Hard gamblers (strugglers, struggling gamblers may be underplanted)
Clear-cutting and conversion	Hard gamblers, strugglers, and some struggling gamblers for minor products
Clear-cutting and natural regeneration	Naturally prevailing are hard gamblers and gambling strugglers; strugglers and others come in later with succession
Intermediate-sized eco-units	
Strip-cutting and plantation	Depends on exposition and width of strip
Group-cutting and natural regeneration	Strugglers, gambling strugglers, struggling gamblers
Small eco-unit	
Selective cutting and natural regeneration	For recruitment, mainly hard strugglers, strugglers, gambling strugglers and struggling gamblers. For ingrowth of higher trees of the future, mainly strugglers and struggling gamblers

trees present. This implies a shift in the natural spectrum of temperaments, which should be compatible with both the vulnerability to abrupt increase of light in trees with more struggler-like temperaments (see above) and the further development of vital eco-units (cf. Figs. 2.1 and 2.9).

These silvicultural techniques can be applied in order to reach different management aims. One example of the correct use of tree species with different temperaments is illustrated in Figure 2.10. Figure 2.10*a* shows large-scale plantations of *Acacia mangium* (foreground, left) and *Eucalyptus* species in East Kalimantan (Indonesia) on sites that had been cleared by forest fires in 1982 and 1983. These trees are hard gamblers, with high wood production rates. The risks implied in creating such huge, homogeneous eco-units are determined by "hidden" heterogeneity, left over from the past history of the site, like surviving sources of pathogens and parasites or soil differences, and by genetic heterogeneity in the planted tree populations. Indeed, this leads to local tree mortality. Wherever this occurs, a smaller young eco-unit is formed (Fig. 2.1). Here, conditions would be propitious for species with other temperaments, such as Dipterocarpaceae (Fig. 2.10*b*).

Human-induced processes at Forest Mosaic level	A. SELECTIVE CUTTING (★= to be cut)	*	B. ALL TIMBER ●	*
Cutting (= to be cut)				
Consequences				
1. ARCHITECTURE	– changed eco-unit distribution	+	– some large eco-units	–
	– lateral gradients eco-units	++	– lateral gradients eco-units	±
	– form eco-units	+	– form eco-units	–
	– changed micro/mesoclimate	+	– change micro/macroclimate	–
2. ENERGY, NUTRIENTS, WATER	– soil compaction, permeability (localized)	±	– soil compaction permeability (all-over)	–
	– erosion	+	– erosion	–
	– leaching (localized)	+	– leaching (generalized)	–
	– energy and matter flows	+	– energy and matter flows	–
3. COMPONENT SPECIES	– richness flora and fauna	++	– richness flora and fauna	±
	– diversity changes	++	– diversity changes	–
	– participants in diversity (species)	+	– participants in diversity (species)	–
	– soil flora and fauna	+	– soil flora and fauna	–

Please note that diagnosis will be different according to site (LMU) and technology applied

* Reversibility of effects ++ very fast + rather fast ± within 1 tree-lifespan – longer than 1 tree-lifespan

Fig 2.9 Selection of tree temperaments in tropical rain forest under two wood harvesting regimes, i.e. selective cutting of two to five marketable timber trees per hectare (A) and cutting of all pulpable wood of most species over a diameter limit, e.g. 25 cm dbh (B). In the first case, rather small gaps give rise to small eco-units, favouring the germination and establishment of trees towards the struggler side of the spectrum of temperaments (Table 2.5). In the second case, large clearings are created, on which large eco-units are built by trees having temperaments towards the gambler side of the spectrum depicted in Table 2.5. Criteria for establishing a **forest diagnosis** before and after such interventions are discussed by Oldeman (1987), which is the source of this figure. The diagnosis given here is a rough approximation because present knowledge does not allow greater precision. "Tree life span", for instance, would average some 50 to 70 years. LMU = Land Management Unit, a concept from land evaluation

Fig. 2.10 Above – Extended plantation of *Acacia mangium* (foreground left) and *Eucalyptus* species on sites homogenized artificially after large-scale forest fires in East Kalimantan (Indonesia) in 1982 and 1983. Note standing burnt trees. The crown canopy pattern in the plantations, less than a year old, shows patches of different tree vitality among the hard gamblers planted here. Such patches become small gaps if these trees die, and give rise to small eco-units in which other temperaments would be favoured, (e.g., the struggling gamblers among the Dipterocarpaceae, if seedlings were present). Below – Such seedlings are underplanted by the Indonesian forest enterprises since the development of appropriate nursery techniques. These struggling gamblers are Dipterocarps and the juveniles display Massart's or Roux's model (shown here). They provide for sustainable wood production either after harvest of pioneers or their untimely death. Photographs Oldeman (1986)

Indeed, underplanting with struggling gamblers such as *Shorea* species (Dipterocarpaceae), grown and inoculated with the right mycorrhizae in nurseries (Smits *et al.* 1987), is now advocated as standard practice in the management of these forests (Soeryono 1987).

Knowledge of tree temperaments is thus considered indispensable in the design of silvicultural systems, because there are no standard recipes for such systems. The above example shows how problems have been addressed in one situation, but each site in each forest region – with its particular floristic composition – demands silvicultural systems which are tailor-made for that situation. The same is true for agro-forestry systems and their tree species.

However, those responsible for designing silvicultural systems in the humid tropics face the difficult problem of transforming a highly heterogeneous forest (cf. Mutoji-A-Kazadi 1977) into a forest that can be managed. Monocrop plantations tend to become more heterogeneous, as illustrated above. Moreover, they are problematic from the viewpoint of sustainability (Jordan, this volume). And polycyclic, multispecies systems (e.g. de Graaf 1986, Jonkers 1987) require the introduction of sufficient homogeneity to make them manageable, and like plantations have to be sustainable as to energy, water and nutrient cycling (e.g. Poels 1987).

Some rules-of-thumb can be put forward in this context.

- The diagnosis of the forest has to be made in terms of trees to be cut and trees to be left over, so as to keep the matrix of the forest (we would say: its eco-unit composition) sustainable and manageable.

- This diagnosis should concern not only marketable trees, but also other species and decaying trees, with the aim of conserving habitats for future wanted trees.

- After harvest according to such a preceding diagnosis, a selection should be made among the juvenile trees left behind, including all species that may conceivably become marketable in the future, and those juveniles should be favoured by silvicultural measures appropriate to their temperament.

- Considerable reduction of the number of species is possible, but is often difficult to handle.

- Forest treatments always, and automatically, select tree species according to their temperaments. With a wrong diagnosis, treatments will favour unwanted trees.

These "rules" can be easily modified in order to fit in with management objectives other than that of wood production. As a matter of fact, such rules tend to optimize the harmonization of trees with a certain temperament and their forest environment. We have seen, in the above paragraphs, that this question is more complicated than it seems at first sight, particularly in the humid tropics. Temperaments are determined by the tree's organs, its organ-bearing shoots, and its overall architecture, and by the interactions between these levels of organization. The immediate environment of a tree, the eco-unit which it inhabits, is determined by the surrounding forest and by abiotic factors, as well as by the interactions between these levels.

The concept of tree temperament is, therefore, crucial in forest management, and the following recommendations are proposed for promoting its further study.

(1) Support or establishment of study groups on life cycles and temperaments of forest trees belonging to the main families in the rain forests of different continents, e.g. *Leguminosae*, *Dipterocarpaceae*, *Lauraceae*, *Vochysiaceae*, *Meliaceae*, *Chrysobalanaceae* and others.

(2) Support or establishment of study groups on life cycles and temperaments of particular forest tree species, determining the rain forest matrix over long time spans, and selected by ecological and economic criteria.

(3) Stimulation of studies on neglected life phases of such trees, particularly the juvenile phase.

(4) Inclusion in tree temperament studies of associated organisms, e.g., mycorrhizal fungi, other microflora and microfauna (Jordan, this volume), epiphytes, lianes, herbs, larger animals.

(5) Inclusion of tree temperament studies in many other rain forest projects, by asking the researchers to unite data files on certain tree species in a standardized and low-cost way, so that a rich data stream may contribute to the building-up of central files.

(6) Creation of a tree temperament periodical, as a forum for publishing new studies and updating the knowledge on many tree species in a regular way, in a form comprehensible to forest managers.

(7) Accumulation of experimental data on tree temperaments (cf. Bongers and Pompa 1988), by growing tropical rain forest patches

(eco-units) *in vivo* or *in vitro*, under controlled conditions and circumstances.

(8) Preparation and formulation of simulation models for the growth of trees with different temperaments, the outputs of which should be both numerical and graphical.

(9) Identification and collation of urgent questions to be answered by researchers, by screening many forest management plans in the humid tropics, with the help of the managers.

(10) Raising of awareness among tropical forest ecologists and silviculturists that tree temperaments are both essential to the understanding of rain forest dynamics and much more complex than is suggested by previously published simple classifications.

REFERENCES

Ashton, P.S. (1978). Crown characteristics of tropical trees. In Tomlinson, P.B. and Zimmermann, M.H. (eds.) *Tropical Trees as Living Systems*, pp. 591–615. (Cambridge University Press: Cambridge)

Bazzaz, F.A. (1984). Dynamics of wet tropical forests and their species strategies. In Medina, E., Mooney, H.A. and Vázques-Yanes, C. (eds.) *Physiological Ecology of Plants of the Wet Tropics*, pp. 233–43. (Junk: The Hague)

Bazzaz, F.A. and Pickett, S.T.A. (1980). Physiological ecology of tropical succession: A comparative review. *Annual Review of Ecology and Systematics*, **11**, 287–310

Begon, M., Harper, J.L. and Townsend, C.R. (1986). *Ecology: Individuals, Populations and Communities*. (Blackwell: Oxford)

Bongers, F. and Popma, J. (1988) *Trees and Gaps in a Mexican Tropical Rain Forest*. Diss. Dr., Rijksuniversiteit, Utrecht.

Brokaw, N.V.L. (1985). Gap-phase regeneration in a tropical forest. *Ecology*, **66**, 682–87

Bruenig, E.F. (1983). Vegetation structure and growth. In Golley, F.B. (ed.) *Tropical Rain Forest Ecosystems*, pp. 49–75. Ecosystems of the World 14A. (Elsevier: Amsterdam)

Budowski, G. (1965). Forest species in successional process. *Turrialba*, **15**, 1, 40–2

Carabias-Lillo, J. and Guevara-Sada, S. (1985). Fenologia en una selva tropical húmeda y en una comunidad derivdada: Los Tuxtlas, Veracruz. In Gómez-Pompa, A. and Amo, R.S. del (eds.) *Investigaciones sobre la regeneración de selvas altas en Veracruz, México*, pp. 27–66. (INIREB: Xalapa)

Chiariello, N. (1984). Leaf energy balance in the wet lowland tropics. In Medina, E., Mooney, H.A. and Vázques-Yanes, C. (eds.) *Physiological Ecology of Plants of the Wet Tropics*, pp. 85–97. (Junk: The Hague)

Cochet, P. (1959). *Etude et culture de la forêt*. (Ecole Nationale du Génie Rural, des Eaux et des Forêts (ENGREF): Nancy)

Coombe, D.E. and Hadfield, W. (1962). An analysis of the growth of *Musanga cecropiodes*. *Journal of Ecology*, **50**, 221–34

Corner, E.J.H. (1952). *Wayside Trees of Malaya*, Vol. I. Second Edition. (The Govt. Printing Office: Singapore)

Denslow, J.S. (1980). Gap partitioning among tropical rain forest trees. *Biotropica*, **12,** (2, supp.), 47–55

Dransfield, J. (1978). Growth forms of rain forest palms. In Tomlinson, P.B. and Zimmermann, M.H. (eds.) *Tropical Trees as Living Systems*, pp. 247–68. (Cambridge University Press,: Cambridge)

Edelin, C. (1977). *Images de l'architecture des Conifères.* Thèse 3ème cycle. (Université des Sciences et Techniques du Languedoc (USTL): Montpellier)

Edelin, C. (1984). *L'architecture monopodiale: l'exemple de quelques arbres d'Asie tropicale.* Thèse Doct. Etat. (Université des Sciences et Techniques du Languedoc (USTL): Montpellier)

Fisher, J.B. (1978). A quantitative study of *Terminalia* branching. In Tomlinson, P.B. and Zimmermann, M.H. (eds.) *Tropical Trees as Living Systems,* pp. 285–320. (Cambridge University Press: Cambridge)

Fisher, J.B. (1984). Tree architecture: relationships between structure and function. In White, R.A. and Dickinson, W.C. (eds.) *Contemporary Problems in Plant Anatomy,* pp. 541–89. (Academic Press: Orlando, Fl)

Florence, J. (1981). *Chablis et sylvigénèse dans une forêt dense humide sempervirente du Gabon.* Thèse 3ème cycle. (Université Louis Pasteur: Strasbourg)

Fournier, A. (1979). *Is Architectural Radiation Adaptive?* (DEA Botanique Tropicale. Université des Sciences et Techniques du Languedoc (USTL): Montpellier)

Givnish, T.J. (1978). On the adaptive significance of compound leaves, with particular reference to tropical trees. In Tomlinson, P.B. and Zimmermann, M.H. (eds.) *Tropical Trees as Living Systems,* pp. 351–80. (Cambridge University Press, Cambridge)

Givnish, T.J. (1984). Leaf and canopy adaptations in tropical forests. In Medina, E., Mooney, H.A. and Vázques-Yanes, C. (eds.) *Physiological Ecology of Plants of the Wet Tropics,* pp. 51–83. (Junk: The Hague)

Givnish, T.J. and Vermeij, G.J. (1976). Sizes and shapes of liane leaves. *American Naturalist,* **110**, 743–78

Gómez-Pompa, A. and Amo R.S. del (eds.) (1985). *Investigaciones sobre la regeneración de selvas altas en Veracruz, México.* Volumen II. (INIREB: Xalapa)

Gómez-Pompa, A. and Vázquez-Yanes, C. (1985). Estudios sobre la regeneración de selvas en regiones cálido-húmedas de México. In Gómez-Pompa, A. and Amo R.S. del (eds.) *Investigaciones sobre la regeneración de selvas altas en Veracruz, México*, pp. 1–25. (INIREB: Xalapa)

Graaf, N.R. de. (1986). *A Silvicultural System for Natural Regeneration of Tropical Rainforest in Suriname.* (Pudoc, Agricultural University: Wageningen)

Granville, J.J. de (1978). *Recherches sur la flore et la végétation guyanaises. Thèse ès Sciences.* (Université des Sciences et Techniques du Languedoc (USTL): Montpellier)

Granville, J.J. de (1984). Monocotyledons and Pteridophytes indicators of environmental constraints in the tropical vegetation. *Candollea,* **39**, 265–9

Hallé, F. (1978). Architectural variation at the specific level in tropical trees. In Tomlinson, P.B. and Zimmermann, M.H. (eds.) *Tropical Trees as Living Systems,* pp. 209–21. (Cambridge University Press: Cambridge)

Hallé, F. and Oldeman, R.A.A. (1970). *Essai sur l'architecture et la dynamique de croissance des arbres tropicaux.* (Masson: Paris)

Hallé, F., Oldemann, R.A.A. and Tomlinson, P.B. (1978). *Tropical Trees and Forests; an Architectural Analysis.* (Springer: Heidelberg)

Hartshorn, G.S. (1978). Tree falls and tropical forest dynamics. In Tomlinson, P.B. and Zimmermann, M.H. (eds.) *Tropical Trees as Living Systems,* pp. 617–38. (Cambridge University Press: Cambridge)

Hartshorn, G.S. (1980). Neotropical forest dynamics. *Biotropica,* **12** (2, supp.), 23–30

Hladik, A. (1978). Phenology of leaf production in Gabon: distribution and composition of food for folivores. In Montgomery, G.G. (ed.) *The Ecology of Arboreal Folivores,* pp. 51–71. (Smithsonian Institution: Washington, D.C.)

Holthuijzen, A.M.A. and Boerboom, J.H.A.. (1982). The *Cecropia* seedbank in the Suriname lowland rain forest. *Biotropica,* **14**(1), 62–8

Horn, H.S. (1971). *The Adaptive Geometry of Trees.* (Princeton University Press: Princeton)

Jeník, J. (1978). Roots and root systems in tropical trees: morphologic and ecologic aspects. In Tomlinson, P.B. and Zimmermann, M.H. (eds.) *Tropical Trees as Living Systems*, 323–49. (Cambridge University Press: Cambridge)

Jonkers, W.B.J. (1987). *Vegetation Structure, Logging Damage and Silviculture in a Tropical Rain Forest in Suriname.* (Pudoc, Agricultural University: Wageningen)

Kahn, F. (1975). *Remarques Sur l'Architecture Végétative Dans ses Rapports Avec la Systématique et la Biogéographie.* (DEA Botanique Tropicale. Université des Sciences et

Techniques du Languedoc (USTL): Montpellier)

Kahn, F. (1982). *La Réconstitution de la Forêt Tropicale Humide: Sud-Ouest de la Côte d'Ivoire.* Mémoires ORSTOM 97. (ORSTOM: Paris)

Kahn, F. (1986). Life forms of Amazonian palms in relation to forest structure and dynamics. *Biotropica,* **18**, 3, 214–18

Kessell, S.R. (1979). *Gradient Modelling: Resource and Fire Management. (*Springer: Heidelberg)

Lanier, L. (1986). *Précis de sylviculture.* (Ecole Nationale du Génie Rural, des Eaux et des Forêts (ENGREF): Nancy)

Lescure, J.-P. (1986). *La reconstitution du couvert végétal après agriculture sur brûlis chez les Wayapi du Haut-Oyapock (Guyane française).* These Doct. Etat. (Université de Paris VI: Paris)

Lipman, E. (1981). *Contribution à l'étude des rythmes et des unités de croissance chez les arbres tropicaux.* (DEA Ecologie Tropicale. Université des Sciences et Techniques du Lànguedoc (USTL): Montpellier)

Mangenot, G. (1958). Les recherches sur la végétation dans les régions tropicales humides de l'Afrique Occidentale. In Unesco (ed.) *Study of Tropical Vegetation. Proceedings of the Kandy Symposium/L'étude de la végétation tropicale. Actes du Colloque de Kandy,* pp. 115–26. (Unesco: Paris)

Mayer, H. (1980). Waldbau. Second Edition. G. Fischer (Verlag: Stuttgart)

Medina, E. (1983). Adaptations of tropical trees to moisture stress. In Golley, F.B. (ed.) *Tropical Rain Forest Ecosystems,* pp. 225–37. Ecosystems of the World 14A. (Elsevier: Amsterdam)

Mooney, H.A., Field, C. and Vázquez-Yanes, C. (1984). Photosynthetic characteristics of wet tropical forest plants. In Medina, E., Mooney, H.A. and Vázques-Yanes, C. (eds.) *Physiological Ecology of Plants of the Wet Tropics,* pp. 113–28. (Junk: The Hague)

Mutoji-A-Kazadi (1977). *Notes de sylvigénèse pour la Guyane: transect et photographies aériennes.* (DEA Botanique Tropicale. Université des Sciences et Techniques du Languedoc (USTL): Montpellier)

Oldeman, R.A.A. (1974*a*). *L'architecture de la Foret Guyanaise.* Mémoires ORSTOM 73. (ORSTOM: Paris)

Oldeman, R.A.A. (1974*b*). Ecotopes des arbres et gradients écologiques en forêt guyanaise. *Revue d'Ecologie (Terre et Vie),* **28**, 487–520

Oldeman, R.A.A. (1978). Architecture and energy exchange of dicotyledonous trees in the forest. In Tomlinson, P.B. and Zimmermann, M.H. (eds.) *Tropical Trees as Living Systems,* pp. 535–60. (Cambridge University Press: Cambridge)

Oldeman, R.A.A. (1979). Quelques aspects quantifiables de l'arborigénèse et de la sylvigénèse. *Oecologia Plantarum,* **14**, 289–312

Oldeman, R.A.A. (1983). Tropical rain forest, architecture, silvigenesis and diversity. In Sutton, S.L., Whitmore, T.C. and Chadwick, A.C. (eds.) *Tropical Rain Forest: Ecology and Management,* pp. 139–50. (Blackwell: Oxford)

Oldeman, R.A.A. (1987). Tropical forest: the ecosystems. In Beusekom, C.F.van, Goor, C.P. van and Schmidt, P. (eds.) *Wise Utilization of Tropical Rain Forest Lands,* pp. 46–67. Tropenbos Scientific Series 1. (Tropenbos: Ede)

Oldeman, R.A.A. (1989). Biological implications of Leguminous tree architecture. In Stirton, C.H. and Zarucchi, J.L. (eds.) *Advances in Legume Biology. Monographs in Systematic Botany of Missouri Botanical Garden*, **29**, 17–34

Oldeman, R.A.A. (1990). *Elements of Silvology.* (Springer Verlag: Heidelberg) (in press).

Peters, T.J. and Waterman, R.H. (1982). *In Search of Excellence: Lessons from America's Best-run Companies.* (Warner Books, Inc: New York)

Pickett, S.T.A. (1983). Differential adaptation of tropical tree species to canopy gaps and its role in community dynamics. *Tropical Ecology,* **24**, 68–84

Pickett, S.T.A. and White, P.S. (1985). *The Ecology of Natural Disturbance and Patch Dynamics.* (Academic Press: New York)

Poels, R.L.H. (1987). *Soils, Water and Nutrients in a Forest Ecosystem in Suriname.* (Pudoc, Agricultural University: Wageningen)

Poncy, O. (1985). Le genre *Inga* (Légumineuses-Mimosoideae) en Guyane Française: systématique, morphologie des formes juvéniles, écologie. *Mémoires du Muséum National d'Histoire Naturelle, (Paris)*, NS, sér. B, **31**, 1–124

Queensland Department of Forestry (1983). *Rain Forest Research in North Queensland.* (Queensland Department of Forestry: Brisbane)

Schermbeek, A.J. van. (1898). *Het bosch* (Dutch transl. of Gayer, K. "Der Waldbau"). (Van Wees Publishers: Breda)

Schimper, A.F.W. (1903). *Plant Geography Upon a Physiological Basis.* (Engl. transl. by W.R. Fischer). (Clarendon Press: Oxford)

Shugart, H.H. (1984). *A Theory of Forest Dynamics. The Ecological Implications of Forest Succession models.* (Springer Verlag: Heidelberg)

Smits, W.T.M., Oldeman, R.A.A. and Limonard, T. (1987). Mycorrhizae and Dipterocarpaceae in East Kalimantan rain forests. *WOTRO Yearbook 1986,* pp. 67–78. (WOTRO: The Hague)

Soeryono, R. (1987). Constraints on the use of tropical rain forest species in silviculture on timber estates. *Duta Rimba,* **83–84/13**, 7–10

Steenis, C.G.G.J. van (1958). Rejuvenation as a factor for judging the status of vegetation types: the biological nomad theory. In Unesco (ed.) *Study of Tropical Vegetation. Proceedings of the Kandy Symposium/L'étude de la végétation tropicale. Actes du Colloque de Kandy*, pp. 212–18. (Unesco: Paris)

Unesco. (1978). *Tropical Forests Ecosystems:* A state-of-knowledge report prepared by Unesco, UNEP and FAO. Natural Resources Research Series 14. (Unesco: Paris)

Vázquez-Yanes, C. and Orozco-Segovia, A.A. (1985). Posibles efectos del microclima de los claros de la selva sobre la germinación de tres especies de árboles pioneros: *Cecropia obtusifolia, Heliocarpus donnell-smithii* y *Piper auritum.* In Gómez-Pompa, A. and Amo, R.S. del (eds.) *Investigaciones Sobre la Regeneración de Selvas Altas en Veracruz, México.* Volumen II: pp. 241–53. (INIREB: Xalapa)

Vooren, A.P. (1986). Nature and origin of tree and branch fall in the Taï Forest (Ivory Coast). *Netherlands Journal of Agricultural Science,* **34**, 112–15

Vooren, A.P. and Offermans, D.M.J. (1985). An ultralight aircraft for low-cost, large scale stereoscopic aerial photographs. *Biotropica,* **17**, 84–8

Whitmore, T.C. (1975). *Tropical Rain Forests of the Far East.* (Clarendon Press: Oxford)

Whitmore. T.C. (1978). Gaps in the forest canopy. In Tomlinson, P.B. and Zimmermann, M.H. (eds.) *Tropical Trees as Living Systems,* pp. 639–55. (Cambridge University Press: Cambridge)

CHAPTER 3

TROPICAL RAIN FOREST DYNAMICS AND ITS IMPLICATIONS FOR MANAGEMENT

T.C. Whitmore

DYNAMICS AT DIFFERENT SCALES OF TIME AND SPACE

Forests are in a continual state of flux, changing all the time and on different spatial scales. We may distinguish changes at several temporal scales which affect tropical rain forests.

Secular changes

The climate of the humid tropics has fluctuated throughout the Quaternary (the last two million years) and probably the Tertiary too, between hot, humid periods and cooler, drier and more seasonal periods (Flenley 1979, Morley and Flenley 1987). These fluctuations coincide with high latitude Interglacials, one of which we are now experiencing, and Glacials. The change from the last Glacial began about 4000 years ago. There were numerous earlier ones. Today's conditions are exceptionally warm and humid and have been of much shorter total duration than drier cooler and more seasonal climates.

Various lines of evidence show that tropical rain forests have expanded and contracted as climate has fluctuated. Today, their natural extent is believed to be at or near the maximum achieved during the Quaternary. In drier times, tropical rain forests of both America (Prance 1983, Whitmore and Prance 1987) and Africa (Hamilton 1976) were of much restricted extent, set as 'islands' in an extensive sea of tropical seasonal (monsoon) forests. Recently, evidence has been found that the Malesian rain forests were also reduced in area (Morley and Flenley 1987).

These secular changes in climate and vegetation have left their imprint on the landscape and on soils. They are also detectable in the geographical ranges of occurrence of many plant and animal species, patterns whose understanding provides important underpinning for the intelligent design of areas for conservation (e.g. Wetterburg and Jorge Padua 1978), but will not be considered further here.

The forest growth cycle

At the other end of the temporal scale, lies dynamic change due to the growth and death of the trees of the forest. The forest canopy is a mosaic of gaps, patches of juvenile trees growing up in former gaps, and mature forest (Watt 1947, Whitmore 1975, 1978, Oldeman 1983, Fig. 1). We may recognize a forest growth cycle of gap, building, and mature phases. What grows up in a canopy gap determines the composition of the forest for a long time, usually at least decades and sometimes centuries. Hence, gaps are in some respects the most important part of the growth cycle. Forest regeneration has come to be called gap-phase dynamics, and is currently the focus of numerous studies. 'Gap' is a general term. Usually, it is taken to mean a space in the above-ground canopy down to, or nearly to, ground-level.

Regeneration defined

It is at the temporal and spatial scale of the forest growth cycle that regeneration occurs. The term has two distinct meanings. First, we may recognize the restoration of biomass and nutrients in a forest gap as the forest canopy builds to maturity. Second, regeneration can refer to the reassembly of floristic and structural diversity back to a self-perpetuating climax state.

Regeneration on small and large surfaces

We may crudely draw distinctions between forest regeneration on small and large surfaces, in small and large gaps respectively.

In a canopy gap created by the death of one or a few trees, seedlings already present in the undergrowth are 'released' and grow up into building-phase forest. In a big gap, by contrast, viz. a gap created by multiple windthrow, landslide, vulcanism, fire (very rare in tropical rain forest), or, at the extreme, by a cyclone (hurricane, tornado), pre-existing seedlings die. There is a major shift in microclimate near and below the ground, and this may sometimes cause death (though we lack hard evidence), otherwise caused by physical disruption. Whatever the cause of death, the gap is filled by new seedlings which were not present below the previous canopy.

Pioneer and climax species

Thus, we have two sorts of tree species, those with shade-bearing (or shade-tolerant) seedlings and those with light-demanding (or shade-intolerant) seedlings (Swaine and Whitmore 1988). The latter cannot regenerate under any shade, including their own. These two species classes are often also known as climax (or primary) and pioneer species respectively, referring to their abilities to perpetuate *in situ* or not. Many other properties are linked to these seedling characteristics, to form two contrasting syndromes (see Table 3.1).

Table 3.1 Character syndromes of the two contrasting ecological classes of tree species

Synonyms	(shade) intolerants, light-demanders, nomads, pioneers, secondary spp.	(shade) tolerants, shade-bearers, dryads, climax spp., primary spp.
Seed	Copious, small, produced continually or continuously annually	less copious, large, produced annually or less than
Dispersal	wind or animals, for considerable distances	diverse, including gravity, sometimes only local
Dormancy	often (?always) present; never (?) recalcitrant	often absent; often recalcitrant
Soil seed-bank[1]	present	absent (? always)
Height growth	fast	slower
Wood	usually pale, low density, not siliceous	often dark, high density, sometimes siliceous
Growth[2]	indeterminate, no resting buds (*viz* sylleptic)	determinate, with resting buds (*viz* proleptic)
Forking[3]	high	low
Leaves[2]	short-lived, one generation present, *viz* high turn-over	long-lived, several generations present, *viz* slow turn-over
Roots[4]	superficial	some deep
Root/shoot ratio[2,4]	low	high
Photosynthesis rate[5]	high	low
Toxic chemicals[6]	low	high
Leaf susceptability to predation[7]	high	low
Geographical range	wide	often narrow
Phenotypeplasticity[8]	high	low

1. Whitmore (1983): 2. Boojh and Ramakrishnan (1982); 3. Whitney (1976); 4. Shukla and Ramakrishnan (1984); 5. Koyama (1978), Oberbauer and Strain (1984); 6. Speculative, evidence slight and equivocal; 7. Coley (1983); 8. Baker (1965)

Succession

If the return-time of large-gap creation is longer than the lifespan of the pioneer trees, then, as these die, small gaps will form in the forest canopy and climax species will invade in a second growth cycle. This process is succession, which is defined as a directional shift with time in floristic composition. It is the second kind of regeneration as defined above.

The pioneer species may be short- or long-lived, and these grow to be small and large trees respectively. In the Eastern tropical rain forests, for example, a *Macaranga* canopy begins to disintegrate at about 20 years age (Kochummen 1966), whereas, on the slopes of Mt Victory in Papua, *Paraserianthes falcataria* and *Octomeles sumatrana* were only beginning to die at 84 years (Taylor 1957), a condition comparable to that in the commercially valuable cyclone-climax forests of the Solomon archipelago dominated by *Campnosperma brevipetiolatum*.

The forest which first colonizes big gaps is known as secondary, and, throughout the tropics, it consists of stands of only one or a few species per hectare, in contrast to most climax forests which have numerous species. Moreover, the total pioneer tree flora is small everywhere, with some of its species very widespread (e.g. *Schefflera (Didymopanax) morototoni* from Brazil to Mexico, *Ochroma lagopus* throughout western tropical America). There are fewer pioneers in America and Africa than the East, where *Macaranga* alone has over 150 pioneer species. The lack of speciation in pioneers relative to climax species is likely to be because their repeated, wide dispersal at every generation prevents evolution of local genetic isolates and eventual speciation. Thus, it is probably no coincidence that the insular East has more pioneers than the continental areas of the rain forest biome, because, there, the islands do create genetic isolation.

The frequency of forest disturbance may also be important. In north-east India, for example, exotics replace native species in secondary succession when disturbance is frequent (P.S.Ramakrishnan *pers. comm.*).

Climax forest

Eventually, probably usually after several centuries, there will no longer be a directional change in floristics; composition at a given spot will change from one tree generation to the next, but, overall, a steady-state will have been achieved. This is the climax forest in which cyclic replacement occurs. In some forests, this condition may never be reached because the return-time of catastrophes is too short.

Rates of turnover

Tropical rain forests develop from a species-poor condition comprising mixtures of one or a few, well dispersed, light-demanding, pioneer tree species existing as a coarse structural mosaic, which progressively breaks up as these trees die, to climax forests of more numerous shade-tolerant, climax species, occurring in a finer structural mosaic. These latter forests are the focus for conservation.

There are now a few plots in rain forest not recently disturbed by man which have been repeatedly observed over several years. The turnover rate of trees is considerable, even in forests believed to be primary: an average figure of one percent of trees dying annually is common (see collations by Whitmore 1975, Lieberman *et al.* 1985, Swaine *et al.* 1987). A Nigerian forest, believed typical of many in West Africa, had not reached steady-state after about 250 years (Jones 1955–56).

Cataclysmic destruction

Where the return time of cataclysmic destruction of the forest and creation of big gaps is frequent, succession to a steady state may never be completed. We have become aware that cataclysms of several kinds, and with a return time of a few centuries, occur in many parts of the humid tropics.

The parts of the tropics lying between 10–20 degrees north and south are prone to cyclones, and forests regrowing after wind destruction have been described from many places. Both foresters planning plantations (e.g. pines in Fiji) and ecologists studying ecosystem processes (e.g. the various long-term investigations in Caribbean island forests) cannot safely disregard cyclones as powerful factors.

Outside the cyclone belts, massive wind destruction is rarer. In Malaya, observations over 12 years in a purportedly primeval dipterocarp lowland rain forest at Sungai Menyala, which was typical of an extensive 'red meranti/keruing' (*Shorea-Dipterocarpus*) forest type, showed that the canopy-top species (including certain dipterocarps) were not regenerating *in situ*. The under-canopy contained species known to be more shade-tolerant (Wyatt-Smith 1966, Whitmore 1975). There are records in Malaya as a whole, of 3–4 windstorms in the past century, each destroying a swathe of forest several kilometres wide and long. It seems likely such storms create sufficient canopy destruction to favour the light-demanding dipterocarps which at present form the Sungei Menyala canopy top. Their return-time is sufficiently frequent for big tracts of red meranti/keruing forest to exist.

Climax tropical rain forest is believed to be fire-resistant (except for the specialized forest formation of limestone hills). New discoveries cast doubt on this belief. Charcoal has recently been found over an extensive area below *terra firme* rain forest in south Venezuela, showing that repeated fires have occurred during the past six millennia. Man is believed to have been present for less than four millennia, and forest destruction by wildfire is deduced to have occurred during drier epochs than at present (of which four have been recognized) (Sanford *et al.* 1985).

In fact, many ecologists have found charcoal in the soil under tropical rain forest, but it had not previously been systematically studied or dated, or its possible significance realized: the idea that primary rain forest does not burn is very firmly entrenched in scientists' minds.

A second example of fire in climax rain forest comes from eastern Borneo. Between about July 1982 and April 1983, this region received only 32 per cent of the usual average rainfall (Lennertz and Panzer 1983, Malingreau *et al.* 1985). Following this exceptional drought, great fires broke out. From aerial reconnaissance, these fires were at first believed to have destroyed 3.5 million hectares of lowland tropical rain forest (an area the size of Holland). The drought also strongly affected Sabah and was similarly followed by fire (Beaman *et al.* 1985, Woods 1987). Later, careful ground survey in Kalimantan revealed big areas in which the canopy-top trees had been killed by drought, with no sign of fire having occurred. There were extensive fires, but these destroyed mainly forest which had recently been logged, and then spread out from roads and cultivated lands only found in the region since commercial logging began in the previous decade (Wirawan 1984). Searches of the climatic records and the recollection of folk memory suggest that similar droughts have occurred several times since the mid nineteenth century. In Sabah, Woods (1987) found 'up to six' previous droughts of similar severity to that of 1982–83 in the 57 years since records began. So, in this case, cataclysmic death after drought is likely to have a return time considerably less than the recovery-time of climax species-stable forest. The 1982–83 drought, latest in a series, also led to extensive destruction by fire, because man was present on that occasion.

Widespread fires also occurred in 1983 in the West African forests. As in Borneo, an unusual climatic perturbation led to unusual drought. Man as agriculturist may have, in the past, caused cataclysmic destruction in both West Africa (the Nigerian forests studied by Jones 1955–56, see above) and Amazonia, by forest-felling and cultivation of the land. In Amazonia, there are now far fewer Amerindians than at the time of European contact, and some of the former occupants were probably shifting cultivators. We still do not know to what extent various structural and floristic facies of the forests reflect this past. Speculations include the big areas dominated by Babassu palm (*Orbignya martiana*). It is known that areas of black soil (*terra preta*) all over Amazonia, which reach a few hectares in extent (and are now sought out for agriculture because of their fertility), are of anthropogenic origin. They are commonly at strategic places and testify to prolonged human occupation and cultivation of the site. It appears that no systematic study has yet been made of the forest over *terra preta* (H.Klinge *pers. comm.*). In the Eastern rain forests, areas of former cultivation may be detected, perhaps for centuries, by their richness in fruit trees (see Ellen 1985 concerning Seram Island), and ritual plants, for example *Areca guppyi* on Malaita in the Solomons, outside its native range.

A third kind of cataclysmic destruction of rain forest has recently been discovered in Peru (Salo *et al.* 1986, Salo and Kalliola this volume). Here, many tributaries, swiftly flowing, heavily loaded with sediments, and with big annual floods, continually alter course by moving laterally several metres a year (the maximum is 170 m). Due to movement of the rivers, little if any of the floodplain is over a century old and the forest consists entirely of belts of primary successional forest communities. Something similar occurs on a smaller scale on the south side of the main cordillera of Papua New Guinea (R.J. Johns 1986).

Conclusion

Primeval tropical rain forest, undisturbed and stable 'since the dawn of time' is a myth. Instability of varying extents occurs on several time-scales. The recovery to a steady-state is likely to take several centuries and is perhaps never achieved in many places. Boreal forests in the oceanic climates of westernmost Europe, Alaska and British Columbia degenerate to a 'post-climax' peat bog bearing heath plants (Spurr and Barnes 1980) instead of forest unless fire intervenes to set the succession back. The only similar succession to non-forest I know in the tropics is 'phasic community 6' in the Sarawak peat swamp forests, which is the last phase and is achieved only on the dome centre of the most highly developed raised bogs. Are we to attribute the lack of such post-climax low stature vegetation in the humid tropics to continual rejuvenation by either cataclysm or by continuing response to secular change? Or have we simply not yet recognized its existence? What is the status of open woodland found mingled with some heath forests in Asia and Amazonia? Is this perhaps either anthropogenic in origin or 'post-climax' vegetation?

GAP-PHASE DYNAMICS

A spectrum of species' gap requirement

Study of the autecology of the tree species in any tropical rain forest quickly reveals that it is too simplistic to recognize just pioneer and climax species.

Table 3.2 Regeneration behaviour of the common big tree species, Kolombangara, Solomon Islands (based on Whitmore 1974, Table 7.6)

Species	*Conditions for seedling establishment*	*Conditions for seedlings to grow up*
Dillenia salomonensis	high forest	high forest
Maranthes corymbosa	high forest	high forest
Parinari salomenensis	high forest	high forest
Schizomeria serrata	high forest	high forest
Calophyllum kajewskii	high forest/small gaps	high forest/small gaps
Calophyllum vitiense	high forest	high forest/gaps
Pometia pinnata	high/disturbed forest	high forest/?small gaps
Elaeocarpus sphaericus	high forest	gaps
Campnosperma brevipetiolatum	high forest/gaps	gaps
Terminalia calamansanai	high forest/gaps	gaps
Endospermum medullosum	mostly gaps	gaps
Gmelina moluccana	mostly gaps	gaps

The forests of Kolombangara Island in the Solomon archipelago have only 12 common canopy-top tree species. Yet, even in such a species-poor forest, four classes can be identified with an increasing requirement for canopy gaps for the two stages of seedling establishment and release (see Table 3.2).

On Barro Colorado Island, Panama, fine discrimination was found in an 8-year study between three pioneer species studied in the range of gap sizes colonized and the year of colonization (Brokaw 1987).

It has been shown recently for a forest in Costa Rica that different species succeed at different parts of small, single tree gaps, namely at the butt (often with upturned soil), where the bole fell, or where the crown fell (Brandani *et al.* 1988). Success of different species on different microsites in a big gap after burning was shown by Uhl *et al.* (1981). By experiment, they created six kinds of surface: bare soil, soil with charcoal, soil with root mat, and these three surfaces plus slash. Then, seeds were planted on these substrata. It is suspected that establishment of some species in tree-fall gaps is better on upturned soil, and this advantage has been demonstrated for *Cecropia obtusa* in French Guyana (Riera 1985). Traditional farmers know well that there are differences between the woody species of secondary forest from big gaps after various degrees of 'soil exhaustion' by agricultural crops. In the Corcovado forest of Costa Rica, different pioneers colonized natural big gaps and abandoned arable land (Herwitz 1981).

Exotic pioneers dominate early succession in some rain forests. For example, *Lantana camara* is such a serious weed in Australia, where it arrests succession, that insect pests have been introduced from its native America to control it. *Piper aduncum*, a treelet from South America, forms pure stands of secondary rain forest in many places in eastern Malesia. *Leucaena leucocephala* forms dense roadside stands over coralline soils in parts of Vanuatu. We do not know what has enabled these weeds to oust indigenous pioneers, nor whether they permanently alter succession. *Cecropia* (from South America) has escaped from plantation trials in a few places in Malesia, and *Paraserianthes falcataria* (from Papuasia) has become naturalized in Johor, Malaysia. Neither species is as abundant as the indigenous pioneers with which it grows mixed. No studies have yet been conducted on the ecological interactions of indigenous and exotic pioneers.

Relay floristics and simultaneous colonization

Secondary succession in plant communities is not so constant and predictable as was thought by early ecologists. Finegan (1984) has given a useful review of forest successions. Two models can be recognized. Sometimes, a set of invading pioneers is later replaced by shade-tolerant climax species. This is the model we have already discussed in connection with succession in big canopy gaps. As an alternative, species with a range of shade-tolerances sometimes all colonize simultaneously, or in the 1st year or 2, but some grow faster and are more *apparent* and, in one sense of the term, are dominant in the early years of the succession. The light-demanding fast growers are commonly shorter lived than the shade-bearers, and, as they die, the latter become more apparent and dominant. The casual observer thus sees different successive dominants, although all species have in fact been present in the forest since its earliest stage. This case was described as a 'competitive hierarchy' by Horn (1976). There is evidence of simultaneous, or nearly simultaneous, colonization at Kade, Ghana (Swaine and Hall 1983) and on Barro Colorado Island (Brokaw 1987).

In reality, the two models are not always mutually exclusive. This was so at Kade, Ghana (Swaine and Hall 1983) and in the South Venezuela *terra firme* forest mentioned earlier, where succession was observed for 5 years (Uhl and Jordan 1984). Forbs dominated in the 1st year then in year 2 *Cecropia ficifolia*. Many *Cecropia* died in year three and the vacant space was invaded by many other pioneer and climax species (viz relay floristics). By year 5, there was little more invasion, 56 species were present, over half of them climax species, but mainly in the under canopy below faster growing pioneers (viz simultaneous colonization).

Besides pioneer and climax species, foresters sometimes distinguish a group of 'late-secondary species' (e.g. Budowski 1970), many of them commercially valuable. It is objectively easy to distinguish between species able to establish on open ground (viz pioneers) and under shade, but not to define different sorts of shade establishment (Swaine and Whitmore 1988). We await detailed elucidation of the autecology and successional status of the so-called 'late-secondary' species guild. I believe that it is likely that it consists of a group of species which occur in forests developing after simultaneous colonization, and that they are the species which become dominant in time between the fast growing pioneers and the slowest growing climax species. This is a topic where 'academic' ecological work is badly needed to underpin silviculture.

Grasses and/or forbs may dominate very early secondary succession. This happens for example on old-field successions where successive weedings have exhausted the soil seed-bank of pioneer tree species. The time to reproductive maturity of many grasses/forbs is so short that they quickly form a soil seed-bank themselves so are able to regrow after weeding. Many also reproduce vegetatively from fragments accidentally left behind at weeding (e.g. *Scleria* in Sarawak, Chin 1985). There is apparently no close study of the additional phenomenon of grass/forb dominance in forest opened for the first time, which sometimes happens. Trees soon take over. Except on the most degraded sites, a tree canopy will be established after a year, and commonly over an even shorter period.

Effects on climate

The albedo (reflectance) of all vegetation is markedly lower than that of bare earth, and the albedo of forest is slightly lower than that of herbaceous vegetation (surface total albedos of forest, grassland and desert are about 0.11, 0.17 and 0.35 respectively, Henderson-Sellers 1980). All humid tropical vegetation has a high potential evapotranspiration rate, especially forest (evapotranspiration rates in mm yr^{-1} of 850 (forest), 550–750 (grass) and 400–500 (crops) are reported by Henderson-Sellers 1980). It follows, therefore, that the published scenarios on putative dramatic changes in regional or global climate which would follow removal of primary tropical rain forests (e.g. Potter *et al.* 1981) are exaggerated because they compare bare earth with primary rain forest, forgetting that nature nowhere abhors a vacuum more strongly than on a patch of bare soil in the humid tropics.

Mother trees

Succession back to climax depends on successful invasion of secondary forest by primary species, which characteristically have poor capacities for dispersal. For example, a secondary forest at Kepong, Malaysia, only received its first dipterocarp after 33 years, though the nearest mother tree was only 180 m distant (Kochummen and Ng 1977, Kochummen 1978). Krakatau, an island 40 km away from rain forests on Java and Sumatra, has received very few climax tree species in the century since it erupted (Flenley and Richards 1982). Jarak, an island 64 km away from Malaya, still has a poor unbalanced flora after 34 000 years and its forests are believed to have been destroyed by the eruption of Mount Toba (Wyatt-Smith 1953). The clear conclusion is that recovery of climax rain forest on extensive areas cleared for pasture may take centuries or millennia in the absence of mother trees, unless we manage actively by planting such climax species.

In Amazonian Peru, *Swietenia*, a long-lived pioneer, colonizes newly deposited riverine alluvium from seed which, although winged, is in fact transported by water. Seedling carpets develop which mature to form monospecific tree stands. It is essential to leave mother trees as seed sources. This requirement is widely realized, but what is not well known is that the only mothers effective to perpetuate these commercially valuable pure communities are those left on the convex banks of river meanders (J. Salo *pers. comm.*).

Because of their poor dispersal abilities, diaspores of climax species commonly mostly fall near the mother tree. This is especially the case with species with diaspores apparently adapted to wind dispersal. For example, Burgess (1970, 1975) found 97% of the fruits of the dipterocarp *Shorea curtisii* fell in an ellipse of long axis 60 m, lying along the direction of the prevailing wind. Within this 'footprint' (sometimes misleadlingly called a 'seed-shadow') diaspores occur at high density. There has been much argument about their fate, for few grow up into trees. Several studies have now been made. For the Central American palm *Scheelea rostrata* fruits were attacked by beetles which failed to find isolated fruits so predation thinned out species-clumping (Janzen 1971). This phenomenon of higher death where the seedling population is densest near to the mother tree has now been found to be very commonly true but it is not always the case. For *Shorea curtisii* ants attacked the seeds, tending to destroy all isolated ones but leaving many within the clumps, so in this instance predation reinforces the clumping so characteristic of the species (Burgess 1969). For recent reviews of diaspore and seedling predation, their density dependence, and hence the importance of predation for clumping in rain forest tree populations, see Howe and Smallwood (1982), Clark and Clark (1984), Becker *et al.* (1985) and Fenner (1985).

Sources of regrowth

Seed-bank, seed-rain

Pioneer seedlings are not present at the time a big canopy gap is created. They develop from seed which was either present in the soil as a 'seed-bank' or which arrived after gap-creation as 'seed-rain'. There have been several studies (reviewed by Whitmore 1983, Vázquez-Yanes and Orozco Segovia 1984) which demonstrate that rain forest soils do have a seed-bank, and that germination is commonly triggered by a change in light quality, or by elevated soil surface temperatures (so-called photoblastic and thermoblastic responses). Much remains to be discovered about spatial and temporal variations in the soil seed-bank on which Saulei (1984) and Young *et al.* (1987) have made valuable contributions, and we should not jump to the conclusion that all primary rain forests have one. None of 18 species at Barro Colorado Island studied by Augspurger (1984) had seeds capable of dormancy. Moreover, there may be other triggers to germination, just as there are in other biomes.

For establishment, the relative importance of variations in the seed-bank or seed-rain, and in microsites need to be analysed before we can explain why, in the Eastern rain forests, different adjacent patches of secondary forest characteristically have different small subsets of the total pioneer tree flora.

Coppice

Coppice-shoots growing from stems or roots to form new trees are utilized by foresters and shifting agriculturists in various parts of the humid tropics (e.g. *Eusideroxylon zwageri* in Borneo, *Swartzia* in south Venezuela). Three studies have all shown that coppicing ability is reduced by fire (Venezuela, Uhl 1982; Queensland, Stocker 1981; Kalimantan, Riswan 1982).

Lack of pioneers

In the East, neither the heath forest formation nor the distinctive forest of some ultrabasic rocks has a pioneer flora. Riswan (1982) found by experiment that regrowth in a Bornean heath forest is nearly all from coppice shoots, hence this formation is easily degraded to an open landscape where fire follows gap-creation. In Amazonia, flooded forest (*igapo*) and heath forest both lack *Cecropia* (H. Klinge, C.Jordan *pers. comm.*).

Gap switch-size

The switch from colonization of small gaps by pre-existing shade-bearers to big gaps by invasive light-demanders is presumed to be correlated with the change in microclimate to open conditions from those of the forest ground layer (relatively cool, constant, highly humid, dimly illuminated by light with far red wavelengths exceeding red, and soil surface layers constant at *c.* 25° C.)

In detail, ecophysiologists have only very recently begun to analyze the characteristics of shade- and sun-seedlings, and we still have much to learn (see Bazzaz, this volume). The question, vital in nature and to silviculturists, concerning why shade-tolerant seedlings fail in gaps bigger than a certain small size remains unanswered. Anecdotal information on dipterocarps (summarized in Whitmore 1984) shows that these succumb to insect attack and not to desiccation. This is a similar finding to that of Langenheim *et al.* (1984) that shade seedlings of five species tested did not die from large increases in photon flux density; they suggested competition instead, but did not carry the analysis further.

Only once in rain forest has a series of canopy gaps of different size been artificially created to study succession against gap size. The experiment was in the lower montane rain forest of west Java. In gaps of 0.1 ha, pre-existing seedlings grew up; in gaps of 0.2 and 0.3 ha, pioneers replaced them (Kramer 1926, 1933). A circular gap of 0.1 ha is 35.7 m in diameter. Casual observation suggests gap-switch size in lowland rain forests is much less, pioneers can often be seen in gaps under 10 m diameter (80 m^2). Probably the cool, cloudy climate of the lower montane zone caused the switch to be at so large a gap diameter in Kramer's study. (The inference from circumstantial evidence that switch size in lowland forest is commonly at 0.1 ha in Whitmore (1982), I now believe to be an artefact). Gap-size is not the only determinant, microsites within the gap also play a role (see above). Pioneers which need soil upheaval to germinate and mineral soil to establish will be favoured at the root-plate sites of fallen trees, and these may be more common in larger gaps.

Numerical models of forest dynamics

The species composition of rain forests is seen, even from the rather few critical studies on gap-phase dynamics, to be immensely complex and to depend on numerous variables. I suggest that the way ahead to a clearer understanding of how species composition and diversity is maintained, and hence how man can manage in a sustainable and conservative manner, lies in further careful ecological observations and experiments of the kind referred to in this paper, to see how these ecosystems work. The outline of what happens seems fairly clear. I do not believe that numerical models will give us much real help in these very complex ecosystems. I believe their role in predicting in detail the response to natural or man-made disturbance is minimal. We shall learn much more from expending an equal amount of effort on good practical ecological enquiry. That is not, however, to deny the role of forest models, to give just one example, in forecasting changes in biomass and productivity in consequence to changes in atmospheric carbon dioxide.

TROPICAL RAIN FOREST MANAGEMENT

Let us now turn to consider how our knowledge of rain forest ecology assists management, and what important gaps there are whose filling would assist management. I define the objectives of tropical rain forest management to be both the sustainable production of industrial or fuel wood, and the conservation of the forest in the broad sense and with its many facets.

Gap-phase dynamics and sustainable timber production

Tropical rain forest silviculture was initiated by foresters steeped, directly or indirectly, in the European tradition. In Europe, foresters as intuitive ecologists have been manipulating forests since the Thirteenth Century to create canopy-gap conditions favourable for regeneration of one species or another. We must not forget that ecological knowledge of rain forest dynamics has sprung from the work of foresters. It is only in the last few years that academic scientists have entered this field of study and begun to ask increasingly finely focused questions about mechanisms and physiology. Nor should we forget that rain forest management fails, as it usually does, not because the silvicultural system is biologically unsound, but because of social, economic and political forces (see Palmer 1975).

Silvicultural systems

Monocyclic and polycyclic types may be recognized. Monocyclic systems remove all timber at one cut and rely on seedlings to form the next crop. They tend to create big canopy gaps and so favour regrowth of pioneer or near-pioneer species with light, pale, eminently marketable timber. Polycyclic systems involve selective removal of a few trees on several occasions per cycle. Theoretically, they yield more timber because use is made of adolescent trees, freed from competition by selective felling. In practice, it is difficult to keep damage sufficiently low to prevent the forest from deteriorating. There may be dysgenic effects by repeated removal of the big (viz fast growing) trees. Where selective felling is at low intensity, a polycyclic system creates small gaps and so favours the more shade-tolerant of the climax species.

In both West Africa and the Eastern rain forests, biologically sound silvicultural systems have been devised (reviewed by Lowe 1978 and Whitmore 1984, respectively), making use of the considerable number of big, long-lived pioneer and relatively light-demanding climax species present in those regions. By contrast, no viable systems have yet been operated in South America. Small scale experiments have proceeded in Suriname for some years (de Graaf 1986, this volume) and by several different groups in Brazil. The problem is that many neotropical rain forests are dominated by very shade-tolerant, heavy-timbered, slow-growing climax tree species. Any modern timber removal operation to be economically viable would be likely to create canopy gaps so big that these climax species would be replaced by faster growing species with paler, softer timber, e.g. *Goupia glabra* and *Laetia procera*, which are at present unknown commercially. There is apparently a difference between great tracts of the South American rain forests and rain forests elsewhere, in that the former have very few of the light-demanders which provide commercially attractive timbers and are easy to grow following modern high-intensity logging.

Logging by machine

Tree felling and log removal creates canopy gaps, and so mimics nature, and the forest responds by its natural processes. Modern logging uses heavy machines and creates conditions never found in nature. Most important is soil compaction, which the flora is not equipped to combat. It seems to be a particularly serious problem in South America. At Jari, in eastern Amazonian Brazil, forest clearance for plantations by machine was abandoned because plantation trees grew so poorly on the compacted soils. Machines also disrupt the soil surface,thus destroying seedlings, the humus layer and superficial feeding roots. The soil seed-bank may also be removed. Thus, the natural sources of building-phase forest are destroyed. It is common, also, for log extraction tracks to dam streams and thereby create swamps. The key to successful forest recovery after logging lies in minimizing damage to seedlings, adolescent trees, the soil surface and drainage pattern. At Bajo Calima in Colombia, excellent forest regrowth occurs, despite total biomass removal, because extraction is aerially by cable and the forest floor is left intact.

Gaps in information and understanding

There are various opportunities for ecologists to do work to underpin forest management.

Climber mats are one example. In many rain forests, mats of climbers are a very serious impediment to recovery after intensive logging of, for example, *Merremia* in the Solomon Islands and some parts of Malesia. There is very little critical information on the ecology of climbers and their role in natural tropical rain forest ecosystems (but see Putz 1984, 1985). It is surprising, for example, that *Merremia* is absent from Gogol in New Guinea even after clear-felling for chipwood (S.Saulei *pers. comm.*), and it was very rare in the Solomons before intensive logging began. Big woody climbers are believed to colonize and grow up in gaps (but see Penalosa (1984) who found vegetative spread more important). Climbers seem much less of a problem in forests recovering from natural cataclysms, though climber towers have been reported from Queensland (deduced to result from cyclone damage, Webb 1958), and carpeted areas of Kalimantan after the great fire of 1982–83 (S.Riswan *pers. comm.*). Are climbers really not a serious impediment to natural succession? If so, why not?

Nutrients and biomass are other subjects on which biologists have much to contribute, including investigation of the consequences of total, or nearly total, utilization of above-ground biomass. The end in view is to devise sustainable systems for this extreme form of rain forest utilization. The harvesting of all or most of the above-ground biomass for wood chips, then turned into paper, is so far practised in several Asian mangrove forests, but in only three dry land rain forests, Bajo Calima (Colombia), Gogol (Papua New Guinea) and Sipitang (Sabah). There are plans to utilize biomass for fuelwood in parts of the Brazilian Amazon (at Carajas to smelt iron, Trombetas to dry bauxite, Manaus to make cement, and widely to fuel power stations). We badly need studies on ecosystem recovery after total harvesting. What is the biomass productivity of natural regrowth of various species, singly or in mixtures, and what are its economics compared to plantation monocultures? Is regrowth from coppice shoots or from seed, and is it sustainable? What happens to plant mineral nutrients under such a drastic management regime? In a lower montane New Guinea forest, the boles of trees ≥ 10 cm in diameter contained about half of all the major plant mineral nutrients, with concentrations in the bark higher than in the wood (Table 10.1 in Whitmore 1984 summarizing data of Grubb and Edwards 1982). Various studies have shown substantial nutrient input in rainfall. In some tropical rain forests, soil erosion is continuously bringing weathered rock into the rooting zone and thereby replenishing ecosystem nutrients (Baillie and Ashton 1983); in others, bedrock may or may not be a closed nutrient cycle. Thus, generalizations cannot be relied upon. We require studies at any location where total biomass harvesting is contemplated to see if, and how, it can be made sustainable and not totally destructive of the habitat.

Sources of big-gap regeneration, whether from soil seed-bank, seed-rain, or coppice is the topic of several recent and current studies (see above). The subject is of clear importance for the understanding of forest recovery on big surfaces, for example after the abandonment of pasture.

Termites are the major group of decomposers in lowland tropical rain forest. In Malesia, drier sites, including logged forests, have a higher proportion of the group which can become serious pests (Macrotermitinae) (Collins 1980). More study is needed.

Species' demographies need further study. By analogy with other kinds of species-rich forest (e.g. of Chile, Veblen 1985) it seems likely that different climax tree species differ in their demography in tropical rain forests. Some are likely to have big seedling populations, frequently replenished, but with rapid loss by death, and others to have fewer, more persistent seedlings. Such differences between species have been discovered in the Solomon Islands (Whitmore 1974). We also have a glimmering understanding of differences in seedling demography for some dipterocarps. For example, there are differences between ecologically similar *Shorea* species (Fox 1972, summarized in Whitmore 1984). *Shorea multiflora* is known to have a sapling/pole 'bank' instead of a seedling-bank (P.F. Burgess *pers. comm.*).

Forest dynamics and conservation

We have shown that tropical rain forests are not stable and unchanging. The objective of conservation cannot be to achieve stability. Instead, it must aim to maintain big enough samples for natural processes to persist at all temporal and spatial scales, thereby maintaining all niches, and hence structural and taxonomic diversity. This objective will permit the continuation of natural adjustments to secular environmental change, and the natural interactions between organisms, and hence continuing evolution. In small areas, active intervention may be necessary.

Management for multiple objectives

Animals in the forest are an important part of the ecosystem. Much has been discovered, especially in the last 2 decades, of the way they are specialized to the many different spatial and temporal niches, and are interdependent with plants in complex ways. To review these studies is beyond the scope of this book, but see Bourlière (1983), Gilbert (1980) and Whitmore (1984).

It is clearly unrealistic to expect that more than a small fraction of the world's tropical rain forests will remain inviolate. Yet, conservation of species diversity, and of the rain forest as habitat for animals, need not be incompatible with management for the sustained production of timber and 'minor forest products' (rattans, resins, latex, drugs, fruits). This is an area where there is still much research needed, first in 'pure science' to learn about forest diversity, maintenance and species' interactions, and second in how these are influenced by timber harvesting. Thus, research is needed into workable multiple-use management systems. Timber production leads to loss of diversity of the forest in many ways.

We still have little knowledge of the extent to which rain forest ecosystems are resilient and recover from logging. The Brazil nut (*Bertholletia excelsa*) is pollinated by a euglossine bee. What does that bee feed on during the numerous months each year when the Brazil nut is not flowering? Some tree species are particularly important for ecosystem maintenance. For example, figs (*Ficus spp.*) provide food for many birds and mammals in both tropical America and South-east Asia. Some of the big banyan and strangling figs are especially important. Collectively, these are in fruit all the time (though individual trees and species have different phenologies), and are periodically augmented by other kinds of tree. It is believed that the abundance of figs determines the numbers of fruit-dependent animals the forest can support. They are the 'base-line' food and Gilbert (1980) called such vital species 'keystone mutualists'. Their demise has disproportionately serious consequences to the ecosystem. At Kutai, East Kalimantan, it has been found that these big figs are significantly associated with some of the major timber species which logging selectively depletes (Leighton and Leighton 1983). At Kutai, therefore, for the forest to continue as a rich animal habitat, logging practice needs to be modified.

A study at Sungai Tekam, Malaya, showed the five resident primate species were fairly resilient to a commercial logging operation, and so were many of the forest-interior birds (though not so the highly specialized groups: cuckoos and trogons and specialized insectivores were absent from the logged forest) (A.D. Johns 1985, 1986). A recent survey has shown that orang-utan occur in logged forest in Sabah. This observation encourages further study to see whether they breed and maintain themselves there, and, if necessary, to modify logging practice (Davies 1986). In South-east Asia generally, it is known that gaur, elephant, rhinoceros and tapir feed selectively on lush herbs and shrubs found naturally along valleys and on landslips. These animals, then, are favoured by the young regrowth vegetation along logging tracks. Pigs, a major protein source for forest-dwelling hunter-gatherers, are more abundant in secondary than in primary forest.

These few examples give glimpses of the complexity of rain forest ecosystems and suggest that logging is not necessarily a complete disaster. There is a vital need for biological research wherever timber production and conservation are planned to coexist so that adequate management systems can be devised. It is likely that a component of management for conservation plus timber production will always be to leave patches of forest untouched. At Sungai Tekam, the animals moved into such patches during the time of logging. Also, it is well known that many climax forest plants disperse effectively over only very short distances, tens of metres rather than hundreds, and their long-term persistence in the forest depends on mother plants surviving logging.

Silviculture for timber production, especially if it has the aim to increase yield in the future rotations, will alter the relative proportions of different tree species and of the age classes of individual species. The forest structure will be altered in many forests from a fine mosaic of the gap, building, and mature phases of the forest growth cycle to a much coarser mosaic, with the next felling timed to take place when the forest has reached the late building phase. These changes are not incompatible with the conservation of high species diversity of plants and animals. There is also always a very strong case for control areas to remain untouched, because only in such areas is the full original structure, the original dynamics, and the original balance within and between species populations maintained. Malaya long ago set aside single compartments of its reserved forests as virgin jungle reserves (Wyatt-Smith 1950) with just that end in view.

Finally, another consequence of timber exploitation is that roads are built into formerly inaccessible tracts and these roads give access to hunters, to collectors of minor forest products, and to people looking for agricultural land. This penetration of previously untouched forest has happened in Brazilian and Ecuadorian Amazonia; East Kalimantan is another example. Massive logging began in the late 1960s, and has been followed by excessive extraction of rattans, all the old social controls and trading networks having been supplanted (Peluso 1983); and by the ingress of farmers from other Indonesian islands (noticeably Bugis from Sulawesi). These are political and social problems, outside the scope of the biologist.

Management for conservation

Conservation is, in essence, the conservation of rare species, not the matrix of common and widespread species in which they are set. The need begins to arise of a new kind of active management for conservation because conserved areas may eventually become rather small relict patches set in a sea of agricultural land or grossly altered forest. Manipulation explicitly for species conservation may then become necessary. Malaya approaches this situation (Ng 1983). In a small relict patch, the dispersers of a tree's fruits may be absent (perhaps because of hunting). Rare species may need to be moved into such patches to conserve them from extinction. The conservation of isolated patches is an acute problem for countries like Malaya with a flora very rich because of many local endemics. Detailed inventories, such as that of the endemic tree species of Malaya (Ng and Low 1982), pin-point the species at risk. Conservation areas may suffer attrition at the margins from neighbouring villagers whose traditional gathering of forest produce becomes too concentrated when the forest has dwindled. This is happening to the Gunung Leuser National Park in Sumatra, Indonesia. There is considerable discussion to combat this depletion by buffer zones, either of intensively managed semi-natural forest or of plantations. As yet, this measure remains largely discussion, though plans are in hand, for example at Kerinci in Sumatra, to establish demonstration schemes.

CONCLUSIONS

All is not yet lost of the world's tropical rain forests. There are powerful economic pressures of many kinds behind their rapid diminishment. On the other side, conservation awareness, as enlightened self-interest, is continuously increasing, and at many different levels. The optimist has to hope that the longer-term view comes to prevail before these richest of all ecosystems are irretrievably devastated. There is scope for closer collaboration between forester and scientist to tackle particular silvicultural problems and, notably, to devise multiple-use management systems.

REFERENCES

Augspurger, C.K. (1984) Light requirements of neotropical tree seedlings: a comparative study of growth and survival. *Journal of Ecology*, **72**, 777–95

Baillie, I.C. and Ashton, P.S. (1983). Some soil aspects of the nutrient cycle in mixed dipterocarp forests in Sarawak. In Sutton, S.L., Whitmore, T.C. and Chadwick, A.C. (eds.) *Tropical Rain Forest Ecology and Management*, pp. 347–56. (Blackwell: Oxford)

Baker, H.G. (1965). Characteristics and modes of weeds. In Baker, H.G. and Stebbins, G.L. (eds.) *The Genetics of Colonizing Species*. (Academic: New York)

Beaman, R.S., Beaman, J.H., Marsh, C.W., and Woods, P.V. (1985). Drought and forest fires in Sabah in 1983. *Sabah Society Journal,* **8**, 10–30

Becker, P., Lee, L.W. and Hamilton., W.D. (1985). Seed predation and the co-existence of tree species: Hubbell's hypothesis revised. *Oikos,* **44**, 382–90

Boojh, R. and Ramakrishnan., P.S. (1982). Growth strategy of trees related to successional status. I, II. *Forest Ecology and Management,* **4**, 359–74, 375–86

Bourliere, R. (1983). Animal species diversity in tropical forests. In Golley F.B. (ed.) *Tropical Rain Forest Ecosystems, Structure and Function,* pp. 77–91. (Elsevier: Amsterdam)

Brandani, A., Hartshorn, G.S. and Orians, G.H.. (1988). Internal heterogeneity of gaps and species richness in Costa Rican tropical wet forest. *Journal of Tropical Ecology,* **4**, 99–119

Brokaw, N.V.L. (1987). Gap-phase regeneration of three pioneer tree species in a tropical forest. *Journal of Ecology,* **75**, 9–19

Budowski, G. (1970). The distinction between old secondary and climax species in tropical Central American lowland forests. *Tropical Ecology,* **ll**, 44–8

Burgess, P.F. (1969). Preliminary observations on the autecology of *Shorea curtisii* Dyer ex King in the Malay Peninsula. *Malayan Forester,* **32**, 438

Burgess, P.F. (1970). An approach towards a silvicultural system for the hill forests of the Malay Peninsula. *Malayan Forester,* **33**, 126–34

Burgess, P.F. (1975). *Silviculture in the Hill Forests of the Malay Peninsula*. Malaysian Forestry Department Research Pamphlet, **66**

Chin, S.C. (1985). Agriculture and resources utilization in a lowland rain forest Kenyah community. Sarawak Museums Journal Special Monograph, **4**.

Clark, D.A. and Clark, D.B. (1984). Spacing dynamics of a tropical rain forest: evolution of the Janzen-Connell model. *American Naturalist,* **124**, 769–88

Coley, P.D. (1983). Rates of herbivory on different tropical trees. In Leigh, E.G., Rand, S.A. and Windsor, D.M. (eds.) *The Ecology of a Tropical Forest: Seasonal Rhythms and Long-term Changes,* pp. 123–32. (Oxford University Press: Oxford)

Collins, N.M. (1980). The effect of logging on termite (Isoptera) diversity and decomposition processes in lowland dipterocarp forests. In Furtado, J.I. (ed.) *Tropical Ecology and Development*, pp. 113–21. (International Society of Tropical Ecology: Kuala Lumpur)

Davies, G. (1986). The orangutan in Sabah. *Oryx,* **20**, 40–5

Ellen, R.F. (1985). Patterns of indigenous timber extraction from Moluccan rain forest fringes. *Journal of Biogeography,* **12**, 559–637

Fenner, M. (1985). *Seed Ecology*. (Chapman and Hall: London)

Finegan, B. (1984). Forest succession. *Nature,* **312**, 109–14

Flenley, J.R. (1979). *The Equatorial Rain Forest: A Geological History*. (Butterworth: London)

Flenley, J.R. and Richards, K. (eds.). (1982). *The Krakatau Centenary Expedition.* Final Report. University of Hull Department of Geography Miscellaneous Series, **25**. (University of Hull: Hull)

Fox, J.E.D. (1972). *Natural Vegetation of Sabah and Natural Regeneration of the Dipterocarp Forests.* Ph.D. thesis, University of Wales.

Gilbert, L.E. (1980). Food web organisation and conservation of neotropical diversity. In Soulé, M.E. and Wilcox, B.A. (eds.) *Conservation Biology*, pp. 11–33. (Sinauer: Sunderland, Mass.)

Graaf, N.R. de (1986). *A Silvicultural System for Natural Regeneration of Tropical Rain Forest in Suriname*. (Agricultural University: Wageningen)

Grubb, P.J. and Edwards, P.J. (1982). Studies of mineral cycling in a montane rain forest in New Guinea. III. The distribution of mineral elements in the above-ground material. *Journal of Ecology,* **70**, 623–43

Hamilton, A. (1976). The significance of patterns of distribution shown by forest plants and animals in tropical Africa for the reconstructions of upper Pleistocene palaeo-environments. In Zinderen Bakker, E.M. van Sr. (ed.) *Palaeoecology of Africa*, pp. 63–97. (Balkema: Cape Town)

Henderson-Sellers A. (1980). The effect of land clearance and agricultural practice on climate. *Studies in Third World Societies,* **14**, 44–86

Herwitz, S.R. (1981). *Regeneration of Selected Tropical Tree Species in Corcovado National Park*, Costa Rica. University of California Pubns Geography, **24**

Horn, H.S. (1976). Succession. In R.M. May (ed.) *Theoretical Ecology*. (Blackwell: Oxford)

Howe, H.E. and Smallwood, J. (1982). Ecology of seed dispersal. *Annual Review of Ecology and Systematics,* **13**, 201–28

Janzen, D.H. (1971). The fate of *Scheelea rostrata* fruits beneath the parent tree: predispersal attack by brucids. *Principes,* **15**, 89–101

Johns, A.D. (1985). Selective logging and wildlife conservation in tropical rain forest: problems and recommendations. *Biological Conservation,* **31**, 355–75

Johns, A.D. (1986). Effects of selective logging on the behavioural ecology of West Malaysian primates. *Ecology,* **67**, 684–94

Johns, R.J. (1986). The instability of the tropical ecosystem in New Guinea. *Blumea,* **31**, 341–71

Jones, E.W. (1955–56). Ecological studies on the rain forest of southern Nigeria. IV. The plateau forest of the Okumo forest reserve. *Journal of Ecology,* **43**, 564–94; **44**, 83–117

Kochummen, K.M. (1966). Natural plant succession after farming in Sungei Kroh. *Malayan Forester,* **29**, 170–81

Kochummen, K.M. (1978). Natural plant succession after farming in Kepong. *Malaysian Forester,* **41**, 76–7

Kochummen, K.M., and Ng, F.S.P. (1977). Natural plant succession after farming at Kepong. *Malaysian Forester,* **40**, 61–78

Kramer, F. (1926). *Onderzoek naar de natuurlijke vergonging in den uitkap in Preanger gebergte-bosch*. Med. Proefst. Boschw. Bogor, **14**

Kramer, F. (1933). De natuurlijke verjonging in het Goenoeng Gedeh complex. *Tectona,* **26**, 156–85

Koyama, H. (1978). Photosynthesis studies in Pasoh forest. *Malaysian Nature Journal,* **30**, 253–78

Langenheim, J.H., Osmond, C.B., Brooks, A. and Ferrar, D.J. (1984). Photosynthetic responses to light in seedlings of selected Amazonian and Australian rain forest tree species. *Oecologia* (Berlin), **63**, 215–24

Leighton, M. and Leighton, D.R. (1983). Vertebrate responses to fruiting seasonality within a Bornean rain forest. In Sutton, S.L., Whitmore, T.C. and Chadwick, A.C. (eds.) *Tropical Rain Forest Ecology and Management*, pp. 181–96. (Blackwell: Oxford)

Lennertz, R. and Panzer, K.F. (1983). *Preliminary Assessment of the Drought and Forest Fire Damage in Kalimantan Timur*. (DFS German Forest Inventory Service Ltd: Samarinda)

Lieberman, D., Lieberman, M., Peralta, R. and Hartshorn, G.S. (1985). Mortality patterns and stand turnover rates in a wet tropical forest in Costa Rica. *Journal of Ecology*, **73**, 915–24

Lowe, R.G. (1978). Experience with the tropical shelterwood system of regeneration in natural forest in Nigeria. *Forest Ecology and Management*, **1**, 193–212

Malingreau, J.P., Stevens, G. and Fellows, L. (1985). Remote sensing of forest fires: Kalimantan and north Borneo 1982–3. *Ambio*, **14** 314–21

Morley, R.J. and Flenley, J.R. (1987). Late Cenozoic vegetational and environmental changes in the Malay archipelago. In Whitmore, T.C. (ed.) *Biologeographical Evolution of the Malay Archipelago*, pp. 50–9. (Clarendon: Oxford)

Ng, F.S.P. (1983). Ecological principles of tropical lowland rain forest conservation. In Sutton, S.L., Whitmore, T.C. and Chadwick, A.C. (eds.) *Tropical Rain Forest Ecology and Management*, pp. 359–75. (Blackwell: Oxford)

Ng, F.S.P. and Low, C.M.. (1982). *Check List of Endemic Trees of the Malay Peninsula*. Malaysian Forestry Department Research Pamphlet, **88**

Oberbauer, S.T. and Strain, B.S. (1984). Photosynthesis and successional status of Costa Rican rain forest trees. *Photosynthesis Research*, **5**, 227–32

Oldeman, R.A.A. (1983). Tropical rain forest, architecture, silvigenesis and diversity. In Sutton, S.L., Whitmore, T.C. and Chadwick, A.C. (eds.) *Tropical Rain Forest Ecology and Management*, pp. 139–50. (Blackwell: Oxford)

Palmer, J.R. (1975). Towards more reasonable objectives in tropical high forest management for timber production. *Commonwealth Forestry Review*, **54**, 273–89

Peluso, N.L. (1983). Networking in the commons: a tragedy for rattan. *Indonesia*, **35**, 95–108

Penalosa, J. (1984). Basal branching and vegetative spread in two tropical rain forest lianas. *Biotropica*, **16**, 1–9

Potter, G.L., Elsasser, H.W., MacCracker, M.C. and Ellis, J.S. (1981). Albedo change by man: test of climatic effects. *Nature*, **291**, 47–50

Prance, G.T. (ed.) (1983). *Biological Diversification in the Tropics*. (Columbia: New York)

Putz, F.E. (1984). The natural history of lianas on Barro Colorado Island, Panama. *Ecology*, **65**, 1713–24

Putz, F.E. (1985). Woody vines and forest management in Malaysia. *Commonwealth Forestry Review*, **64**, 559–64

Riera, B. (1985). Importance des buttes de racinement dans la regeneration forestiere en Guyana francaise. *Revue d'Ecologie* (*Terre Vie*), **40**, 321–9

Riswan, S. (1982). *Ecological Studies on Primary, Secondary and Experimentally Cleared Mixed Dipterocarp Forest and Kerangas Forest in East Kalimantan, Indonesia*. Ph.D. thesis, Aberdeen University.

Salo, J., Kalliôla, R., Häkkinen, I., Mäkinen, Y., Niemalä, P., Puhakka, M. and Coley, P.D. (1986). River dynamics and the diversity of Amazon lowland forest. *Nature*, **322**, 254–8

Sanford, R.L. Jr., Saldarriaga, J., Clark, K.R. Uhl, C. and Herrera, R. (1985). Amazon rain forest fires. *Science*, **227**, 53–5

Saulei, S.M. (1984). Natural regeneration following clear-fell logging operations in the Gogol Valley, Papua New Guinea. *Ambio*, **13**, 351–4

Shukla, R.P. and Ramakrishnan, P.S. (1984). Biomass allocation strategies and productivity of tropical trees related to successional status. *Forest Ecology and Management*, **9**, 315–24

Spurr, S.H. and Barnes, B.V. (1980). *Forest Ecology*. Third Edition. (John Wiley: New York)

Stocker, G.C. (1981). Regeneration of a north Queensland forest following felling and burning. *Biotropica*, **13**, 86–92

Swaine, M.D. and Hall, J.B. (1983). Early succession on cleared forest land in Ghana. *Journal of Ecology*, **71**, 602–27

Swaine, M.D. and Whitmore, T.C. (1988). On the definition of ecological species groups in tropical forests. *Vegetatio*, **75**, 81–6

Swaine, M.D., Lieberman, D. and Putz, F.E. (1987). The dynamics of tree populations in tropical forests, a review. *Journal of Tropical Ecology,* **3**, 359–66

Taylor, B.W. (1957). Plant succession on recent volcanoes in Papua. *Journal of Ecology,* **45**, 233–43

Uhl, C. (1982). Recovery following disturbances of different intensities in the Amazon rain forest of Venezuela. *Interciencia,* **7**, 19–24

Uhl, C. and Jordon, C.F. (1984). Succession and nutrient dynamics following forest cutting and burning in Amazonia. *Ecology,* **65**, 1476–90

Uhl, C., Clark, K., Clark, H. and Murphy, P. (1981). Early plant succession after cutting and burning in the upper Rio Negro region of the Amazon basin. *Journal of Ecology,* **69**, 631–49

Vázquez-Yanes, C. and Orozco Segovia, A. (1984). Ecophysiology of seed germination. In Medina, E. Mooney, H.A. and Vázquez-Yanes, C. (eds.) *Physiological Ecology of Plants of the Wet Tropics*, pp. 37–50. (Dr. W. Junk Publishers: The Hague)

Veblen, T.T. (1985). Forest development in tree-fall gaps in the temperate rain forests of Chile. *National Geographic Research*, Spring 1985, pp. 162–83

Watt, A.S. (1947). Pattern and process in the plant community. *Journal of Ecology,* **35**, 1–22

Webb, L.J. (1958). Cyclones as an ecological factor in tropical lowland rain forest, north Queensland. *Australian Journal of Botany,* **6**, 220–8

Wetterberg, G.B. and Jorge Padua, M.T. (1978). *Preservacao da natureza na Amazonia brasiliera. Situacao em* 1978. PRODEPEF Serie Tecniica (PNUD/FAO/IBFD/BRA-545), **13**, IBDF, Brasilia.

Whitmore, T.C. (1974). *Change With Time and the Role of Cyclones in Tropical Rain Forest on Kolombangara, Solomon Islands*. Commonwealth Forestry Institute, Paper 46.

Whitmore, T.C. (1975). *Tropical Rain Forests of the Far East.* (Clarendon: Oxford)

Whitmore, T.C. (1978). Gaps in the forest canopy. In Tomlinson, P.B. and Zimmermann, M. H. (eds.) *Tropical Trees as Living Systems*, pp. 639–55. (Cambridge University Press: Cambridge)

Whitmore, T.C. (1982). On pattern and process in forests. In Newman, E.I. (ed.) *The Plant Community as a Working Mechanism*, pp. 45–59. (Blackwell: Oxford)

Whitmore, T.C. (1983). Secondary succession from seed in tropical rain forests. *Forestry Abstracts,* **44**, 767–79

Whitmore, T.C. (1984). *Tropical Rain Forests of the Far East.* Second Edition. (Clarendon: Oxford)

Whitmore, T.C. and Prance, G.T. (eds.) (1987). *Biogeography and Quaternary History in Tropical America.* (Clarendon: Oxford)

Whitney, G.G. (1976). The bifurcation ratio as an indicator of adaptive strategy in woody plant species. *Bulletin of the Torrey Botany Club,* **103**, 67–72

Wirawan, N. (1984). *Good Forests Within the Burned Forest Area in East Kalimantan.* World Wildlife Fund Indonesia Project No. 1687 Field Report. (WWF: Bogor)

Woods, P. (1987). Drought and fires in tropical forests in Sabah – an analysis of rainfall patterns and some ecological effects. In Kostermans, A.J.G.H. (ed.) *Proceedings of the Third International Round Table Conference.* Samarinda, 16–20 April 1985. pp. 367–87. (Unesco: Jakarta)

Wyatt-Smith, J. (1950). Virgin jungle reserves. *Malayan Forester,* **13**, 40–5

Wyatt-Smith, J. (1953). The vegetation of Jarak island, Straits of Malacca. *Journal of Ecology,* **41**, 207–25

Wyatt-Smith, J. (1966). *Ecological Studies on Malayan Forests.* I. Malayan Forestry Department Research Pamphlet, **52**

Young, K.R., Ewel, J.J. and Brown, B.J. (1987). Seed dynamics during forest succession in Costa Rica. *Vegetatio,* **71**, 157–76

CHAPTER 4

REGENERATION OF TROPICAL FORESTS: PHYSIOLOGICAL RESPONSES OF PIONEER AND SECONDARY SPECIES

F.A. Bazzaz

INTRODUCTION

In recent years, scientists have called repeatedly for more research into the dynamics of tropical forests. They want to study these complex ecosystems before the extensive clearing in all tropical regions of the world reduces these forests to small, isolated fragments. Because of their unusually high levels of biological diversity, tropical forests challenge the understanding of ecosystem structure and function, community organization, population dynamics, speciation, and the fundamental physiological, demographic and behavioural attributes that underlie these processes. Furthermore, tropical forests are still a potentially important source of timber, food, and medicine for human populations.

Foresters have long tried to base forest management practices on a solid understanding of the basic biology of trees and their environment. Because of its importance to forest management, the dynamics of regeneration after exploitation has received particular attention. Most pioneer, and many secondary, species are fast growing and therefore are important in agro-forestry. Because they retain nutrients, and preserve watersheds, these species also help the recovery of forest structure after disturbance. Therefore, this review will concentrate on regeneration, with special reference to pioneer species and secondary species.

Bazzaz and Pickett (1980) and Mooney *et al.* (1980) reviewed aspects of the physiological ecology of tropical plants and identified research directions and needs. Major contributions to the field since 1980 can be found in *Physiological Ecology of Plants in the Wet Tropics*, Medina, Mooney, and Vázquez-Yanes (1984) and *Tropical Rain Forest Ecology and Management*, Sutton, Whitmore and Chadwick (1983). The new edition (1984) of Whitmore's book *Tropical Rain Forests of the Far East* contains much information about regeneration of tropical forests in South-east Asia and Australia. In my article (Bazzaz 1984*a*), I related the physiological attributes of tropical trees to gap dynamics, created a table (Table 4.1) of the physiological and demographic attributes of pioneer species, and recommended future research needs. The table was based on available literature and my own work on temperate succession, and is now modified in the light of new evidence (Table 4.1). The present article is, in part, an update of my previous reviews, with special reference to forest management.

THE FOREST ENVIRONMENT

Forest regeneration begins with the dispersal of seeds to sites suitable for germination. The dispersed seeds must be viable; they must escape predators; and they must encounter the light, moisture, and temperature conditions required to germinate. These factors, together with nutrient relations and herbivory, control growth and reproduction. I begin with an

Table 4.1 Physiological characteristics of pioneer and secondary species

1.	Often have seed banks.
2.	Possibly long seed and seedling dormancy.
3.	Germination is enhanced by light, decreased by high Fr/R ratios, increased by temperature fluctuations and by higher nutrient concentrations.
4.	Mostly epigeal germination; photosynthetic cotyledons.
5.	High rates of photosynthesis, respiration, transpiration, high conductances, high N content.
6.	Continuous production of leaves; fast leaf turnover rates; leaves arranged in flat crowns; not multilayered.
7.	Rapid growth; low density wood; large leaves.
8.	Highly branched, intensive deep root system; mostly NO_3 users (as opposed to NH_4).
9.	Early and long flowering time.
10.	Rapid response to changes in resource levels.
11.	High acclimation potential.
12.	High susceptibility to herbivores and pathogens.

overview of the tropical forest environment and then discuss how this environment affects germination, photosynthesis, and growth.

The light environment

The upper canopy of the forest in wet tropical regions receives less radiant energy than the canopy of other tropical forests because water molecules in the rain forest's humid air absorb this energy. Much attenuation occurs as the light passes through the vegetation; the amount of radiant energy available near the forest floor can be extremely low. The slope of the light attenuation curve varies greatly and is determined by leaf area density and leaf absorbance characteristics. The strata that affect vertical light distribution are more recognizable in forests dominated by one or a few species (Richards 1983). Kira *et al.* (1969) found that, in a secondary forest in peninsular Thailand, the leaf area index was highest at 20–35 m. In a young mixed dipterocarp forest at Pasoh in Malaysia there were two maxima: one at 20–35 m and one below 5 m above ground (Kira 1978). Detailed analysis of the light energy profiles in tropical forests in the Far East can be found in Yoda (1974), Aoki *et al.* (1975), and Sasaki and Mori (1981).

Chazdon and Fetcher (1984a) compared the light environments in a 0.5 ha clearing, a 400 m^2 gap, and a heavily shaded understorey at La Selva, Costa Rica (Fig. 4.1). Photosynthetic photon flux density (PPFD) reached a peak of over 1000 $\mu mol\ m^{-2}\ s^{-1}$ in the clearing, but only 90 $\mu mol\ m^{-2}\ s^{-1}$ in the understorey, and that for only a short time. The percentage of incident PPFD reaching the understorey was higher on cloudy and overcast days than on sunny days. However, in a dipterocarp forest, Sasaki and Mori (1981) showed that diffuse light remains relatively stable in spite of significant fluctuations in light levels in the open. Thus, the understorey plants very likely function best when sunflecks contribute significantly to total daily PPFD. Total daily PPFD in clearings was generally higher in the dry than in the wet season. Calculations of potential photon flux densities for the site show peaks of approximately 1400 W m^2 in April and September and troughs of approximately 1200 W m^2 in January.

Another aspect of the forest's light environment is the shift in spectral quality as the light passes through the canopy. The foliage absorbs red over far red light, so the light near the forest floor may have little red wavelength light (Fig. 4.2). This filtering has been repeatedly shown in several rain forests in the neotropics, Africa, Australia, and Malaysia (Chazdon and Fetcher 1984b, Sasaki and Mori 1981, Pearcy 1983, Bazzaz and Pickett 1980, and references therein). Lee (1987) has shown that the ratio of red to far red light is closely correlated to the percent of full sunlight at different points in the forest.

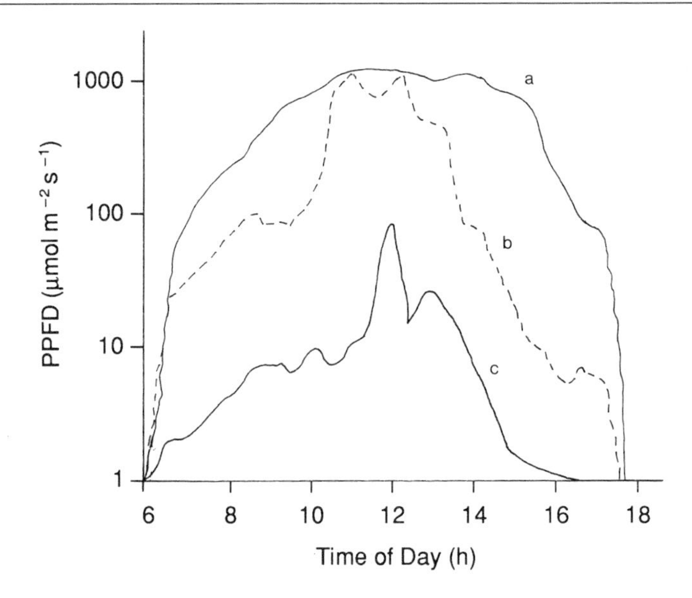

Fig. 4.1 Daily pattern of photosynthetic photon flux density in tropical rain forest in Costa Rica for a sunny day in (a) 5000 m^2 clearing, (b) 400 m^2 gap, and (c) understorey. From Chazdon and Fetcher (1984*a*)

Despite different spatial variations in the average understorey PPFD of different tropical forests (Pearcy 1983, Chazdon and Fetcher 1984*a*, Chazdon 1986), certain patterns are evident. The substantial three-dimensional variation in the light environment is largely due to the distribution of the vegetation. Different parts of individuals may simultaneously experience vastly different light environments. Because of sunflecks, a shoot or a single leaf may experience rapid shifts from very low to very high light levels. This shift affects carbon-gain capabilities, and the growth and resource allocation of seedlings (Pearcy 1983).

The temperature environment

The light environment dictates the temperature of air, plants, and soil in tropical forests. Although the temperature of emergents and outer parts of the forest canopy may rise during mid-day, there are no detailed or accurate measurements of this rise. In the understorey, daily variation in leaf temperatures may be small and leaf and air temperatures remain close except during sunflecks. Air temperature profiles have been made of several locations in tropical forests. Representative profiles, as well as profiles of irradiance, relative humidity, windspeed, and saturation deficits,

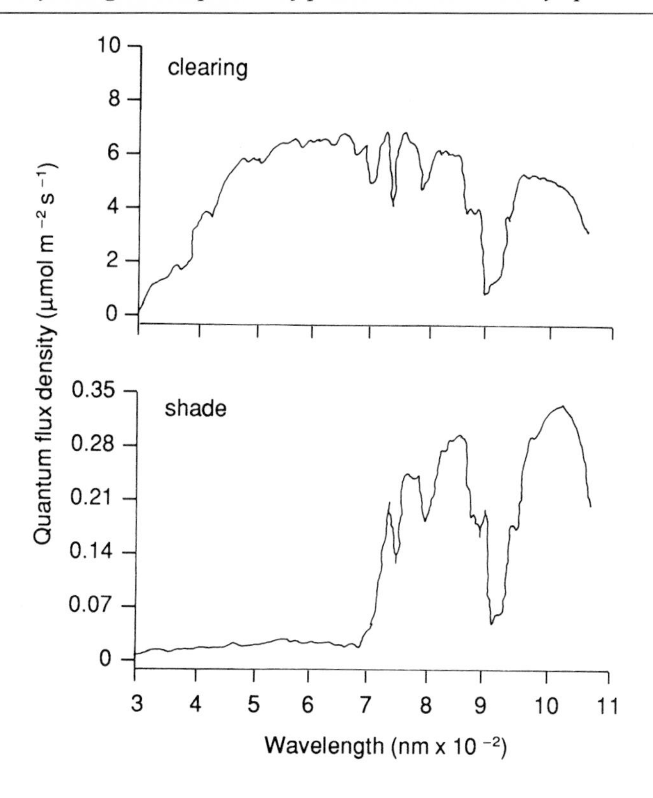

Fig. 4.2 The spectral distribution of radiant energy in a clearing and in the shade of a tropical rain forest in Costa Rica. From Chazdon and Fetcher (1984*b*)

are presented in Chiariello (1984). At Pasoh, Malaysia, the canopy air temperature at approximately 45 m was considerably higher than air temperature above and below the canopy (Aoki *et al.* 1975). Soil temperature may differ little from air temperature in the understorey, but may be much higher than air temperature in gaps and clearings (Schulz 1960). The differences are most pronounced near the soil surface, where the germination and early seedling growth usually occurs.

Fetcher *et al.* (1984) measured temperature and atmospheric humidity in a single tree gap, a 0.5 ha clearing, the canopy, and the understorey of the intact forest in a moist tropical forest in La Selva, Costa Rica. They showed how the degree of exposure influences the microenvironment and how the change in stature of the vegetation influences the patterns of temperature and atmospheric humidity. Predictably, temperatures and vapour pressure deficits were higher and more varied in the clearing, and were lower and less varied in the understorey. After 2 years of growth, the microclimate of the gap at seedling height was very similar to that of the understorey.

Water economy

Landsberg (1984) has discussed in detail the water economy of tropical rain forests in terms of the standard hydrological equation. He showed that, when the canopy is thoroughly wet, significant stemflow occurs and the soil at the base of large emergent and canopy trees may become considerably wetter than the rest of the forest. This effect may contribute to spatial heterogeneity of resources in the forest. Evapotranspiration from the canopy may be large and is driven by the energy balance of the canopy, resistances to water flow, and soil moisture levels. Incoming radiation and the albedo are the most important controllers of evaporation; the latter is high in clearings with exposed soil surfaces or dry vegetation, lower in gaps, and lowest in intact forest.

Energy in the forest canopy is dissipated by evapotranspiration (LE), and sensible heat transfer (H), which includes advection and convection. The ratio H/LE (Bowen ratio) ranges between 0.45 in the wet season and 6.4 during the dry season (Landsberg 1984). Low evaporation from soil and low air mixing in dense tropical forests lead to high relative humidity, which in turn reduces evapotranspiration. Of course, the amount and distribution of rainfall may vary considerably among locations. This affects forest dynamics by changing the morphology, physiology, and behaviour of forest organisms. In gaps and large clearings, relative air humidity may be low and may limit the establishment and growth of some forest species. However, soil moisture content may be high in these locations.

Nutrients

Nutrient dynamics have received much attention in Amazonia, Hawaii, and, recently, at La Selva (Costa Rica) and Pasoh (Malaysia). Vitousek (1984) reviewed available literature on patterns of nitrogen, phosphorus and calcium cycling, using data from 62 tropical forests. He concluded that, in general, lowland tropical forests have more nitrogen and a lower dry mass to nitrogen ratio in litterfall than do most temperate forests. Phosphorus return, however, is very low. Vitousek concluded that phosphorus, but not nitrogen, is usually limiting in these systems. Cuevas and Medina (1986) showed differences in nitrogen, phosphorus, calcium, and magnesium concentrations and in nutrient-use efficiencies by plants growing in different soils in the Amazon. They concluded that different vegetation types in the area may be limited by different elements. *Tierra firme* is phosphorus limited, *caatinga* is nitrogen limited, while *bana* appears to be limited by both nitrogen and phosphorus. The authors related differences in the degree of sclerophylly and leaf duration among the three vegetative types to nutrient availability. On a smaller scale, Vitousek and Denslow (1986) measured the pattern of nutrient concentrations in various parts of gaps and found significant differences among some locations in a gap. This heterogeneity may influence recruitment patterns of species in gaps.

FOREST REGENERATION

It is now clear that tropical forests are very dynamic and that the process of gap creation and filling controls the structure and function of the forest, heavily influencing plant evolution. Recent discussions of this process include Hubbell (1979), Orians (1982), Alexandre (1982), Pickett (1983), Hartshorn (1980), Bazzaz (1984*a*), Brokaw (1985*a*), and Martinez-Ramos (1985). Regeneration of the forests involves recruitment, survivorship, and growth of a very large number of species that may differ in their modes of life and the roles they play in regeneration. Changes in spatial and temporal patterns of vegetation in a given location will be influenced by the interactions of resource levels, colonization patterns, and each species' ecological properties.

The physiological ecology of tropical plants has been treated in Medina *et al.* (1984). Specific topics include: C. Vázquez-Yanes and A. Orozco Segovia, seed germination; T. J. Givnish, leaf and canopy adaptation; N. Chiariello, leaf energy balance; J. J. Landsberg, water regime; R. H. Robichaux *et al.*, tissue water deficit; H. A. Mooney *et al.*, photosynthesis; E. Medina, nutrients; J. H. Langenheim, R. Dirzo, P. A. Morrow, herbivory; F. A. Bazzaz, physiological ecology in relation to the dynamics of tropical forests.

These reviews cover the literature to 1983. Since then, there has been much work in the physiological ecology of tropical plants, especially in the neotropics. Unfortunately, other than seed germination and seedling growth, physiological data are still woefully lacking for the tropical forests of the Far East and Africa.

Germination and recruitment

Seed germination characteristics, requirements, and patterns have been extensively studied in many rain forest locations. Especially prominent and comprehensive among these are studies in Mexico by C. Vázquez-Yanes and in Malaysia by F.S.P.Ng (e.g. Ng 1983). The literature published prior to 1980 was summarized by Bazzaz and Pickett (1980). Vázquez-Yanes and Orozco Segovia (1984) summarized the work to 1983. Whitmore (1984) discusses seed germination in tropical forests of the Far East. These studies, together with others such as Augspurger (1984) confirm earlier predictions and suggest that our present knowledge of seed germination requirements is sufficient to make certain predictions about the regeneration of the humid tropical forests. Adaptation of seed germination to the process of gap creation and filling in tropical forests has been previously discussed by Bazzaz (1984*a*). The following statements about seed germination and seedling establishment rely heavily on the sources mentioned above.

(1) Early successional and pioneer species flower early in life and usually produce seeds annually. In areas with mild dry seasons, plants tend to fruit at the end of the wet season. Where dry seasons are severe, plants concentrate their fruiting at the beginning of the wet season.

(2) Seed longevity is generally low in tropical trees. Advanced regeneration (suppressed seedlings) may be more important than the seed-bank as a source of regeneration of some tropical trees. However, seed longevity is usually higher for pioneer than for late successional species and, in pioneers, the seed-bank may be a major source of regeneration. In contrast to pioneers, seeds of most primary species have no dormancy and a short life span.

(3) Resprouting is common in tropical trees, but severe fire reduces it substantially. Severe fire and erosion destroy seed-banks as well, and regeneration will depend on immigrants.

(4) Seed germination of the many secondary forest species is enhanced by increased irradiance.

(5) Germination is generally rapid in tropical trees. But there is also within-species variation in the speed of germination. Seeds with harder coats generally have a lower moisture content, are longer-lived, and take longer to germinate.

(6) The germination of many pioneer and secondary species is triggered by disturbance. Shifts in red/far red ratios and the temperature fluctuations that result from the removal of vegetation enhance germination. In contrast, seeds of many primary species, except for emergents, are able to germinate in the shade. It must be remembered, however, that tropical tree species vary widely in their germination responses to different light environments. Primary and pioneer species are probably not the only groups that are sensitive to changes in the light environment.

Most disturbances create highly heterogeneous habitats that recruit different species and play out different growth scenarios. In treefall gaps, tips and mounds have different nutrient contents (Bazzaz 1983, Vitousek and Denslow 1986), which could recruit different species (Putz 1983). Experiments by Uhl *et al.* (1981) in Amazonia showed that the cutting and burning of the forest created a number of microhabitats which differed in soil temperature, plant diversity, and water loss potential (Fig. 4.3). Most of the species tested showed distinct micro-habitat preferences. Recruitment from various sources of propagules may vary depending on gap size and severity of disturbance (Fig. 4.4). The seedling-bank is most important in small gaps, while seed-banks become more important in larger gaps and/or moderately severe disturbances. Birds, bats, and other animals, as well as wind, are the major sources of recruited seeds. However, Uhl *et al.* (1981) found that seed-banks, although they may be reduced after fire, contribute to recruitment. Furthermore, they found that seeds of the primary forest species arrived rather early in succession. The pioneers grow fast initially, but their growth rates decline later while the growth of primary species is generally slower but continues at a steady rate (Uhl and Jordan 1984). It is not known, however, whether this pattern of recruitment is common in other tropical forests. Brokaw (1987) found in Barro Colorado Island (BCI), Panama, that *Trema micrantha* and *Cecropia insignis* were mostly recruited in the first year, while *Miconia argentea* was recruited mostly in years two and three, and that some individuals were recruited as late as 7 years after gap formation. *Trema* grew very fast (up to 7 m yr^{-1}) and was persistent in large gaps. *Cecropia* grew less fast (up to 4.9 m yr^{-1}) and persisted in large and somewhat small gaps. *Miconia* grew slowest (2.5 m yr^{-1}) and persisted in the widest range of gap size.

Expected patterns of the early dominance and rapid demise of pioneers and the continued growth and later dominance of the primary species have been observed for several tropical forests (e.g. Brokaw 1985*b*, C.Uhl unpublished).

Energy and material exchange with the environment

Variation in the light environment in tropical forests affects plant germination, growth, and reproduction. The interaction between the vegetation and the environment generates a very complex and continually changing ecological setting. The three-dimensional environmental pattern influences the plants and is influenced by them. It is important to recognize that plants as individuals may experience much variation in the levels of resources as they grow from small seedlings to mature, reproducing individuals, and that parts of the same individual may simultaneously experience different environments.

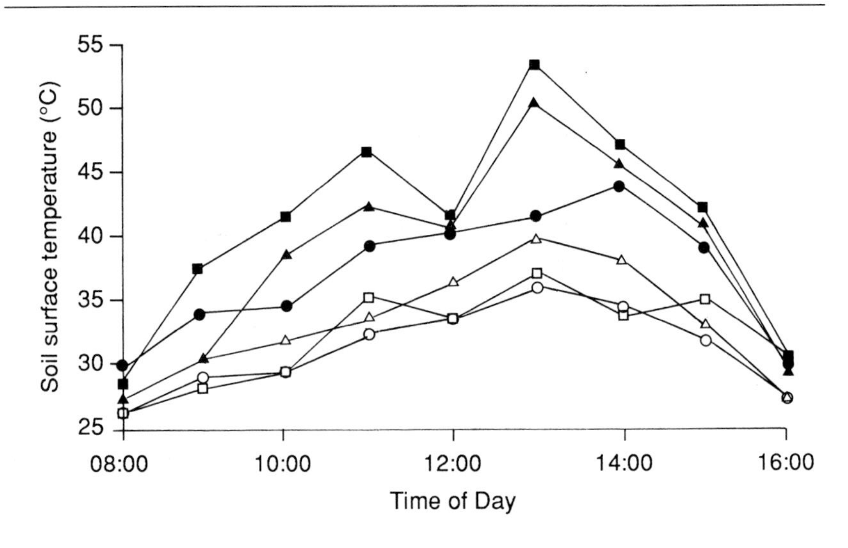

Fig. 4.3 Soil surface temperature in each of six microhabitat types. Bare soil (circles), Chorcol (squares), root mat (triangles), with slash (open symbols) and without slash (closed symbols). From Uhl *et al.* (1981)

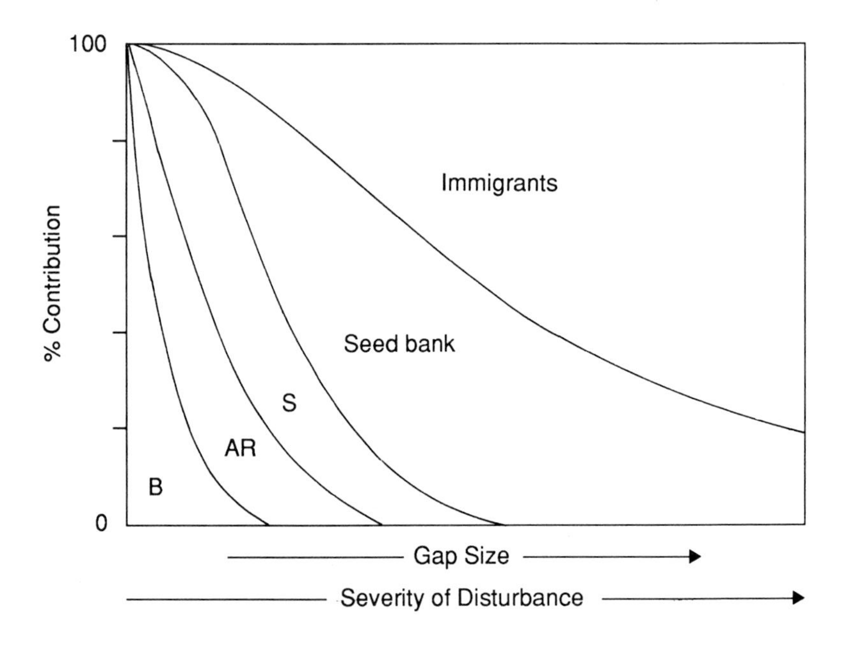

Fig. 4.4 Relationship between gap size and the relative contribution of various guilds to gap filling. Increased severity of disturbance during gap creation moves the time axis to the right. B = branches, AR = advance regeneration, S = sprouts. From Bazzaz (1984*a*)

Leaf-level considerations

Vertical microclimatic gradients govern, to a large extent, the energy balance of individual leaves and their carbon-gain capacities. Leaf energy balance measurements for tropical trees are very limited. Chiariello (1984) used a modelling approach to examine the effects of covarying trends in irradiance, humidity and air temperature on leaf temperature. She showed that a typical leaf should be warmer than air under open conditions and near or slightly above air temperature in shaded conditions. Measurements have shown that leaves overheat in situations of high radiation, low conductance, and low wind speed. Temperatures up to 15° C higher than the air temperature have been observed in attached leaves in several species. Boundary layer resistance in the large-leaved *Tectona grandis*, *Gmelina arborea*, and *Triplochiton scleroxylon* were studied by Grace *et al* (1980). The values they obtained are comparable to those of temperate large-leaved species. These two studies suggest that there may be no fundamental differences in energy balance between tropical and temperate forest trees.

A wide range of stomatal conductances have been documented in tropical plants. Under some field situations conductance may be as high as 1 mol m^{-2} s^{-1} (Chiariello 1984). Grace *et al.* (1982) measured 1.4 mol m^{-2} s^{-1} for *Tectona grandis* in the wet season. Values about 1 mol m^{-2} s^{-1} have been measured in *Heliocarpus*, *Urera alata*, and *Gmelina*. Whitehead *et al.* (1981) measured stomatal conductance in *Tectona* and *Gmelina* in the field in Nigeria. The highest value for *Gmelina* was 12 mm s^{-1} and for *Tectona* 35 mm s^{-1}, though the first was at 400 μE $m^{-2} s^{-1}$ and the second was at 1000 μE $m^{-2} s^{-1}$ PPFD. However stomatal conductance is low in some understorey plants. Robichaux and Pearcy (1980) measured 1–2 mm s^{-1} in *Euphorbia*, a value which did not change much throughout the day.

Givnish (1984) summarized data on leaf width in relation to annual rainfall, soil depth and water holding capacity. He concluded that leaf width increases with increased rainfall and decreases with decreased soil moisture-holding capacity and soil nutrient levels along edaphic gradients (but see Fetcher 1981). Leaf size increases from the sunlit canopy to the lower strata of tropical forests. Some early successional trees also have larger leaves than do mature phase species. Younger trees that are lower in the canopy have larger leaves than do mature trees, and leaves of shade-grown trees are larger than leaves in the sun. In tropical forests in Ghana, Hall and Swaine (1981) found that macrophylls and megaphylls were present only in the wettest forests and that microphylls became increasingly rare as moisture increased. Steep leaf angle is characteristic of some species exposed to sun (Hall and Swaine 1981) and may produce substantial reduction in leaf temperature (Medina *et al.* 1978). Upper canopy leaves are steeply inclined and leaves may change orientation diurnally to reduce their projected area along the solar beam. *Piper auritum* in open situations shows severe mid-day wilting and reduced laminar area exposed to direct light (Chiariello *et al.* 1987).

Plant–water relations

Patterns of rainfall interact with forest structure and dynamics to generate complex patterns of water economy for different individuals. Robichaux *et al.* (1984) discussed water relations in tropical forests with special attention to Hawaiian euphorbs of the genus *Dubautia*. They show that tropical trees may experience severe water stress, which may not be comparable to water stress observed in temperate forests. Measurements by Robichaux (1984) showed a wide range of leaf water potentials, especially near the end of the dry season. Fetcher's data (1979) show very low leaf water potential at the end of long dry season (-3.9 MPa in *Trichilia*). Oberbauer *et al.* (1987) measured -1.7 MPa and understorey leaves were 0.5 MPa higher. Medina and Klinge (1983) reported diurnal water potentials of -0.1 to -1.0 MPa during a clear day in mountain wet forest in Puerto Rico.

In the genus *Dubautia*, there was much variation among species in the course of daily water potential. Turgor pressure declined faster in the species from shaded habitats than from sunny habitats. Turgor pressure was lower at any given water potential in the shade than in the sun species (Robichaux 1984). In *Metrosidoros polymorpha*, maximum tissue turgor pressure was 0.35 to 0.40 MPa higher in varieties that colonize exposed dry sites than varieties in wet forest.

P.W. Rundel (cited in Robichaux *et al.* 1984) compared tissue osmotic potentials and total water potentials in several species at BCI, Panama. He found that osmotic potentials at full hydration differ among species and are generally more negative for canopy trees than for understorey plants. In gaps, species with the same water potential differed in their turgor potential. During the severe droughts of 1983 in BCI, the leaves of the three understorey shrubs remained at or close to zero turgor for a 24 hour period. But leaves of the gap species *Leuhea seemanii* and *Cordia alliodora* remained under positive turgor throughout the day and exhibited moderately high rates of leaf conductance, perhaps because of higher soil moisture in the gap. Pre-dawn water potentials were less negative in the successional species than in those of the intact forest (Fig. 4.5).

During the dry season, differences in vapour pressure deficit between a gap and the intact forest may be large and also vary during the day (Fetcher *et al.* 1984). Fetcher (1979) compared xylem pressure potential and leaf conductance in five species of the semi-evergreen forest of BCI, Panama. Xylem pressure potentials of approximately -2.0 MPa were recorded during the mid-day in *C. alliodora*, corresponding to -20 mbars vapour pressure deficit and stomatal conductance of 0.15 s cm^{-1}. Xylem pressure potentials

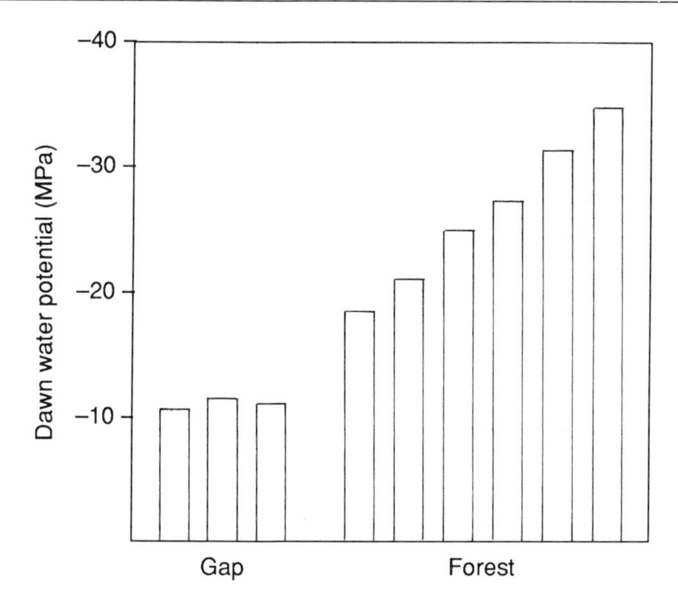

Fig. 4.5 Pre-dawn water potentials of species in a large gap and intact forest habitats during March 1983 on Barro Colorado Island, Panama (P.W. Rundel, unpublished). The species from left to right are *Luehea semanii, Cordia lliodora, Miconia argentea, Hybanthus prunifolias, Piper cordulatum, Psychotria marginata, Mouri myrtilloides, Trichilia tuberculata* and *Faramea occidentalis*

and leaf conductance varied widely during the day. *Trichilea cipo* growing at the edge of the clearing had an even more dramatic negative xylem pressure during the dry season, but recovered appreciably after rain. Pre-dawn water potentials were as low as -2.6 MPa for *Trichilea* during the dry period.

Within-tree variation in leaf morphology, structure, environmental relations, and physiological activities has been examined in great detail by Oberbauer and Strain (1985, 1986). Understorey, mid-canopy and canopy leaves were compared in *Pentaclethra macroloba*. Canopy leaves had the smallest leaflets and the highest stomatal density, the lowest specific leaf area and the greatest thickness.

Leaf osmotic potentials at full turgidity decreased with height but turgor pressures were similar for leaves from the three levels. Bulk tissue elastic modulus increased with height and therefore understorey leaves remained turgid to lower water content while mid-canopy and canopy leaves did not. Leaf water potentials were lowest in the canopy and highest in the understorey. Consequently minimum turgor pressure was lowest in the canopy and approached zero in full sunlight on clear days. There were also no significant adjustments in osmotic potential between wet and dry seasons.

Photosynthesis

There has been a significant increase in the number of studies on photosynthesis of tropical plants. Summaries of previous works presented in Bazzaz and Pickett (1980) and in Mooney *et al.* (1980) can now be significantly augmented. A series of studies by B.R. Strain and associates, a paper by J.H. Langenheim and one by R. Koyama are among the most detailed recent studies of photosynthesis of tropical trees. There is also work in progress on shrubs of genus *Piper* by C.B. Field and H.A. Mooney in Los Tuxtlas, Mexico, on herbaceous plants by A. Smith, and on trees in gaps by P. Rundel in BCI, Panama. Our laboratory has studies underway on secondary forest species from Mexico and dipterocarps from Malaysia. Data previous to 1980 are somewhat less reliable than more recent data. This is largely due to the more advanced technology that allows for more precise measurement of the exchange of water and CO_2 between leaves and the surrounding air.

Mean photosynthetic rates reported in 1980 are slightly lower than the more recent measurements (Table 4.2). In general, species preferring open and sunny habitats have higher maximum photosynthetic rates than do species preferring shady habitats. Furthermore, within a species, shade-grown seedlings generally have lower light saturated photosynthetic rates than do sun-grown seedlings. But there are exceptions, e.g. *Hymenea parviflora* and *H. courbaril* (Langenheim *et al.* 1984). This is also true for leaves on the same individual (Oberbauer and Strain 1986). Recent data also confirm previous conclusions that understorey species have lower photosynthetic rates than do primary canopy species and those in turn have lower rates than pioneers (Bazzaz and Pickett 1980, Bazzaz 1984*a*). These trends in photosynthesis are similar to those in temperate forest (Bazzaz 1979) and the rates are not very different between these two systems. There are not as many measurements of dark respiration, but available data suggest that the ratio of photosynthesis/dark respiration (PS/Rd) is between 7 and 10 and that understorey plants and shade leaves generally have a lower PS/Rd ratio. There is need for more research on this relationship.

Sunflecks are an important part of the light environment and carbon gain of understorey plants. Rapid response of photosynthesis to changes in light levels may be important for these plants. In the understorey, saplings of *Claoxylon sandwicense* demonstrated rapid response to changes in the light environment (Pearcy and Calkin 1983). Plants saturated at 10% of full sunlight had a light compensation point less than 2 μmol m^{-2} s^{-1}. Diffuse light was approximately 1 μmol $m^{-2}s^{-1}$ but sunflecks exceeded 200 μmol $m^{-2}s^{-1}$, the saturation point of this species. The stomata increase their conductance to near maximum early in the morning and remain steady throughout the day even when light levels change dramatically. Pearcy and Calkin concluded that 60% of daily carbon gain is attained during sunflecks

Table 4.2 Photosynthetic rates (in µmol $m^{-2}s^{-1}$) published since 1980 of some tropical plants grown under near optimal conditions

Species	Rate	Species	Rate
Ochroma lagopus	27.7	*Virola*	6.3
Ochroma lagopus	24.6	*Pentaclethra macroloba*	6.0
Hampea	13.9	*Carapa*	6.0
Bursera	11.8	*Hymenea courboril*	5.0
Gnetum	8.5	*Scheelia*	5.0
Diptryx	8.1	*Agathis microstachya*	4.5
Heliocarpus appendiculatus	8.0	*Simarouba amara*	4.0
Copaifera venezuelana	7.5	*Minquartia*	4.0
Agathis robusta	7.3	*Hymenea parviflora*	3.5
Cordia alliodora	6.8	*Socratea*	2.8

and that overall seedling growth is strongly correlated with total daily sunfleck duration.

Acclimation to the light environment has been studied in some tropical plants. Langenheim *et al.* (1984) chose species that germinate under extremely dense shade, but grow to become canopy dominants or emergents with crowns exposed to full sunlight. This strategy is common to numerous species in all tropical forests. Seedlings of *Hymenaea courbaril*, *H. parviflora* and *Copaifera venezuelana* from Amazonia and *Agathis microstachya* and *A. robusta* from the Australian forests were grown at 100% and 6% of full sunlight. The photosynthetic responses of these plants were typical of sun versus shade plants: shade-grown seedlings had lower compensation points, higher quantum yields and lower respiration rates per unit area than did those grown in full sun. Light-saturated photosynthetic rates of the seedlings grown in sun were slightly, but not significantly, different (except for *Agathis robusta*) from seedlings grown in the shade. Furthermore, intercellular CO_2 concentrations were similar in sun and shade leaves and assimilation was limited by intrinsic mesophyll factors (r_m) rather than by stomatal conductance. Individuals grown in the sun were taller and produced more leaves. *Agathis robusta* grew most under all conditions and *Copaifera venezuelana* grew least.

Fetcher *et al.* (1983) compared the acclimation to the light environment of the pioneer, large-gap species *Heliocarpus appendiculatus* with the small-gap species *Dipteryx panamensis* in Costa Rica. Seedlings were grown in full sun, partial shade (20% sun), and full shade (2% sun) and were switched between environments. Growth of the pioneer species was more plastic than that of the small-gap species in response to changes in the light environment. In high light, vertical growth was much more increased in *Heliocarpus* than in *Dipteryx*, but survival was higher in *Dipteryx* than in *Heliocarpus*. Changes in leaf thickness, specific leaf weight and stomatal density were also higher for *Heliocarpus* than for *Dipteryx*. These results confirm previous findings about small- and large-gap species.

In the forest, both sun and shade leaves may be found on the same individual. Their carbon-gain capacities and contribution to the total carbon budget of that individual are probably different. There is, however, little information about differential response of leaves of single individuals. Oberbauer and Strain (1986) compared seedling leaves to leaves from three heights in the canopy of *Pentaclethra macroloba*. They found that changes in leaf characteristics along the canopy gradient paralleled changes that could occur in seedlings grown on a light gradient. Individuals of the same species grown in full and 25% sun had equal biomasses; switching seedlings between these two light environments had no effect on growth. However individuals switched from full and 25% sun to 1% sun had a negative CO_2 exchange rate and suffered leaf abscission resulting in negative growth. Plants switched from 1% to full sun showed severe photoinhibition and leaf damage. There was little acclimation of photosynthesis to changes in the light environment. It was concluded that seedlings of this species will respond quickly and positively to the formation of small or medium gaps, but not large ones.

Oberbauer and Strain (1984) studied the photosynthetic capacity of seven tree species in a Costa Rican rain forest and related photosynthetic rates to successional status (Table 4.2). They found that light-saturated photosynthetic rates were related to preferred light environment in the field. Plants preferring heavy shade had a mean photosynthetic rate of 6.8 μmol m^{-2} s^{-1}, those in canopy gaps 11.3 μmol m^{-2} s^{-1} and the species from large clearings 27.7 μmol m^{-2} s^{-1}. Furthermore, light saturation of plants from the clearings occurred at PPFD greater than 1000 μmol m^{-2} s^{-1}, while those from shadier environments reached light saturation at much lower PPFDs. Their findings confirmed earlier predictions about photosynthetic rates and successional status (Bazzaz 1979, Bazzaz and Pickett 1980).

Leaf-level and whole-plant carbon gain are influenced by leaf life span which is estimated to be 6 to 9 months in deciduous species and little more than 1 year in evergreen ones. Phenology of leaf expansion ranges from almost continuous in some pioneer species (e.g. *Cecropia*) to discontinuous flushes in mature phase species (Coley 1983). There are both obligatory deciduous (Medina 1984) and facultative deciduous (Whitmore 1984) trees. The behaviour of the latter depends on water availability during the dry season. Deciduous leaves have lower mean specific leaf weights and higher N and P content per unit dry weight. P and N contents of leaves are positively correlated. Some N and P are translocated out of leaves before they fall, approximately 50% of N, 60% of P (Medina 1984). It is assumed that most tropical tree species have low nitrogen-use efficiency, perhaps because of allocation of some N to defense. It is not known how these factors influence photosynthetic rates in tropical trees, and how this relates to fast-growing trees, which may be defended by C-based rather than N-based chemicals.

Plant architecture

The rates of photosynthesis in individual leaves are only one factor in the growth and production of tropical plants. Allocation of photosynthate to competing structures and functions and the display of these structures, the plants' architecture, play a significant role as well. The general growth form and branching patterns of tropical trees have been treated using the Hallé and Oldeman growth models (see Hallé *et al.* 1978). De Foresta (1983) attempted to relate growth models to successional status in tropical trees of French Guyana. He suggests that the Roux and Rauh models are common in pioneer species, while the Troll model is more common in primary forest. Similar trends were obtained for pioneer vegetation in Kalimantan, Indonesia, by Huc and Rosalina (1981). Architectural configurations which reduce self-shading and allow the development of monolayered canopies would enhance carbon-gain capacities and may be common in secondary species.

Chazdon (1985) has worked on leaf display, canopy structure and the pattern of light interception in two understorey palms, *Asterogyne maritima* and *Geonoma cuneata* in Costa Rica. She examined leaf size and display, and crown size, in considerable detail. She found that seedlings of the two species did not differ in light interception efficiency. However, adults of *G. cuneata* had higher light interception efficiency than did those of *A. maritima*, despite the fact that the latter had longer leaves, more leaves per plant, and, therefore, a greater total leaf area than the former. The decreased efficiency of adult *A. maritima* crowns was largely due to increased proportion of pendent leaves relative to the adults of *G. cuneata*. In both species, there was relatively little self-shading because of the even distribution of leaves around the main axis. The leaves of the two species also possess other morphological attributes that maximize light-interception capacities in the light-limited environment of the understorey. Studies of this kind, together with photosynthetic measurements and growth analysis, will be essential in the understanding of the mechanisms of adaptation of species to tropical forest environments.

The high photosynthetic rates and fast growth of pioneer species are generally associated with the development of low density wood; this is one of the distinguishing characteristics of pioneers. Energetically cheap branches can produce a thin, open canopy that exposes leaves to the high light levels to which they respond. Shukla and Ramakrishnan (1986) showed that fast growth over a longer period of time results in sparse branch arrangement and greater leaf exposure in two pioneer tree species in India. Allocation to stilt roots, which is common among the palms, may contribute to their success. These roots enable the palms to produce an extending axis early in their development. Individuals are able to attain greater heights per unit of diameter earlier in their development than do non-stilt root species. The latter, however, may allocate a greater amount of biomass to underground parts than do the former (Schatz *et al.* 1985).

The higher proportion of far red light in the understorey influences plant architecture as well. Sasaki and Mori (1981) showed that far red stimulated internode elongation of some dipterocarp seedlings and restricted their root growth. Under low light conditions, internode elongation was greater relative to diameter growth. Thus, the ratio of height to diameter was higher than under high light conditions.

Growth

Because of the length of the growing season, most tropical trees grow faster (especially above ground) than do temperate deciduous trees. It is also well established that early successional pioneer and secondary forest species grow faster than do primary forest species. Reports of extremely fast growth of the early successional species are common. Growth has been measured most commonly as height extension because of the ease and non-destructive nature of the measurements. Girth growth has also been recorded for a large number of species, usually of large individuals. There are technical difficulties with this type of measurement because deciding where to take the trunk measurement may be arbitrary and because bole–taper relationships may vary in different species.

The range of possibilities for timing growth is best viewed as a spectrum from continuous growth to annual rhythmic growth (Tomlinson and Longman 1981). Annual growth rings are uncommon in most tropical trees. On the other hand, many species have clear zonations of wood that resemble annual rings but are not necessarily the result of annual growth (Bormann and Berlyn 1981). Periodicity of growth has been observed in several tropical trees. These flushes of girth growth and branch extension may result because of seasonality in resources, especially moisture (Borchert 1978) and temperature, or because of endogenous rhythms. In pioneers, the growth seems to be strongly influenced by environmental factors such as temperature and photoperiod (Longman 1978). Mature-phase species may have more or less continuous growth, but emergents may have intermittent growth (Ashton 1978). Mariaux (1981) gives an extensive list of tree species with various patterns of growth. Borchert (1980) and Reich and Borchert (1982, 1984) have studied growth and phenology in relation to soil moisture in several species of tropical trees in Guanacaste Province, Costa Rica. A decline in soil moisture, which changes the tree water status, affects the timing of leaf fall, bud break, and (in many species) anthesis. Interestingly, rehydration occurred on wet sites after leaf fall, perhaps because of reduced transpiration by the trees. Continuous growth was more common in seedlings and saplings than in canopy individuals.

Height extension

Height growth of pioneers and commercially important species has been studied in several tropical forests. Some of the results are astonishing (Table 4.3). In a forest clearing in Nigeria, Ross (1954) measured *Trema guianeensis* that grew approximately 10 m in 5 years. In 14 years *Musanga cecropioides* formed a canopy at 23 m. In Malaysia, a stand of *Macaranga gigantia* grew to approximately 20 m in 15 years. In Guatemala, Ewel (1977) reported height growth of 15 to 20 m in *Ochroma lagopus* and *Cecropia peltata* in 14 years. In Rio Negro, Uhl *et al.* (1981) found 5 m tall *Cecropia* spp. 22 months after the forest had been cut and burned.

A detailed study of tree growth after forest clearing was carried out by Swaine and Hall (1983) in Ghana. Some secondary species reached a height growth of up to 4 m yr^{-1} (Fig. 4.6). Forty-five individuals, mostly *Harungana madagascariensis* and *Trema orientalis*, exceeded 10 m in 5 years with one individual of *Trema* reaching 17 m. Another individual of the same species reached 50 cm girth at breast height (gbh). Primary species such as *Albizzia adianthifolia*, *A. zygia*, and *Antiaris toxicaria* achieved 5 m or more in 5 years. Brokaw (1985*b*) reported even higher values from BCI. He measured 13.5 m height extension in 2 years in *Trema micrantha* and 4.9 and 5.0 m in one year by *Cecropia insignis* and *Zanthoxylum* species respectively. Some individuals of the primary species also grow rapidly in gaps. *Simarouba amara* grew 3.0 m yr^{-1}, *Virola sebifera* 2.4 m yr^{-1}, and *Protium panamensis* 2.3 m yr^{-1}.

Table 4.3 Growth rates of some tropical trees

Species	*Height extension* (m yr^{-1})	*Source*
Trema micrantha	6.75	Brokaw 1985
Zanthoxylum sp.	5.0	Brokaw 1985
Cecropia insignis	4.9	Brokaw 1985
Trema orientalis	3.4	Swaine and Hall 1983
Simarouba amara	3.0	Brokaw 1985
Cecropia spp.	2.7	Uhl *et al.* 1981
Harungana madagascarensis	2.4	Swaine and Hall 1983
Virola sebifera	2.4	Brokaw 1985
Protium panamensis	2.3	Brokaw 1985
Trema guianensis	2.0	Ross 1954
Musanga cecropioides	1.4	Ross 1954
Ochroma lagopus	1.4	Ewel 1977
Cecropia peltata	1.04	Ewel 1977

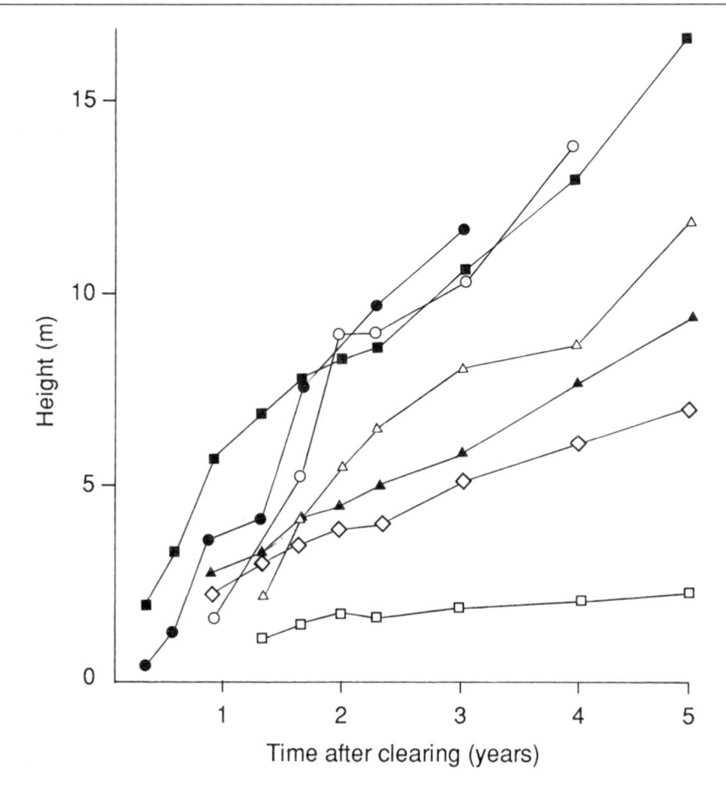

Fig. 4.6 Height growth of some of the largest trees of selected secondary species at Atewa Range Forest Reserve, Ghana. Symbols: ● *Musanga cecropioides*; ○ *Terminalia ivorensis*; ■ *Trema orientalis*; △ *Harungana madagascariensis*; ▲ *Margaritaria discoidea*; ◊ *Discoglypremna caloneura*; □ *Nauciea diderrichii*. From Swaine and Hall (1983)

A comprehensive long-term study of tree growth in tropical forests has been reported by Lieberman *et al.* (1986). Growth of 46 common species in a Costa Rican tropical forest was measured on plots established in 1969–1970. The authors used both canopy and subcanopy species and measured annual diameter at breast height (dbh), age, and life span. They made trajectories of maximum, minimum and median growth rates and life span. The results of the study show that: a) shade-intolerant canopy and subcanopy species have maximum growth rates and are short-lived; b) shade-tolerant subcanopy trees live about twice as long as understorey trees and grow at approximately the same maximum rate; c) canopy and subcanopy trees that are shade tolerant but respond opportunistically to increased light levels have long life spans and high maximum growth rates; and d) understorey species have slow maximum growth rates and short life spans.

Girth growth

Lang and Knight (1983) studied growth in a large number of tropical trees in BCI, Panama. They found that diameter growth was highly variable among species and among size classes. Individuals of the pioneer species *Cordia alliodora*, *Leuhia seemenii*, and *Spondias radikoferi* of 30 to 50 cm dbh grew at 0.67, 1.09, and 0.25 cm yr^{-1} respectively. Deciduous species grew as rapidly or more so than evergreen species. Using long-term forestry data from Sarawak, Primack *et al.* (1985) estimated mean annual diameter increments of between 0.4 and 3 mm yr^{-1} in several *Artocarpus* and two *Ficus* species. Growth rates increased most in the first and second year following selective logging but began to decline later. Growth rates varied considerably among species and among individuals: most trees showed little or no growth and a few showed high growth rates. Growth of *Artocarpus* saplings increased as fast as that of adults. There are, of course, site-to-site differences in growth rates of the same species. For example, Dilmy (1980), cited in Primack *et al.* (1985), measured the annual growth increment of three *Artocarpus* species and found it to be approximately 2.4 mm yr^{-1}. In ideal conditions, growth may be several times higher (Ng and Tan 1974).

The information on allocation to roots and on root morphology in tropical trees is very limited. A study of two early and two late successional species in northeast India by Shukla and Ramakrishnan (1986) showed that the former have a more shallow-feeding root system and that they allocate less to roots than do late successional species. The relationship between soil moisture and nutrient availability and between soil moisture and allocation to roots should be investigated.

Growth analysis has not been frequently used in the study of tropical trees. Okali (1971) reported relative growth rates (RGRs) of 0.36, 0.68, 0.28, and 0.18 g g^{-1} w^{-1} respectively for *Ceiba*, *Terminalia*, *Chlorophora* and *Musanga* seedlings in Ghana. Net assimilation rates (NAR) were respectively 54.1, 41.6, 14.3, and 20.6 g m^{-2} w^{-1}. Whitmore and Wooi-Khoon (1983) reported a net assimilation rate of 33 g m^{-2} w^{-1} for *Ochroma lagopus*. Some investigators (e.g. Oberbauer and Strain 1986, Fetcher *et al.* 1983) have combined growth analysis with measurements of photosynthetic rates. These approaches will undoubtedly result in a better understanding of the control of productivity in tropical forests and the strategies of individual species. Whitmore and Bowen (1983) also used growth analysis to study seedlings of *Agathis macrophylla* and *A. robusta*. They showed that relative growth rate (RGR) declined by only 30% under shade of 12% of full light. They showed that these species can maintain growth under reduced irradiance and can establish in small gaps.

Herbivory

Herbivores play an important role in forest regeneration through the removal of photosynthetic and support tissue. This may result in changes in plant architecture and, in extreme cases, the death of individuals. Herbivory may indirectly increase mortality by providing entry for pathogens. Insects seem to be the most important herbivores in tropical forests (Janzen 1981, Dirzo 1984). Massive defoliation does occur, but more commonly only parts of leaves or of individual plants are consumed. The average value of leaf area consumed by herbivores in many tropical forests seems to be between 10 and 20% (Coley 1983, Dirzo 1984, Lowman 1984). The removal of tissue at the seedling stage is especially crucial for plant growth and reproduction. The contribution a damaged leaf makes to the carbon economy of the plant depends on when in its life-span the leaf was damaged (Crawley 1983, Bazzaz 1984*b*, Hartnett and Bazzaz 1984). There is good evidence that young leaves are more susceptible to herbivory than old leaves and that leaves of early successional species are more susceptible than late successional species (e.g. Coley 1983). Removal of seed tissue by herbivores may be even more critical for the fate of the resulting seedlings. Removal of only 5% of seed tissue resulted in much reduced height growth in *Omphalea oleifera* (Dirzo 1984), a primary forest species.

The consequences of herbivory to photosynthesis and water and nutrient use efficiencies have not been studied in any detail. Furthermore, the costs (lower photosynthetic capacity) and benefits (herbivore deterrence) of secondary compounds, which occur in high diversity and concentrations in tropical forests (Langenheim 1984) are unknown. How maintenance, growth, reproduction, and defense control the allocation of plant resources, as well as the kinds of defense, is also unknown (see Coley *et al.* 1985).

CONCLUSIONS AND RECOMMENDATIONS

Successful management of tropical forests and their continued use as a renewable resource must be based on a good understanding of the basic biology of the component species and their role in the process of regeneration. The levels of detail and sophistication of the knowledge required for management decisions vary considerably and it would not be prudent to define these in rigid terms. We have a reasonable base of biological data to make some management decisions, but we certainly do not have the data to make all the required decisions. Clearly, there is an urgent need for research relevant to forest management as the human pressures on these forests accelerate.

One of the most critical areas related to forest regeneration is that of seed biology. Though it is well-established that seed-banks may be quite different in composition from that of the forest above, we do not know how the composition of the seed-bank changes throughout the season and the rates of input and output that govern their dynamics. We can safely assume, however, that seeds of pioneers and many secondary species form a major portion of seed-banks and that seeds of primary species have rather short life spans. Cues for seed germination of early successional species and pioneers are also well known: germination requires a disturbance that makes light more available at the forest floor and reduces the high proportion of inhibitory far-red light.

We also know that seedling-banks (large numbers of suppressed seedlings in the understorey) are common for many late successional and primary forest species and may be the most important means of regeneration of primary forests. But we do not know enough about seedling persistence, demography, causes of mortality, and carbon gain capacities and nitrogen metabolism. There is now good evidence that drought, pathogens, and herbivores play major roles in seedling-bank dynamics and that each by itself, under some circumstances, can lead to the destruction of entire seedling populations. Species reponse to these factors and their interactions remains unknown. For a large number of species we need: demographic studies; detailed investigations of the physiology of seedlings in relation to drought, high and low light intensities, and nutrient availability; studies of the effects of pathogens on seedlings; and research into the consequences of tissue removal on carbon-gain capacities and competitive interactions.

The timing of disturbances by natural causes or by deliberate removal will influence the structure and composition of the forest. Furthermore, the level of environmental heterogeneity created by disturbances needs to be studied. It is now clear that heterogeneity and the differential distributions and availability of resources favour the growth of certain species in different locations. But even if we know which species perform better in which sites, we know virtually nothing about the physiological basis for these preferences. Clumping of individuals in a species may result because of this heterogeneity and the patterns of seed rain and seedling shadows. One implication of this clumping (which is common for many species, Hubbell and Foster 1983) is that genetically related seedlings grow near the sources of dispersal (see model of Bazzaz 1983). Clumping of closely related individuals may also enhance the local spread of pathogens that are able to breach defenses of certain genotypes. The spatial and temporal organization of genetic variation in tropical forest trees is being investigated in a number of sites, especially at BCI, Panama, by J. Hamrick and L. Loveless, and in Asia by K. Bawa.

Differential species response to canopy gaps of different sizes and with different levels of PPFD is just beginning to be examined; these studies will yield very useful information both for management purposes and for basic biology about the diversity of forest. The effect that sudden changes in environmental resources and controllers have on water-use efficiency, carbon-gain capacity, growth, and allocation of resources in seedlings and saplings will yield much useful information about differential species response to the creation of gaps or selective cutting of large trees. Causes of seedling and sapling death after canopy opening are not well researched. Photoinhibition, drought, and overheating of leaves are probable causes, but increased insect herbivory and pathogens may also be important. Studies of some species' ability to acclimate to changing environments are just beginning and undoubtedly will be important to forest management.

Allocation in seedlings of carbon, nitrogen, and other resources, and the flexibility of this allocation is also unknown. Trade-offs between height extension, girth growth, and allocation to stems and roots, as well as how these allocations are affected by changes in resource levels, also need further investigation.

Most individuals in tropical forests spend much of their life suppressed under the canopy. The physiological ecology of these individuals is virtually unknown. The trade-offs – in terms of carbon, nitrogen, etc. – between limited growth, maintenance respiration, and defense against herbivores and pathogens have yet to be investigated. Response surfaces of species on a number of resource gradients, especially light quantity and quality, soil moisture, and nutrients are needed. These data must be obtained for individuals of different ages because age is likely to greatly affect their response.

A comparative approach to the physiological ecology of regeneration is necessary. Tropical forests occur on soils of different fertility and the physiological attributes of the component species are likely to differ. Furthermore, tropical forests occur in three main regions that have few species but many guilds in common. In order to learn enough to manage the tropical rain forests of the world, we need an internationally supported plan that identifies the major research needs, suggests uniform methodology and data analysis, and gives an overall synthesis.

REFERENCES

Alexandre, D.Y. (1982). Aspects de la regeneration naturelle en forêt dense de Côte Ivoire. *Candollea,* **37**, 579–88

Aoki, M., Yabuki, K. and Koyama, H. (1975). Micrometeorology and assessment of primary production of a tropical rain forest in West Malaysia. *Journal of Agricultural Meteorology* (Tokyo), **31**, 115–24

Ashton, P.S. (1978). Crown characteristics of tropical trees. In Tomlinson, P.B. and Zimmermann, M.H. (eds.) *Tropical Trees as Living Systems*, pp. 591–615. (Cambridge University Press: Cambridge)

Auspurger, C.K. (1984). Light requirements of neotropical tree seedlings: a comparative study of growth and survival. *Journal of Ecology,* **72**, 777–95

Bazzaz, F.A. (1979). Physiological ecology of plant succession. *Annual Review of Ecology and Systematics,* **10**, 351–71

Bazzaz, F.A. (1983). Characteristics of populations in relation to disturbance in natural and man-modified ecosystems. In Mooney, H.A. and Godron, M. (eds.) *Disturbance and Ecosystems – Components of Response*, pp. 259–75. (Springer-Verlag: Berlin)
Bazzaz, F.A. (1984*a*). Dynamics of wet tropical forests and their species strategies. In Medina, E., Mooney, H.A. and Vázquez-Yanes, C. (eds.) *Physiological Ecology of Plants of the Wet Tropics*, pp. 233–43. (Dr. W. Junk Publishers: The Hague)
Bazzaz, F.A. (1984*b*). Demographic consequences of plant physiological traits: some case studies. In Dirzo, R and Sarukhan, J.R (eds.) *Perspectives in Plant Population Ecology*, pp. 324–46. (Sinauer Publishers: Sunderland, Massachusetts)
Bazzaz, F.A. and Pickett, S.T.A. (1980). The physiological ecology of tropical succession: a comparative review. *Annual Review of Ecology and Systematics*, **11**, 287–310
Borchert, R. (1978). Feedback control and age–related changes of shoot growth in seasonal and nonseasonal climates. In Tomlinson, P.B. and Zimmermann, M.H. (eds.) *Tropical Trees as Living Systems*, pp. 497–515. (Cambridge University Press: Cambridge)
Borchert, R. (1980). Phenology and ecophysiology of tropical trees: *Erythrina poeppigiana* Cook O.F. *Ecology*, **61**, 1065–74
Bormann, F.H., and Berlyn, G (eds.) (1981). *Age and Growth Rate of Tropical Trees: New Directions for Research*. Bulletin No. 94. (Yale University School of Forestry and Environmental Studies: New Haven)
Brokaw, N.V.L. (1985*a*). Treefalls, regrowth, and community structure in tropical forests. In Pickett, S.T.A. and White, P.S. (eds.) *The Ecology of Natural Disturbance and Patch Dynamics*, pp. 53–69. (Academic Press: New York)
Brokaw, N.V.L. (1985*b*). Gap–phase regeneration in a tropical forest. *Ecology*, **66**, 682–7
Brokaw, N.V.L. (1987). Gap-phase regeneration of three pioneer tree species in a tropical forest. *Journal of Ecology*, **75**, 9–19
Chazdon, R.L. (1985). Leaf display, canopy structure, and light interception of two understorey palm species. *American Journal of Botany*, **72**, 1493–1502
Chazdon, R.L. (1986). Light variation and carbon gain in rain forest understorey palms. *Journal of Ecology*, **74**, 995–1012
Chazdon, R.L. and Fetcher, N. (1984*a*). Photosynthetic light environments in a lowland tropical rain forest in Costa Rica. *Journal of Ecology*, **72**, 553–64
Chazdon, R.L. and Fetcher, N. (1984*b*). Light environments of tropical forest. In Medina, E., Mooney, H.A. and Vázquez-Yanes, C. (eds.) *Physiological Ecology of Plants of the Wet Tropics*, 27–36. (Dr. W.Junk Publishers: The Hague)
Chiariello, N. (1984). Leaf energy balance in the wet lowland tropics. In Medina, E., Mooney, H.A., and Vázquez-Yanes, C. (eds.) *Physiological Ecology of Plants of the Wet Tropics*, pp. 85–98. (Dr. W.Junk Publishers: The Hague)
Chiariello, N.R., Field, C.B. and Mooney, H.A. (1987). Midday wilting in a tropical pioneer tree. *Functional Ecology*, **1**, 3–11
Coley, P.D. (1983). Herbivory and defensive characteristics of tree species in a lowland tropical forest. *Ecological Monographs*, **53**, 209–33
Coley, P.D., Bryant, J.P. and Chapin, F.S. III. (1985). Resource availability and plant antiherbivore defense. *Science*, **230**, 895–9
Crawley, M.J. (1983). *Herbivory, the Dynamics of Animal-Plant Interactions*. (University of California Press: Berkeley, Calif.)
Cuevas, E. and Medina, E. (1986). Nutrient dynamics within Amazonian forest ecosystems. I. Nutrient flux in fine litterfall and efficiency of nutrient utilization. *Oecologia*, **68**, 466–72
Dilmy, A. (1980). *Dynamics of Forest Trees in Lowland Dipterocarp Forest in University Forest at Lempake, East Kalimantan*. Dissertation. University of Mulawarman, Samarinda, Indonesia.
Dirzo, R. (1984). Insect-plant interactions: some ecophysiological consequences of herbivory. In Medina, E., Mooney, H.A. and Vázquez-Yanes, C. (eds.) *Physiological Ecology of Plants of the Wet Tropics*, pp. 209–24. (Dr. W. Junk Publishers: The Hague)
Ewel, J.J. (1977). Differences between wet and dry successional tropical ecosystems. *Geo-Eco-Trop.*, **1**, 103–17
Fetcher, N. (1979). Water relations of five tropical tree species on Barro Colorado Island, Panama. *Oecologia*, **40**, 229–33
Fetcher, N. (1981). Leaf size and leaf temperature in tropical vines. *American Naturalist*, **117**, 1011–14
Fetcher, N., Oberbauer, S.F. and Strain, B.R. (1983). Effects of light regime on the growth, leaf

morphology, and water relations of seedlings of two species of tropical trees. *Oecologia,* **58**, 314–19

Fetcher, N., Oberbauer, S.F. and Strain, B.R. (1984). Vegetation effects on microclimate in lowland tropical forest in Costa Rica. *International Journal of Biometeoerology,* **29**, 145–55

Fetcher, N., Oberbauer, S.F., Rojas, G. and Strain, B.R. (1987). Efectos del régimen de luz sobre la fotosíntesis y el crecimiento en plántuals de árboles de un bosque lluvioso tropical de Costa Rica. *Revista de Biologia Tropical* **35**, Supplemento, 1, 97–110

Foresta, de H. (1983). Le spectre architectural: Application a l'étude des relations entre architecture des arbres et écologie forestiere. *Bulletin du Muséum national d'histoire naturelle* **4**, série 5, 295–302

Givnish, T.J. (1984). Leaf and canopy adaptations in tropical forests. In Medina, E., Mooney, H.A. and Vázquez-Yanes, C. (eds.) *Physiological Ecology of Plants of the Wet Tropics*, pp. 51–84. (Dr. W. Junk Publishers: The Hague)

Grace, J., Fasehun, F.E. and Dixon, M. (1980). Boundary layer conductance of the leaves of some tropical timber trees. *Plant, Cell and Environment,* **3**, 443–50

Grace, J., Okali, D.U.U. and Fasehun, F.E. (1982). Stomatal conductance of two tropical trees during the wet season in Nigeria. *Journal of Applied Ecology,* **19**, 659–70

Hall, J.B. and Swaine, M.D. (1981). *Distribution and Ecology of Vascular Plants in a Tropical Rain Forest. Forest Vegetation in Ghana.* (Dr. W. Junk Publishers: The Hague)

Hallé, F., Oldeman, R.A.A. and Tomlinson, P.B. (1978). *Tropical Trees and Forests – an Architectural Analysis*. (Springer: New York)

Hartnett, D.C. and Bazzaz, F.A. (1985). Leaf demography and plant–insect interaction: Goldenrods and phloem-feeding aphids. *American Naturalist,* **124**, 137–42

Hartshorn, G.S. (1980). Neotropical forest dynamics. *Biotropica,* **12**, 23–30

Hubbell, S.P. (1970). Tree dispersion, abundance and diversity in a tropical dry forest. *Science,* **203**, 1299–309

Hubbell, S.P. and Foster, R.B. (1983). Diversity of canopy trees in a neotropical forest and implications for conservation. In Sutton, S.L., Whitmore,T.C. and Chadwick, A.C. (eds.) *Tropical Rain Forest Ecology and Management*, pp. 25–42. (Blackwell: Oxford)

Huc, R. and Rosalina, V. (1981). *Aspects of Secondary Forest Succession on Logged Over Lowland Forest Areas on East Kalimantan.* Tropical Forest Research Paper **2(2)**. Mimeographed. (SEAMEO Regional Center for Tropical Biology (BIOTROP): Bogor)

Janzen, D.H. (1981). Patterns of herbivory in a tropical deciduous forest. *Biotropica,* **13**, 271–82

Jordan, C.F. (1984). Nutrient regime in the wet tropics: physical factors. In Medina, E., Mooney, H.A. and Vázquez–Yanes, C. (eds.) *Physiological Ecology of Plants of the Wet Tropics*, pp. 3–12. (Dr. W. Junk Publishers: The Hague)

Kira, T. (1978). Community architecture and organic matter dynamics in tropical lowland rainforests of Southeast Asia with special reference to Pasoh Forest, West Malaysia. In Tomlinson, P.B. and Zimmermann, M.H. (eds.) *Tropical Trees as Living Systems*, pp. 561–90. (Cambridge University Press: Cambridge)

Kira, T., Shinozaki, K. and Hosumi, K. (1969). Structure of forest canopies as related to their primary productivity. *Plant and Cell Physiology,* **10**, 129–42

Landsberg, J.J. (1984). Physical aspects of the water regime of wet tropical vegetation. In Medina, E., Mooney, H.A. and Vázquez-Yanes, C. (eds.) *Physiological Ecology of Plants of the Wet Tropics,* pp. 13–26. (Dr. W. Junk Publishers: The Hague)

Lang, G.E. and Knight, D.H. (1983). Tree growth, mortality, recruitment, and canopy gap formation during a 10-year period in a tropical moist forest. *Ecology,* **64**, 1075–80

Langenheim, J.H. (1984). The roles of plant secondary chemicals in wet tropical ecosystems. In Medina, E., Mooney, H.A. and Vázquez-Yanes, C. (eds.) *Physiological Ecology of Plants of the Wet Tropics*, pp. 189–208. (Dr. W. Junk Publishers: The Hague)

Langenheim, J.H., Osmond, C.B. and Ferrar, P.J. (1984). Photosynthetic responses to light in seedlings of selected Amazonian and Australian rain forest tree species. *Oecologia,* **63**, 215–24

Lee, D.W. 1987. The spectral distribution of radiation in two neotropical rain forests. *Biotropica,* **19**, 161–6

Lieberman, D., Lieberman, M., Hartshorn, G. and Peralta, R. (1986). Growth rates and age-size relationships of tropical wet forest trees in Costa Rica. *Journal of Tropical Ecology*, **1**, 97–109

Lowman, M.D. (1984). An assessment of techniques for measuring herbivory: Is rain forest defoliation more intense than we thought? *Biotropica,* **16**, 264–8

Longman, K.A. (1978). Control of shoot extension and dormancy in external and internal factors. In Tomlinson, P.B. and Zimmermann, M.H. (eds.) *Tropical Trees as Living Systems*, pp.

465–95. (Cambridge University Press: Cambridge)
Mariaux, A. (1981). Past efforts in measuring age and annual growth in tropical trees. In Bormann, F.H. and Berlyn, G. (eds.) *Age and Growth Rate of Tropical Trees: New Directions for Research*. pp. 20–30. Bulletin No.94. Yale University School of Forestry and Environmental Studies, New Haven.
Martinez-Ramos, M. (1985). Claros, ciclos vitales de los arboles tropicales y la regeneracion natural de las selvas atlas perennifolias. In Gómez-Pompa, A., Vázquez-Yanes, C., del Amo R, S. and Butanda, A. C.(eds.) *Regeneracion de Selvas* (Second edition), pp. 191–240. (Campania Editorial: Mexico City, Mexico)
Medina, E. (1984). Nutrient balance and physiological processes at the leaf level. In Medina, E., Mooney, H.A. and Vázquez-Yanes, C. (eds.) *Physiological Ecology of Plants of the Wet Tropics*, pp. 139–54. (Dr. W. Junk Publishers: The Hague)
Medina, E. and Klinge, H. (1983). Tropical forests and tropical woodlands. In Lange,O.L., Nobel, P.S., Osmond, C.B. and Ziegler, H.J. (eds.) *Encyclopedia of Plant Physiology*, New Series, Vol. **12D**, Physiological plant ecology IV. (Springer-Verlag: Berlin)
Medina, E., Mendozo, A. and Montes, R. (1978). Nutrient balance and organic matter production in the Trachypogon savannas of Venezuela. *Tropical Agriculture* (Trinidad), **55**, 243–52
Medina, E., Mooney, H.A. and Vázquez-Yanes, C. (eds.) (1984). *Physiological Ecology of Plants of the Wet Tropics*. (Dr. W. Junk Publishers: The Hague)
Mooney, H.A., Bjorkman, O., Hall, A.E., Medina, E. and Tomlinson, P.B. (1980). The study of the physiological ecology of tropical plants – current status and needs. *BioScience*, **30**, 22–6
Mooney, H.A., Field, C. and Vázquez-Yanes, C. (1984). Photosynthetic characteristics of wet tropical forest plants. In Medina, E., Mooney, H.A. and Vázquez-Yanes, C. (eds.) *Physiological Ecology of Plants of the Wet Tropics*, pp. 113–28. (Dr. W. Junk Publishers: The Hague)
Mooney, H.A., Field, C., Vázquez-Yanes, C., and Chu, C. (1983). Environmental controls on stomatal conductance in a shrub of the humid tropics. *Proceedings of the National Academy of Science*, **80**, 1295–7
Ng, F.S.P. (1983). Ecological principles of tropical lowland rain forest conservation. In Sutton, S.L., Whitmore, T.C. and Chadwick, A.C. (eds.) *Tropical Rain Forest Ecology and Management*, pp. 359–76. (Blackwell: Oxford)
Ng, F.S.P. and Tan, H.T. (1974). Comparative growth rates of Malaysian trees. *Malaysian Forester*, **37**, 2–28
Oberbauer, S.F. and Strain, B.R. (1984). Photosynthesis and successional status of Costa Rican rain forest trees. *Photosynthesis Research*, **5**, 227–32
Oberbauer, S.F. and Strain, B.R. (1985). Effects of light regime on the growth and physiology of *Pentaclethra macroloba* (Mimosaceae) in Costa Rica. *Journal of Tropical Ecology*, **1**, 303–20
Oberbauer, S.F. and Strain, B.R (1986). Effects of canopy position and irradiance on the leaf physiology and morphology of *Pentaclethra macroloba* (Mimosaceae). *American Journal of Botany*, **73**, 409–16
Oberbauer, S.F., Strain, B.R and Reichers, G.H. (1987). Field water relations of a wet-tropical forest tree species, *Pentaclethra macoloba* (Mimosaceae). *Oecologia*, **71**, 369–74
Okali, D.U.U. (1971). Rates of dry-matter production in some tropical forest-tree seedlings. *Annals of Botany*, **35**, 87–97
Orians, G.H. (1982). The influence of tree-falls in tropical forests on tree species richness. *Tropical Ecology*, **23**, 255–79
Pearcy, R.W. (1983). The light environment and growth of C_3 and C_4 tree species in the understorey of a Hawaiian forest. *Oecologia*, **58**, 19–25
Pearcy, R.W. and Calkin. H.C. (1983). Carbon dioxide exchange of C_3 and C_4 tree species in the understorey of a Hawaiian forest. *Oecologia*, **47**, 99–105
Pickett, S.T.A. (1983). Differential adaptation of tropical species to canopy gaps and its role in community dynamics. *Tropical Ecology*, **24**, 68–84
Primack, R.B., Ashton, P.S., Chai, P. and Lee, H.S. (1985). Growth rates and population structure of Moraceae trees in Sarawak, East Malaysia. *Ecology*, **66**, 577–88
Putz, F.E. (1983). Treefall pits and mounds, buried seeds, and the importance of soil disturbance to pioneer trees on Barro Colorado Island, Panama. *Ecology*, **64**, 1069–74
Reich, P.B. and Borchert, R. (1982). Phenology and ecophysiology of the tropical tree, *Tabeuia neochrysantha* (Bignoniaceae). *Ecology*, **63**, 294–9
Reich, P.B. and Borchert, R. (1984). Water stress and tree phenology in a tropical dry forest in the

lowlands of Costa Rica. *Journal of Ecology,* **72**, 61–74
Richards, P.W. (1983). The three-dimensional structure of tropical rain forest. In Sutton, S.L., Whitmore, T.C. and Chadwick, A.C. (eds.) *Tropical Rain Forest Ecology and Management*, pp. 3–10. Special Publication Number 2 of the British Ecological Society. (Blackwell Scientific Publications: Oxford)
Robichaux, R.H. (1984). Variation in the tissue water relations of two sympatric Hawaiian Dubautia species and their natural hybrid. *Oecologia,* **65**, 75–81
Robichaux, R. and Pearcy, R.W. (1980). Environmental characteristics, field water relations, and photosynthetic responses of C_4 Hawaiian *Euphorbia* species from contrasting habitats. *Oecologia,* **47**, 99–105
Robichaux, R.H., Rundel, P.W., Stemmermann, L., Canfield, J.E.,Morse, S.R. and Friedman, W.E. (1984). Tissue water deficits and plant growth in wet tropical environments. In Medina, E., Mooney, H.A. and Vázquez-Yanes, C. (eds.) *Physiological Ecology of Plants of the Wet Tropics*, pp. 99–112. (Dr. W.Junk Publishers: The Hague)
Ross, R. (1954). Ecological studies on the rain forest of southern Nigeria. II. Secondary succession in the Shasha Forest Reserve. *Journal of Ecology,* **42**, 259–82
Sasaki, S. and Mori, T. (1981). Growth responses of Dipterocarp seedlings to light. *Malaysian Forester,* **44**, 319–45
Schatz, G.E., Williamson, G.B., Cogswell, C.M. and Stam, A.C. (1985). Stilt roots and growth of arboreal palms. *Biotropica,* **17**, 206–9
Schulz, J.P. (1960). *Ecological Studies on Rain Forest in Northern Surinam. (*Amsterdam, North Holland)
Shukla, R.P. and Ramakrishnan, P.S. (1986). Architecture and growth strategies of tropical trees in relation to successional status. *Journal of Ecology,* **74**, 33–46
Shulka, R.P. and Ramakrishnan, P.S. 1984. Biomass allucation strategies and productivity of tropical trees related to successional status. *Forest Ecology and Management*, **9**, 315–24
Sutton, S.L., Whitmore, T.C. and Chadwick A.C. (eds.) (1983). *Tropical Rain Forest Ecology and Management.* Special Publication Number 2 of the British Ecological Society. (Blackwell Scientific Publications: Oxford)
Swaine, M.D. and Hall, J.B. (1983). Early succession on cleared forest land in Ghana. *Journal of Ecology,* **71**, 601–28
Tomlinson, P.B. and Longman, K.A (1981). Growth phenology of tropical trees in relation to cambial activity. In Bormann, F.H. and Berly, G. (eds.) *Age and Growth Rate of Tropical Trees: New Directions for Research,* pp. 7–19. Bulletin No. 94. (Yale University School of Forestry and Environmental Studies: New Haven)
Uhl, C. (1987). Factors controlling succession following slash and burn agriculture in Amazonia. *Journal of Ecology,* **75**, 377–407
Uhl, C., Clark, K., Clark, H. and Murphy, P. (1981). Early plant succession after cutting and burning in the upper Rio Negro region of the Amazon Basin. *Journal of Ecology,* **69**, 631–49
Uhl, C. and Jordan, C. (1984). Succession and nutrient dynamics following forest cutting and burning in Amazonia. *Ecology,* **65**, 1476–90
Vázquez-Yanes, C. and Orozco Segovia, A. (1984). Ecophysiology of seed germination in the tropical humid forests of the world: a review. In Medina, E., Mooney, H.A. and Vázquez-Yanes, C. (eds.) *Physiological Ecology of Plants of the Wet Tropics*, pp. 37–50. (Dr. W. Junk Publishers: The Hague)
Vitousek, P. (1984). Litterfall, nutrient cycling, and nutrient limitation in tropical forests. *Ecology,* **65**, 285–98
Vitousek, P.M. and Denslow, J.S. (1986). Nitrogen and phosphorus availability in treefall gaps of a lowland tropical rain forest. *Journal of Ecology,* **74**, 1167–78
Whitehead, D., Okali, D.U.U. and Fasehun, F.E. (1981). Stomatal response to environmental variables in two tropical forest species during the dry season in Nigeria. *Journal of Applied Ecology,* **18**, 571–87
Whitmore, T.C. (1984). *Tropical Rain Forests of the Far East.* (Clarendon Press: Oxford)
Whitmore, T.C. and Bowen, M.R. (1983). Growth analyses of some *Agathis* species. *Malaysian Forester,* **46**, 186–96
Whitmore, T.C. and Wooi-Khoon, G. (1983). Growth analysis of the seedlings of balsa, *Ochroma* lagopus. *New Phytologist,* **95**, 305–11
Yoda, K. (1974). Three-dimensional distribution of light intensity in a tropical rain forest of West Malaysia. *Japanese Journal of Ecology,* **24**, 247–54

CHAPTER 5

REPRODUCTIVE BIOLOGY AND GENETICS OF TROPICAL TREES IN RELATION TO CONSERVATION AND MANAGEMENT

K.S. Bawa and S.L. Krugman

INTRODUCTION

Proper forest management practices are essential components for maintaining and improving the quality of life of the general population of any country. The term management carries different connotations to different people. In our view, management of tropical rain forests often has at least the following four major goals:

(1) Selective removal and regeneration of commercially important species with a minimum impact on the ecology and species-richness of the community.

(2) Variation in species composition with a view to increasing the frequency of commercially more valuable species.

(3) Restoration of degraded lands with native or exotic species.

(4) Conservation of the original biota for both immediate and long-term benefits to the national population.

In order to have responsible forest management, it is essential that there be a level of knowledge concerning the reproductive biology and genetic structure of the forest tree species under management. Although our knowledge of these subjects for tropical forests may be limited, there is still information and experience that can be shared and can be helpful in deciding management strategies. In this paper, we first emphasize the differences between the temperate zone and tropical forest trees with respect to reproductive modes and then evaluate the existing knowledge about the reproductive biology and genetics relevant to the management of tropical rain forests. We then point out constraints to our knowledge in the application of genetics to management practices, and finally present recommendations to overcome these constraints.

Are the genetics of tropical trees different from that of the temperate zone trees? It is difficult to answer this question. In a broad sense, there is no reason to assume that the two groups differ with respect to the mechanisms of inheritance of major traits or in their genetic systems. Superficially, at least, both the tropical and the temperate zone trees are long-lived and outcrossed. However, tree species in tropical lowland rain forests have unique demographic features and diverse modes of reproduction. For example, most tropical forest trees in the humid forests occur in low densities. Hubbell and Foster (1983) found that more than 50% of species in a tropical lowland forest in Panama have only 2–3 individuals over 20 cm dbh per hectare. Ashton (1984) reports similar densities for most species in South-east Asian dipterocarp forests. Further, the species display complex flowering patterns (Bawa 1983). Pollination is almost exclusively by animals, ranging from tiny wasps to large bats (Appanah 1981, Bawa, Bullock, Perry, Grayum and Coville 1985); sexual systems are equally diverse (Ashton 1969, Bawa, Perry and Beach 1985) and seeds of many (if indeed not most) species are dispersed by a wide variety of animals. In contrast, the northern temperate zone trees typically occur in high densities, are largely wind pollinated, monoecious (bisexual trees with separate male and female flowers) and have seeds either wind or gravity dispersed. Thus, as compared to their north-temperate zone counterparts, tropical trees possess complex and varied modes of reproduction. The consequences of low density and diverse, but largely unknown, mating systems on the genetic structure of populations in tropical trees are poorly understood. We know very little about the patterns of genetic variation within and between populations. Finally, for many species, there is a lack of information about the basic reproductive biology, from the initiation of flowering to the formation of seeds. Thus, the extent to which the utilization of genetic principles in management and conservation requires detailed knowledge of reproductive modes, mating system and population genetic structure, the available information is woefully inadequate. Whether or nor the genetics of tropical trees are fundamentally different is an interesting, though a moot, issue.

REPRODUCTIVE MODES AND GENETICS OF TROPICAL RAIN FOREST TREES

Below, we review the existing knowlege about such basic features as flowering and fruiting phenology, models of pollination, sexual systems, gene flow and genetic structure of populations, emphasizing the gaps in our knowledge that are of critical importance in the management and conservation of tropical rain forest trees.

Flowering and fruiting phenology

Much of our knowledge about flowering and fruiting periodicity is derived from community-wide studies (e.g. Frankie *et al.* 1974, Bawa 1983, Appanah 1985). However, it is the behaviour of individuals in a species that is of interest to a forest manager, and there are very few detailed studies of individual species based on large sample sizes.

Flowering

Tree species in a tropical rain forest display much variation in the timing, duration and frequency of flowering (Bawa 1983).

At the community level, even in aseasonal humid forests, there is some seasonality in flowering, though there is hardly a month when several species are not in flower (Frankie *et al.* 1974, Medway 1972). Species vary considerably in the duration of flowering, which extends from a few days in some species to several months in others (Frankie *et al.* 1974). Individuals are usually highly synchronous in their flowering in those cases where flowering lasts a few days or a few weeks. In species with extended blooming, asynchronous blooming is common, but not of universal occurrence.

Flowering may occur annually, supra-annually, or several times a year. In those species that flower more than once a year, quantitative variation in the intensity of flowering among episodes has been documented (Bullock *et al.* 1983). Species that bloom annually may also show variation in flowering (and fruiting) intensity among years (Bullock and Bawa 1981).

The supra-annual species constitute a heterogeneous group. The duration of fruiting intervals varies among species (Janzen 1974, Frankie *et al.* 1974, Ashton 1989). In general, the intervals between fruiting are better documented than the intervals between flowering. It is not clear if each flowering episode is followed by a fruiting episode. In South-east Asian dipterocarp forests, supra-annual flowering and fruiting is a community-wide phenomenon in some cases, but not in others (Janzen 1974, Ashton 1989).

At the community level, flowering of congeners, or related species, is often sequential if the taxa are pollinated by the same guild of pollinators.

This sequence has been demonstrated in both the Central American (Stiles 1978) and South-east Asian rain forests (Appanah 1985). Pollinators probably switch from one species to another as the floral resources of one species decline and that of the other increase.

The temporal variation in flowering should be of considerable interest to managers. The number of seeds (and fruits) produced are often positively correlated with the number of flowers (Bullock and Bawa 1981, Bullock *et al.* 1983), but we do not know how the quantity influences the genetic quality of the seeds. For example, it is possible that, in poor seed years, fewer trees produce a small crop of flowers, thereby increasing the likelihood of inbreeding. The reverse is possible too. Fewer resources may increase competition among pollinators, and hence induce them to increase their foraging ranges and inter-tree visits. It has been suggested that temporal variation in recruitment may be one of the mechanisms maintaining species coexistence in species-rich communities (Chesson 1986). If this is true, then it is not inconceivable that episodes of recruitment coincide with the episodes of bursts in genetic variability.

The relationship between seed quantity and seed quality is of particular interest in species that bloom supra-annually. In this group, many species do produce a small number of seeds in off years. Whether the quality of the seed is comparable to that produced in mast years is not known.

The staggered blooming periods of related species also have management implications. Some of the sequentially blooming species may provide resources to the pollinators at critical times when other food items are in short supply (cf. Gilbert 1980). If so, their removal from the community could influence the pollinator guilds, and consquently, the plant guilds that depend upon these pollinators.

Fruiting

As in the case of flowering, there is considerable variation among species with respect to timing, duration and frequency of fruiting.

At the community level, fruiting in neotropical rain forests is strongly seasonal (Frankie *et al.* 1974, Foster 1982, Terborgh 1986). For example, in both Panama and Peru, peaks in fruit production straddle the start and the later part of the rainy season; in between, the fruit is scarce (Foster 1982, Terborgh 1986). When fruit is scarce, frugivores are believed to sustain themselves on a handful of "keystone" plant species that provide nectar and nuts (Terborgh 1986). At first glance, such community level patterns may not be of much interest to a resource manager interested in a particular species, but the particular species may be dependent upon certain frugivores for seed dispersal. These frugivores, in turn, may rely on other species during lean periods of fruit abudance. The example also emphasizes the fundamental difference between tropical and temperate zone forest tree species in the type of information required for management purposes. The wide variety of ways in which the life cycle of a tropical forest tree is linked with the ecology of other plant and animal species in the community makes it difficult to design a sound management strategy on the basis of biological knowledge of the subject species alone.

How long do various species remain in fruit? Surprisingly, very little is known about the duration of fruiting in tropical tree species. One might imagine selection to favour rapid dispersal of seeds after the ripening of the fruit. However, individual trees may be asynchronous in their fruit ripening schedules, thereby extending the period over which seeds can be collected. Even if the flowering is synchronous, individuals may mature seeds at different times, as is the case in *Pithecellobium pedicellare*, a large canopy tree in the rain forests of central America (Bawa, unpublished observations). Very little is known about the time over which seeds remain viable after they land on the ground. The phenology of events between the time seed matures and the time it germinates is a largely unexplored area in tropical forest ecology.

The comments on the frequency of flowering in the preceding section also apply to the frequency of fruiting. Janzen (1978) has an excellent discussion of fruiting patterns of tropical trees in a dry deciduous forest and Ashton (1989) sums up the information for dipterocarps.

Pollination modes

There is a tremendous diversity of pollination modes among tropical rain forest trees (Appanah 1981, Bawa, Perry and Beach 1985). The pollinators range from tiny wasps in the case of figs, to large bats in the case of *Ceiba pentandra* and other Bombacaceae. Although several large non-flying mammals have also been implicated in the pollination of some trees (Sussman and Raven 1978), the evidence for their role as effective pollinators remains weak.

From the management perspective, pollination biology of individual species is of the greatest interest. Yet, even for most prominent species, such information is largely anecdotal, scant and incomplete. At the level of individual species, it is especially important to know the extent to which the various species are dependent upon particular pollinators. In cases of extreme specialization, the management of the pollinator population becomes as important as the management of the tree species serviced by these pollinators. The extreme specialization between figs and fig wasps is well-known (Wiebes 1979). In many oil palm plantations, seed and fruit set is severely restricted by the absence or low frequency of its beetle pollinators (Syed 1979, Henderson 1986). On the other hand, *Gmelina arborea*, a native of South-east Asian tropics, sets abundant seeds in Costa Rica where it is artifically planted and pollinated by European honey bees (Bawa, unpublished data). Apparently, in this case, honey bees are as effective in pollinating as the native South-east Asian tropical bees that normally pollinate *Gmelina arborea*. Note that flowers of *Gmelina arborea* are specifically adapted for pollination by relatively large bees. However, the specialization is not as great as in the case of figs. Similarly, the nutmeg of commerce, *Myristica fragrans*, was found to set abundant seeds in south India, outside its native range. In this case, the pollination is achieved by a beetle, *Formicomus braminus*, which pollinates related species of *Myristica* in the forests surrounding the plantations of the nutmeg of commerce; the beetle is apparently a strong flier (Armstrong and Drummond 1986).

An adequate management strategy must also take into account the way the populations of pollinators might be influenced by changes in the frequency and composition of forest species. Tropical rain forests pose a tremendous challenge in this regard because, as mentioned earlier, groups of species often share the same pollen vectors. Because we know virtually nothing about the phenology and population dynamics of most pollinators, we cannot assess the impact of changes in frequency of various plant species on the populations of their pollinators. The population dynamics of pollinators in a seasonal dipterocarp forests are particularly intriguing. In these communities, the dipterocarps as well as other species, flower and fruit every few years (Ashton 1989), and certain groups of trees are serviced by pollinators that migrate from the secondary forests during the gregarious flowering (Appanah 1985). The fate of pollinators during other years is generally unknown. They could subsist on other species that are relatively uncommon but very significant in providing resources to the pollinators at critical periods. Alternatively, one may speculate that the life cycles of the pollinators may be so closely tied to the flowering of their hosts that these species remain in diapause when their hosts are not in flower.

In summary, the effective management of tropical rain forests may require a knowledge of plant pollinator interactions at the level of individual species as well as at the level of communities.

Sexual systems

The diversity of pollination mechanisms in tropical rain forest trees is matched by the diversity in sexual systems. Most species bear bisexual flowers, but are self-incompatible. Many are dioecious, that is, characterized by the presence of separate male and female plants (Bawa 1974, 1979, Bawa and Opler 1975, Bawa, Perry and Beach 1985). Essentially, then, most tropical forest trees are obligately outcrossed. There is, however, considerable variation in the strength of the self-incompatibility barrier. In several species and within the same population, some individuals may be entirely self-incompatible, but others may display self-compatibility to varying degrees (Bawa 1974, 1979, Chan 1981, Bawa, Perry and Beach 1985). In other species, such as *Cordia alliodora*, almost all individuals may display a certain degree of self-compatibility (Bawa 1974 and references therein). The range of quantitative variation in self-incompatibility and whether or not such variation has a genetic basis remains to be investigated. In some dipterocarp forests, facultative apomixis (development of a fraction of embryos from somatic cells) appears to be quite common (Ashton 1989).

In terms of management, the degree of inbreeding displayed by the individual is a critical issue in assessing the genetic quality of the seed crop. The amount of inbreeding is not only dependent upon the genetic propensity towards selfing of an individual, but also on the spatial configuration of the relatives. To the extent that the management involves alteration of dispersion patterns, the effect of changing spatial patterns on the amount of inbreeding becomes an important issue.

In dioecious species, optimal spacing of male and female trees is an even more complex issue and cannot be ignored, especially in those cases where the product of economic interest is seed or fruit. The optimal yield requires an appropriate number and mixture of male and female plants. However, in many cases, pollinator attraction to a particular species is maintained by the great preponderance of male flowers, which alone offer floral rewards to the pollinators. In such cases, the female flowers are infrequent, devoid of floral rewards and mimic male flowers in order to get pollinated (Baker 1976, Bawa 1980). Thus, for example, in Papaya and its wild relatives, there are many more male flowers than female flowers. The male, but not the female, flowers offer nectar to the pollinators. The female flowers, however, mimic male flowers by having petaloid stigmas that resemble the corolla lobes of the male flowers (Bawa 1980). Pollinators visit female flowers by "mistake" (Bawa 1976). The example illustrates that, in some dioecious species, a determination of an "appropriate number and mixture of male and female plants" for optimal fruit and seed set is not a simple matter.

Mating systems, pollen flow and effective population size

A convenient way to estimate the amount of outcrossing (or inbreeding) is to analyze the mating system by means of genetic markers (Clegg 1980). The mating system is determined by the degree of self-fertilization, dispersion pattern of related individuals (family structure) and the characteristics of gene flow. The parameters that determine the mating system also define the effective population size, which in essence describes the boundaries of genetic neighbourhoods within which the individuals mate freely with each other.

The mating system has been analyzed quantitatively so far in two species of tropical rain forest trees: *Bertholletia excelsa* (Brazil-nut) and *Pithecellobium pedicellare*, a mimosoid legume (Bawa and O'Malley 1987, O'Malley and Bawa 1987, O'Malley *et al.* 1988). The data indicate that outcrossing rate in both species is very high. (Outcrossing rate is measured by $\hat{t}$ its value typically ranges from 0 to 1 as one moves towards a gradient of increasing outcrossing rate; the multilocus estimate of $\hat{t}$ for *Pithecellobium pedicellare* is 0.951 ± 0.0422, while that for *Bertholletia excelsa* is 0.849 ± 0.033). In the population sample, *P. pedicellare* occurs at the density of 1–2 reproductively mature individuals per hectare. It seems that very wide outcrossing can tie individuals over a wide area by means of pollen flow in this low density population. The data also revealed considerable variation among individuals in outcrossing rates. This variation could be due to differences among individuals in consanguineous mating or in selfing rates.

Despite a very high outcrossing rate, the potential for inbreeding in *Pithecellobium pedicellare* is not negligible. A high frequency of albino seedlings in progeny arrays and the structuring of populations is a further indication of inbreeding (O'Malley and Bawa 1987). Albino seedlings have also been reported in other tropical trees (Venkatesh and Emmanuel 1976).

There are absolutely no data on gene flow in tropical forest tree populations. However, the identification of allozyme markers in tropical trees and paternity exclusion procedures offer a unique opportunity to measure accurately the movement of genes via pollen and seed (Ellstrand 1984, Meagher 1986, Bawa and O'Malley 1987, Hamrick and Loveless 1986*a*).

In summary, most tropical forest trees are certainly widely outcrossed, but we know very little about the extent of gene flow or subdivision within the populations, or the effective size of populations.

The importance of knowledge about mating systems, gene flow and population structure in management is quite obvious. It is important to know the amount of inbreeding (or outcrossing) within a population and whether or not individual trees significantly differ in outcrossing rates. Selective removal of trees can alter the mating patterns with unknown consequences on the (genetic) quality and quantity of the seed crop. If populations are structured on a local scale, and if neighbours are more related to each other than to individuals farther away, the thinning of the stand may reduce inbreeding by consanguineous matings and may actually result in greater outcrossing. But if the selective removal were to increase the distance among conspecifics to an extent that it could not be bridged by means of pollen flow, the removal may result in lower fecundity.

In the case of tree-breeding programmes, information about the pollinator flight distances and the extent of pollen flow is vital. On the one hand, one might wish to locate seed orchards near the native forest to draw pollinators, but, on the other hand, such pollinators may "contaminate" the pollen reaching the stigmas of flowers within the orchards.

Genetic variation within and between populations

The most widely used method to determine the amount of genetic variation in natural populations is to estimate the amount of polymorphism for loci coding for specific proteins by means of gel electrophoresis. Recently, this method has been employed to measure genetic diversity in tropical forest trees (Hamrick and Loveless 1986*b*, Loveless and Hamrick 1987, Bawa and O'Malley 1987). The two most commonly used measures of genetic diversity are H, the average level of heterozygosity, and P, the proportion of polymorphic loci. Preliminary analyses reveal that, for tropical trees, the values of H and P, respectively, range from 0.00 to 0.216 and 0.111 to 0.542 (Hamrick and Loveless 1986*b*, Loveless and Hamrick 1987). There is a pitfall in comparing levels of genetic variability if the same loci are not sampled for all species (Simon and Archie 1985). Nonetheless, it seems that the levels of genetic variation in some species are as high as those reported for the temperate zone conifers. In others, there is little genetic variability. Considering the diversity of reproductive modes and other features of biology, it is not surprising that the tropical tree species vary considerably with respect to the amount of genetic diversity.

Much less is known about the spatial organization of genetic variability. In particular, we do not know the spatial scale over which allele and genotype frequencies change. Nor do we have enough information about the extent to which populations are genetically differentiated. However, the use of allozyme markers is now beginning to provide first estimates of genetic differentation among populations. Using Nei's (1972) methods, total genetic diversity can be partitioned into, within, and between population components. For the large canopy species *Bertholletia excelsa*, the statistic G_{st}, which represents the proportion of genetic diversity accounted for by differences among populations, has been found to be 0.0375 for two populations separated by approximately 900 km in Brazil (Buckley *et al.* 1988). On a local scale, for *Swartzia simplex* var. *ochnacea*, a small understorey tree, G_{st} values ranged from 0.018 among four collections on a 50-hectare plot at the Barro Colorado Island (BCI) Panama, 0.022 among three sites separated by 1 km or more at BCI, and 0.107 among four sites in Panama and Costa Rica (Hamrick and Loveless, unpublished data). In comparison, the G_{st} value was found to be 0.016

among 17 populations of the temperate zone *Pinus contorta*, collected from an area roughly the size of BCI island (Hamrick and Smith, unpublished data). All in all, the preliminary data indicate little genetic differentiation among populations of tropical rain forest trees.

In terms of management, high levels of heterozygosity suggest that populations of each species may carry considerable genetic load. A reduction in population size in those species could result in increased homozygosity and lowered reproductive output (Ledig 1986). The alloyzme data and the analysis of mating systems also indicate that many tropical species are very widely outcrossed with extensive gene flow among populations (O'Malley and Bawa 1987, Hamrick and Loveless, unpublished data). Conservation of such species may therefore require preservation of individuals scattered over a wide area.

FOREST GENETICS AND TREE IMPROVEMENT

In many tropical zones, dry and moist, reforestation of degraded lands is the most important management objective. If there is going to be a reforestation programme, why not plant the best trees possible? The same is true whenever artificial regeneration is an important component of the management of natural forests. A proven method for increasing forest productivity is the careful application of basic genetic knowledge. When properly established and carried-out, such genetic activities are a standard forest management practice for increasing productivity for fibre or fuel. This has been the basic philosophy of tree improvement programmes for the last 30 years.

There are probably as many tree improvement strategies as there are forest geneticists. Here, we offer a brief commentary on the basic criteria for selecting a successful strategy. Once the key criteria of a programme are understood, tree improvement activities can be tailored to the needs of forest management. It is worth noting that selecting or even developing a strategy for tropical tree species is a continuous process. Tree improvement efforts are not static, but a process that evolves as we learn more about the biological systems we are attempting to manipulate.

The first step, of course, is the selection of species to be improved. Obviously, species priorities are initially based on their commercial importance or high value in meeting other management needs, i.e. species selection for shelterbelts and windbreaks can be a priorty need in addition to fibre and fuel production. In other words, the effort placed on different species should be proportional to the potential or relative importance of each species to provide the necessary mix of forest products. With a limited financial and technical support base, it is still possible to make a useful contribution by concentrating the effort just on priority species.

Once the array of important commercial species has been identified, priority selection is then based on the amount of genetic variability within the species. The greater the genetic variability, the greater is the opportunity to make useful improvements in desirable characteristics. Finally, the degree of heritability of the desired characteristics to be improved must be considered. There is little value in attempting to improve those desired features of a tree that have a low heritability. However, this can be a serious constraint since there are only a few important commercial tropical species in which some knowledge of the genetic variability has been demonstrated. There is considerable genetic information related to the semi-tropical pines, but very little useful genetic information has been reported for the tropical hardwoods (Loveless and Hamrick 1987, Buckley *et al.* 1988).

The next step in the tree improvement programme is the identification of a base population to provide the source for seeds. Adequate documentation of seed origins is a critical requirement at this stage. No matter if a native or exotic species is to be planted, seed of all species must come from an identified source. Many a failure has been repeated because the same wrong seed source has been employed. Equally as bad is the inability to repeat a successful planting because a seed source origin was not known. To control seed source carefully, identified seed-collection or seed-production units should be established in natural stands or in plantations of known seed sources. In such special units, collection should be made only from phenotypically superior trees. When it is necessary or desirable to collect from plantations of unknown seed sources, only collect from those plantations at least one-third rotation age. To maintain the integrity of the collection or seed production areas, regeneration of such areas should be only from their own seed sources. In the base programme, the forested areas need to be divided into appropriate planting zones based on available ecological and biological data. Seedlings are then planted only in their own planting zones.

Frequently, there is a need to have a more intensive genetics programme. If there is sufficient biological and reproductive information and the species are of high priority, a superior tree selection testing and screening programme can be initiated. In a typical programme, phenotypically superior trees are selected from natural stands and the selected trees are established in seedling or grafted orchards. In this manner, phenotypically superior trees are permitted to mate and produce seeds. A tree may be superior phenotypically but not genetically. While seed from such orchards are, on the average, superior to wild seed collections, it is normal practice to begin testing the offspring from the orchards to ensure the genetic superiority of the parents. In other words, individual trees are selected that have characteristics more desirable than the original base population. The initial selection will form the production plantations, but, following additional testing and further selection, they are used to develop the next generation of improvement, sometimes called the expanded breeding population. Normally we attempt to have the base population as large as possible, while the initial selected breeding population should consist of at least 50 unrelated individuals. For each stage in the selection of genetically superior parents, steps must be taken to maintain a large genetic base. Care must be taken to ensure that the selected parents are unrelated.

There are many different ways to make selections, to establish seed orchards, to test the progeny and to continue to increase the genetic base of the programme. Such details need to be carefully developed for the species involved and the conditions of the selection. Many reports, plans and some books have now been written to provide these details. The point of this discussion is merely to call attention to many of the essential elements that must be considered. The key issue is the level of knowledge of the species to be improved, and an appropriate organization to carry out a designed programmme. There is no one right way to design a tree improvement programme. The programme should be tailored to the individual situations. There is a large adequate knowledge and experience base in tree improvement available that can assist new programmes to get established.

As noted earlier, there are genetic field tests and population studies for only a relatively few of the important tropical tree species (Zobel and Talbert 1984). In spite of this lack of actual field experience in tree improvement, the basic genetic principles still apply (Namkoong *et al.* 1980). As more countries decide to grow their native species, this list of tropical species undergoing genetic improvement will increase.

CONSTRAINTS ON PROGRESS

More than 10 years ago, Bawa (1976) emphasized the importance of information about reproductive modes and genetic structure of populations in the development of tropical tree breeding programmes. Indeed, in recent years, our knowledge about reproductive modes and genetics has begun to increase steadily, especially in the case of lowland tropical rain forest trees (Bawa, Perry and Beach 1985, Bawa, Bullock, Perry, Grayum and Coville 1985, Bawa and O'Malley 1987, Kageyama and Souza Diaz 1985, Ashton 1986, Hamrick and Loveless 1986*a*). However, this information is not being gathered as part of a management strategy directed towards a particular ecosystem or a species. Rather, the goal of these studies is to gather basic data to elucidate ecological and evolutionary processes in the tropics. Progress in management oriented research has unfortunately been limited to the collection of seeds and the establishment of a few species trials. It is therefore pertinent to explore constraints that limit progress. We suggest below that constraints are largely administrative and organizational rather than scientific.

First, the scientific constraints are obvious ones. In many cases, we know virtually next to nothing about even such basic features as when seeds mature, how seeds germinate, and how to grow and plant seedlings in large numbers. In most simple and brutal terms, if one cannot plant and grow a tree species, the application of forest genetics (or some other management strategy) will be of limited value.

Second, the science of tropical forest management itself has not progressed much during the last 40 years or so, in our view. This stagnancy can be attributed to two developments. One is the replacement of heterogeneous natural forests by even-aged homogeneous plantations, and the other is the use of exotic species in afforestation and reforestation programmes. It is only now, when attention is being shifted back to the sustained management of multiple resources of tropical forests, that one can expect to see the incorporation of modern concepts and findings of ecology and genetics of natural forest ecosystems into management practices.

Third, in many instances, geneticists have not been a part of the management teams responsible for afforestation and reforestation programmes. There are many cases where, for example, the seed source or the genetic base of the plantations is unknown. Very often, in the case of tropical tree plantations or species trials, we do not know if the inadequate performance of a species or a provenance is due to site factors, poor experimental design, or the genetic base and the origin of the material.

Fourth, most genetics research on tropical forest trees is narrowly focussed on the genetic improvement of a particular species. While this is useful for artificial regeneration purposes, it is uninformative about such processes as flowering, pollination, mating systems and micro-differentiation in natural populations. Such an approach also provides no information about the interaction of a particular tree species with other tree species.

Fifth, tropical forest tree geneticists often work in isolation from forest managers. It is not unusual to find a plant physiologist practising tissue culture for artificial propagation on one species, a geneticist undertaking a tree improvement programme aimed at another species, and the forester interested in planting a different species. It sounds elementary and somewhat trivial, but a geneticist must be aware of regional or national forestry goals.

CONCLUSIONS AND RECOMMENDATIONS

In conclusion, we offer the following comments and recommendations:

First, information about flowering and fruiting phenology, plant–pollinator interactions, sexual systems and mating systems at the level of individual species and communities is critical for success in management practices that impinge on variation in species composition. Such information is also vital for forest tree breeding programmes. It is true that many tropical forests in the past have been managed without basic information about reproductive modes and genetic consequences of alterations in population structure. But, in many cases, failure to manage tropical forests due to lack of regeneration may be due to ignorance about the basic features of the reproductive biology of the trees. Furthermore, in the past, the extensive forest stands surrounding the managed areas could serve as sources of pollinators as well as genes, and therefore buffer the managed areas against reproductive failure or loss of allelic diversity. But this is no longer true for the contemporary tropical forest "islands" in most areas.

Second, for conservation and long-term management purposes, it is imperative that we begin to gather data on the spatial organization of genetic variability on both a local and geographical scale. Conservation of genetic diversity will remain empty talk until we begin to understand how the diversity which we wish to conserve is distributed in space. In particular, we need to know the consequences of forest fragmentation and reduction in population size on the distribution of genetic variability. Furthermore, there is a need to determine the relationship between genetic variation and fitness; the study of genetic diversity by means of allozyme markers must be linked with the study of morphological variation in those traits that directly influence fitness.

Third, keeping in view that genetics research is expensive and that tropical forest ecosystems are diverse, we suggest that a management approach incorporating genetic principles be focused on particular ecosystems or groups of species in a given geographical area. Even within an ecosystem, we should expect diverse reproductive modes and equally diversified population structures in various species.

Fourth, genetics work must be an integral part of the overall management effort. Apart from forest tree breeding programmes, geneticists can contribute in many ways to the sustained management of natural forests and reforestation of degraded lands.

Fifth, geneticists working with tropical forest trees must shed the temperate zone bias accrued through experience with the temperate zone conifers. The richness of biotic interactions encountered by tropical forest trees is greater by several orders of magnitude when compared to the temperate zone conifers. Consequently, the genetic outcomes are more complex and diverse in tropical forest trees.

Finally, we offer a number of recommendations concerning forest tree improvement programmes:

(1) The justification, technical direction, and resources committed to a

tree improvement effort must be closely related to the national or regional forest policy. The national or regional policy to a sizeable degree determines the very nature of a tree improvement effort. For example, the decision as to whether exotic tree species should be introduced, or whether only native tropical trees, or both, should be planted, determines the very nature of a tree improvement effort and should be addressed in the policy statements. There is obviously little value in proposing or developing a programme that does not fit national or regional goals.

(2) It should be clearly understood at the outset that a successful tree improvement programme is an element of forest management and should not be isolated from the more traditional forestry activities. To be efficient, tree improvement programmes need to be an integral part of the artificial regeneration programme, since the size, nature, and type of artificial regeneration programme determines the scope of the tree improvement effort.

(3) Another criterion essential for a successful tree improvement activity is the careful evaluation of technical skills available to implement an appropriate level of activity. Not infrequently, complex and sophisticated programmes are destined for failure because of a lack of staff training and education. Experience has clearly suggested that it is far better to start with a simple programme and let it evolve as staff education, training, and experience improve.

(4) Tree improvement programmes are of little value if the seeds and/or pollen cannot be collected, stored, and grown in a nursery. A nursery operation needs to be a major element of the tree improvement programme. Without a successful nursery programme, there is no way a tree improvement activity can succeed. The nursery will have the responsibility for handling large as well as small seed lots, storage of seeds and, frequently, pollen. The nursery will have the main responsibility for growing the individual seed lots and keeping them separate. And finally, the nursery will be expected to provide quality seedlings at the appropriate quantities.

(5) A tree improvement programme is only as good as the field tree planting programme. Good professional plantation establishment and management is essential to the success and continual acceptance of improved material. There must be a matching of seed source to sites, and this is best accomplished by joint co-operation of the tree improvement staff with those responsible for plantation establishment.

ACKNOWLEDGEMENTS

Supported in part by grants from the U.S.National Science Foundation, Healey Grant and Faculty Development Grants from the University of Massachusetts. Frank Crome, Jim Hamrick and G. Harrington offered many valuable comments on the first draft of the paper.

REFERENCES

Armstrong, J.E. and Drummond, B.A. III. (1986). Floral biology of *Myristica fragrans* Houtt. (Myristicaceae), the nutmeg of commerce. *Biotropica,* **18**, 32–8

Appanah, S. (1981). Pollination in Malaysian primary forests. *Malaysian Forester,* **44**, 37–42

Appanah, S. (1985). General flowering in the climax rain forests of South-east Asia. *Journal of Tropical Ecology,* **1**, 225–40

Ashton, P.S. (1969). Speciation among tropical forest trees: some deductions in the light of recent evidence. *Biological Journal of the Linnean Society*, London, **1**, 155–96

Ashton, P.S. (1984). Biosystematics of tropical woody plants: a problem of rare species. In Grant, W.F. (ed.) *Plant Biosystematics*, pp. 497–518. (Academic Press: New York)

Ashton, P.S. (1989) Dipterocarp reproductive biology. In Leith, H. and Werger, M.J.A. (eds.) *Tropical Forest Ecosystems. B. Biogeographical and Ecological Studies*, pp. 219–40 Ecosystems of the World 14b (Amsterdam: Elsevier)

Baker, H.G. (1976). "Mistake pollination" as a reproductive system with special reference to Caricaceae. In Burley, J. and Styles, B.T. (eds.) *Tropical Trees: Variation, Breeding and Conservation*, pp. 161–9. (Academic Press: London)

Bawa, K.S. (1974). Breeding systems of tree species of a lowland tropical community. *Evolution,* **28**, 85–92

Bawa, K.S. (1976). Breeding of tropical hardwoods: an evaluation of underlying bases, current status and future prospects. In Burley, J. and Styles, B.T. (eds.) *Tropical Trees: Variation, Breeding and Conservation*, pp. 43–59. (Academic Press: London)

Bawa, K.S. (1979). Breeding systems of trees in a tropical wet forest. *New Zealand Journal of Botany,* **17**, 521–4

Bawa, K.S. (1980). Mimicry of male by female flowers and intrasexual competition for pollinators in *Jacaratia dolichaula* (Smith, D.) Woodson (Caricaceae). *Evolution,* **34**, 467–74

Bawa, K.S. (1983). Patterns of flowering in tropical plants. In Jones, C.E. and Little, R.J. (eds.) *Handbook of Experimental Pollination Biology*, pp. 394–410. (Van Nostrand and Reinhold Co: New York)

Bawa, K.S. and O'Malley, D.M. (1987). Estudios geneticos y de sistemas de cruzamiento en algunase especies arboreas de bosques tropicales. *Review of Tropical Biology,* **35** (Supplement I), 177–88

Bawa, K.S. and Opler, P.A. (1975). Dioecism in tropical forest trees. *Evolution,* **29**, 167–79

Bawa, K.S., Perry, D. and Beach, J.H. 1985. Reproductive biology of tropical lowland rain forest trees. I. Sexual systems and incompatibility mechanisms. *American Journal of Botany,* **27**, 331–45

Bawa, K.S., Bullock, S.H., Perry, D., Grayum, M.H. and Coville, R.E. (1985). Reproduction biology of tropical lowland rain forest trees. II. Pollination systems. *American Journal of Botany,* **72**, 346–56

Buckley, D., O'Malley, D.M., Apsit, V. and Prance, G.T. (1988). Genetics of Brazil–nut (*Bertholletia excelsa* Humb. and Bonpl.: Lecythidaceae). I. Genetic variation in natural populations. *Theoretical and Applied Genetics,* **76**, 923–8

Bullock, S.H. and Bawa, K.S. (1981). Sexual dimorphism and the annual flowering pattern in *Jacaratia dolichaula* (Smith D.) Woodson (Caricaceae) in a Costa Rican rain forest. *Ecology,* **62**, 1494–504

Bullock, S.H., Beach, J.H. and Bawa, K.S (1983). Episodic flowering and sexual dimorphism in *Guarea rhopalocarpa* in a Costa Rican rain forest. *Ecology,* **64**, 851–61

Chan, H.T. (1981). Reproductive biology of some Malaysian dipterocarps. III. Breeding systems. *Malaysian Forester,* **44**, 28–34

Chesson, P.L. (1986). Environmental variation and the coexistence of species. In Diamond, J. and Case, T.J. (eds.) *Community Ecology*, pp. 240–56. (Harper and Row Publishers: New York)

Clegg, M.T. (1980). Measuring plant mating systems. *Bioscience,* **30**, 814–18

Ellstrand, N.C. (1984). Multiple paternity within fruits of the wild radish, *Raphanus sativus*. *American Naturalist,* **123**, 819–28

Foster, R.B. (1982). Famine on Barro Colorado Island. In Leigh, E.G., Jr., Rand, A.S. and Windsor, D.M. (eds.) *The Ecology of a Tropical Rain Forest: Seasonal Rhythms and Long–term Changes*, pp. 151–71. (Smithsonian Institution Press: Washington, D.C.)

Frankie, G.W., Baker, H.G. and Opler, P.A. (1974). Comparative phenological studies of trees in tropical wet and dry forests in the lowlands of Costa Rica. *Journal of Ecology,* **62**, 881–919

Gilbert, L. (1980). Food web organization and conservation of neotropical diversity. In Soulé, M.E. and Wilcox, B.A. (eds.) *Conservation Biology*, pp. 11–34. (Sinauer Associates: Sunderland, Mass.)

Hamrick, J.L. and Loveless, M.D. (1986*a*). The influence of seed dispersal mechanisms on the genetic structure of plant populations. In Estrada, A. and Fleming, T.H. (eds.) *Frugivores and Seed Dispersal*, pp. 211–23. (Dr. W. Junk Publishers: The Hague, Netherlands)

Hamrick, J.L. and Loveless, M.D., (1986*b*). Isozyme variation in tropical trees: Procedures and preliminary results. *Biotropica,* **18**, 201–7

Henderson, A. (1986). A review of pollination studies in the Palmae. *Botanical Review*, **52**, 221–59

Hubbell, S.P. and Foster, R.B. (1983). Diversity of canopy trees in a neotropical forest and implications for conservation. In Sutton, S.L., Whitmore,T.C. and Chadwick, S.C. (eds.) *Tropical Rain Forest Ecology and Management*, pp. 25–41. (Blackwell Scientific Publications: Oxford)

Janzen, D.H. (1974). Tropical black–water rivers, animals and mast fruiting by the Dipterocarpaceae. *Biotropica,* **6**, 69–103

Janzen, D.H. (1978). Seeding patterns of tropical trees. In Tomlinson, P.B. and Zimmermann, M.H. (eds.) *Tropical Trees as Living Systems*, pp. 83–128. (Cambridge University Press: Cambridge)

Kageyama, P.Y. and Souza Dias, I. (1985). The application of genetic concepts to native forest species in Brazil. *Forest Genetics Resources Information* **No. 13**, 2–11. (FAO: Rome)

Ledig, T.H. (1986). Heterozygosity, heterosis and fitness in outbreeding plants. In Soule, M.E. (ed.) *Conservation Biology: The Science of Scarcity and Diversity*, pp. 77–104. (Sinauer Associates: Sunderland, Mass.)

Loveless, M.D. and Hamrick, J.L. (1987). Distribucion de la variacion genetica en especies arboreas tropicales. *Revista de Biologia Tropical*, **35** (Supplement 1), 165–75

Meagher, T.R. (1986). Analysis of paternity within a natural population of *Chamaelirium luteum*. 1. Identification of most likely parents. *American Naturalist,* **128**, 119–215

Medway, L. (1972). Phenology of a tropical rain forest in Malaya. *Biological Journal of the Linnean Society,* **4**, 117–46

Namkoong, G., Barnes, R.D. and Burley, J. (1980). *A Philosophy of Breeding Strategy for Tropical Forest Trees*. Tropical Forestry Papers **No.16**. (Commonwealth Forestry Institute, University of Oxford: Oxford)

Nei, M. (1972). Genetic distance between populations. *American Naturalist,* **106**, 283–92

O'Malley, D.M. and Bawa, K.S. (1987). Mating system of a tropical rain forest tree. *American Journal of Botany,* **74**, 1143–9

O'Malley, D.M., Buckley, D., Prance, G.T. and Bawa, K.S. (1988). Genetics of Brazil-nut (*Bertholletia excelsa* Humb. Bonpl.: Lecythidaceae). II. Mating system. *Theoretical and Applied Genetics,* **76**, 929–32

Simon, C. and Archie, J. (1985). An empirical demonstration of the liability of heterozygosity estimates. *Evolution,* **39**, 463–7

Stiles, F.G. (1978). Temporal organization of flowering among the hummingbird food plants of a tropical wet forest. *Biotropica,* **10**, 194–210

Sussman, R.W. and Raven, P.H. (1978). Pollination by lemurs and marsupials: an archaic coevolutionary system. *Science,* **200**, 731–6

Syed, R.A. (1979). Studies on oil palm plantation by insects. *Bulletin of Entomological Research,* **69**, 213–24

Terborgh, J. (1986). Keystone plant resources in tropical forests. In Soulé, M.E. (ed.) *Conservation Biology: The Science of Scarcity and Diversity*, pp. 330–44. (Sinauer Associates: Sunderland, Mass.)

Vankatesh, C.S. and Emmanuel, C.J.S.K.. (1976). Spontaneous chlorophyll mutations in *Bombax* L. Slivae. *Genetica,* **25**, 137–9

Wiebes, J.T. (1979). Coevolution of figs and their insect pollinators. *Annual Review of Ecology and Systematics,* **10**, 1–12

Zobel, B.J. and Talbert, J.T. (1984). *Applied Forest Tree Improvement.* (John Wiley and Sons: New York)

CHAPTER 6

ASPECTS OF TROPICAL SEED ECOLOGY OF RELEVANCE TO MANAGEMENT OF TROPICAL FORESTED WILDLANDS

D.H. Janzen and C. Vázquez-Yanes

INTRODUCTION

There is already an enormous data base on tropical seed biology in the literature and in human memory. This data base is growing rapidly while its source is shrinking even more rapidly. The relevance of any particular part of this data base to tropical wildland forest management is extremely situation dependent, just as in the case with the management of other agro-ecosystem components. The components of the management process are the site (what particular forest), the management goal(s), the raw materials (plant and animal forest occupants), and the competence of the managers. The guiding principles are 'know thy organisms' and 'know thy habitat'. The habitat determines the relevance of any particular aspect of seed ecology. A thorough knowledge of the literature on tropical seed biology is unlikely to provide specific answers to any particular management problem, but, on the other hand, the seed biology literature often suggests relevant possibilities.

The word 'seed' is deceptively simple. It appears to denote a boring pill that is seemingly removed from environmental impact and xeroxed in enormous numbers, and out of which appears a plant (at which point things get interesting). But the seeds in a laboratory bag have been stripped of their appendages. At one end, a seed is normally attached to a large and complex physiological history – the timing of fruit (seed) production, size of reserves given by the parent, numbers of sibs, duration of development, chemical defense properties, genetic programme, etc. At the other end, a seed is attached to a large and complex future – dispersers, germination conditions, seed predators, dormancy, microdemography and microgeography of deposition and survival, etc. The appendages are extremely responsive to environmental conditions. A seed lacking its two major appendages is like a car battery that has been removed from its vehicle; it can be used, managed and manipulated, but only a small part of its potential is realized. Furthermore, if it is subsequently installed in a truck, its performance may be poor, to say the least.

Therefore, this contribution concerns itself more with the way seeds interact with their pre- and post-dispersal surroundings in tropical forest than with seeds themselves. Some generalizations about tropical seeds themselves are a necessary part of the framework, but are only distantly useful to a person actually attempting to grow or manage a particular forest on a particular site. Of far greater use are the natural history facts of interaction in particular circumstances, facts that are far too numerous to list here.

We concern ourselves not with plantations but rather with regeneration and management (manipulation) of wildland forest. We also do not dwell on the evolutionary biology of most of the traits discussed; for the present time, regeneration and management has quite enough problems in the pragmatics of ecological interactions without spinning its wheels in the mud of how those traits came to be.

This book (and the workshop on which it is based) is focused on regeneration and management of 'rain forest'. However, since rain forest originally covered less than half of the tropics, since more than half of the tropics' forests to be regenerated or managed occur(red) in strongly seasonal tropical habitats, since a great deal of what we know about tropical forest reproductive biology comes from strongly seasonal habitats, and since what is called 'rain forest' in one part of the world is called moist or dry forest in other parts of the world, our contribution is concerned with tropical forest, rather than restricted to tropical 'rain forest' (however that is defined).

Four processes interact to generate the seed shadow that finally produces a seedling shadow: seed production, predation, dispersal and dormancy. The processes that determine subsequent production of adult tree recruitment from that seedling shadow are the same, except for the deletion of dispersal and the addition of growth. Here we concern ourselves almost entirely with seeds, but recognize that, in forest management, seedling and sapling biology may be of equal or greater importance.

SEED PRODUCTION

What do we know?

Flowering dates and seasons are often very different from seed maturation dates and seasons; many tropical trees flower in one season and bear mature seeds in a quite different season, or at least many months later. The causes and consequences are multiple and centre on the time course of the physiological and allocation demands of accumulating reserves to nourish growing seeds, the availability of dispersal agents (including wind), the absence of pre- and post-dispersal seed predators (absences induced by seasonal and mast cycles), and cue availability and perception. Most trees either have immature dormant fruits for a long time, and then abruptly mature them, or have relatively mature-sized fruits throughout seed development but display various patterns of filling the seeds. Seed provisioning may occur through some photosynthesis by the embryo during the seed development stage (especially in dry forest), as well as through parental subsidy.

Individual- and species-specific seed production patterns vary within the individual, population, year, season and habitat. As a general rule, species of the primary forest canopy wait longer (up to many years) between seed crops, and tend to be more synchronous at the level of the population and habitat than are the species that colonize early primary and secondary successional stages. Many species of bamboos are, however, conspicuous exceptions in that they may broadly colonize disturbed sites but wait decades between seeding events. Within a species of tree, an individual's pattern of seed production among years is very situation-dependent rather than locked into a genetically-fixed cuing system (except for bamboos). Individuals in arboreta and other isolated circumstances are notorious for fruiting in years when the population at large does not fruit. For example, adults left standing in pastures (and thus with fully insolated crowns) may make large fruit crops almost every year while conspecifics in nearby forest fruit synchronously at intervals of 2 or more years.

Forest tree seeds are usually produced by outcross pollination. Wind pollination is almost non-existent in tropical forests (there is, for example, only one species of tropical legume tree that is known to be wind pollinated). Outcross pollination by animals in tropical habitats is extremely variable with respect to how far and where the pollen moves, but, in general, animals produce more far-flung and irregular pollen rain than does wind. Species with seeds dispersed by wind increase in proportion in drier habitats, and in intact forest canopies. However, in most species-rich tropical forests, animals disperse the seeds of more than 75% of the species of woody plants. Even in forests characterized by (wind dispersed) Dipterocarpaceae, it is commonplace for the majority of woody plants to be animal dispersed. Since animal pollinators often move long and irregular distances, and since seeds dispersed by animals do also, it is likely that the seed crop of an individual tropical tree is more genetically mixed than is the seed crop of an extra-tropical tree (most of which are wind pollinated, wind dispersed, or both).

There is enormous inter- and intra-specific variation in the size of individual tree's seed crops. A large crop may range from only a few dozen huge seeds to several million small seeds. Likewise, it is commonplace for a given individual to vary as much as 100-fold in the size of its seed crop among years. Within an individual tree's seed crop, it is commonplace for the lightest viable seeds to weigh as little as less than half of the heaviest viable seeds.

Despite our ability to make reasonable generalizations such as those above, when it comes to any particular species of tropical tree, the details of the natural history of its seed production are either totally unknown or unknown to a significant fraction of the potential users of that information. There is no centralized data pool or publication system that accumulates and files what what we already know about the natural history of tropical tree reproductive biology.

Implications for management

While seed production is often a critical part of the process that leads to restoration of a particular forest, its study in the abstract is of little assistance in management. In general, habitat disturbance increases seed yields for the surviving individuals *if* the pollinator services have not been depressed (or altered to where pollinators produce detrimentally inbred genotypes through their patterns of pollen movement) and if the new environment is not detrimental to the tree's reproductive physiology. The increase in seed production comes about directly through increased resources for the tree and indirectly through decreased seed predation by specialist insects and certain forest-loving vertebrates. The increased seed yields may be pleasant for the surviving frugivores and seed predators (as well as those persons that are collecting seed), but they will also severely alter the proportional demography of the seed rain on to the site. Whether this is prejudicial or beneficial to management depends on the management goals.

The present-day climate changes generated by agriculturization will definitely influence the interaction between seed production and forest regeneration. For example, if a large primary forest tree is left standing as a seed tree in a clearing operation, the outcome will depend very much on whether the tree subsequently becomes surrounded by bushy low-grade pasture, fields, or wildland secondary succession.

Flowering is often conspicuous but is not a good indicator of where and when mature seeds will be available either for collection or dispersal by natural agents. Likewise, many tropical trees are functionally males in many years, and therefore their density and location as flowering individuals is a poor indicator of the density and location of seed bearing individuals.

Since seed production varies strongly among and between years, the time of year and the year in which a reforestation scheme or a forest alteration scheme begins will strongly affect the subsequent outcome. Likewise, as long appreciated and worried over by Malaysian foresters, the number of years since the last mast seeding will influence the number of seedlings that are available to generate new trees at the time that the forest is perturbed. As mentioned earlier, trees that are on strong individual, population-wide or habitat-wide seed production cycles are likely to lose their synchronization when the habitat is removed around them. With seed production coming at intervals, the seed (and hence seedling) dynamics of a given year should not be taken as necessarily representative of the subsequent years. These statements apply most strongly to primary forest, but even young secondary succession can have years of high and low seed production.

The selection of trees as seed trees in tropical forests is substantially more complicated than simply leaving a nice large tree. Many tropical trees are dioecious; males have to be left, yet will produce no seed. Many hermaphrodite or female trees never set seed, and the presence of a seed crop on a tree at the time of lumbering (or in pre-cutting inventory) may be the only reliable indicator of a functional seed tree. Seed trees must be accompanied by appropriate habitat for pollinators and seed dispersers (as well as appropriate habitat for seedlings) if pollination and seeding is to be natural. However, substitutions of one agent by another are quite possible and likely (even though, in any specific case, the surrogates are likely to generate a different pollination or seed dispersal pattern than did the originals); whether this will occur in a given habitat with a given species can only be determined by trial and error. It is tempting to view isolated trees left standing on roadsides, in pastures, near houses, etc. as seed trees and as members of the population. However, owing to the conditions on the ground below and the general absence of pollinators and dispersers in such habitats, such trees are in fact 'living dead', physiologically alive but making no contribution to the breeding population in still forested areas.

In sum, about the only kind of positive manipulation that can occur with seed crop production is species-specific reduction in environmental constraints to resources for the adult tree, and then pray that it is not followed by a concomitant or worse increase in seed predator impact. While trees may be bred for high seed yield, the desirability of releasing such trees into a managed habitat will depend on goals. If seed crops are harvested, the harvester is just another kind of seed predator; harvest impact will depend on what would have been the fate of the seeds that are harvested.

SEED PREDATION

What do we know?

On tropical mainlands, greater than 90% of all tree species have more than 50% of their seeds killed by animals and fungi between fruit set and seed germination (and nearly all seeds land in places where the seedling has no chance of survival). It is often very difficult to find crops of seeds that are free of pre-dispersal seed predators (a 'seed predator' is an organism that kills a seed by eating it). However, this applies more to dry forest than to rain forest, and more to intermediate-age and primary forest than to the very first stages of succession. On islands lacking most vertebrates (and many insects), and in mainland habitats in which the animals have been extinguished, the dense lawns of seedlings (and their subsequent death through competition and disease) stand in silent testimony to the importance of seeds as animal food and animals as designers of tropical forest structure.

The degree of pre-dispersal seed predation is extraordinarily variable and very dependent on the circumstances of the tree, year and habitat. Species range from suffering essentially no pre-dispersal seed predation anywhere, to species in which pre-dispersal seed predation intensifies changes as the habitat changes, to species that suffer very high levels of pre-dispersal seed predation almost anywhere (except when the seed predators have been eliminated by habitat destruction). The same gradient occurs in post-dispersal predation on seeds and young seedlings, but there is a poor (if any) correlation between the percentage of predation in one and the percentage in the other.

The degree of seed predation of any given tree species does not correlate well with the abundance of individual adults of that species; tree species that normally occur at very low adult density may suffer 0–99% seed predation. However, if a given tree species suffers a given regime of pre-dispersal seed predation, then a change in that regime is likely to lead to a subsequent change in other demographic and microgeographic traits of recruitment by that tree.

Just as different species of seeds have different dispersal and germination properties, different species of seeds have different susceptibilities to different species of seed predators. Acorns (*Quercus* seeds) are highly edible to many vertebrates, but rejected by some small rodents. Mahogany seeds (*Swietenia*) are rejected by *Liomys* mice but agoutis (*Dasyprocta*) eat them readily. An agouti can gnaw through a palm nut seed coat that a *Liomys* mouse cannot penetrate. A particular species of bruchid beetle larva can eat a *Lonchocarpus* seed that kills a *Liomys* mouse. While seed predators range from extremely host-specific to very generalized, even the very generalized ones reject, or cannot find or kill a substantial fraction of the species of seeds in the habitat.

Implications for management

While it has not been the subject of much explicit study, it is clear that the intensity of pre- and post-dispersal seed predation declines as the scale and intensity of habitat destruction increases. There are multiple causes. First, the pioneer species with very small seeds seem to have seeds sufficiently small that they are often not fed on by either generalist or specialist seed predators. The dynamics of seed predation are quite size-dependent. Furthermore, once a seed has been dispersed, the smaller it is the greater is the chance that it will be totally free of animate seed predators in any habitat. Fungi, on the other hand, appear to be more successful at killing small rather than large seeds. Second, isolated trees in open fields often bear crops that have no contact with the seed predators that kill large portions of conspecific seed crops within the forest. Much seed predation in forest is committed by animals that are quite unwilling to move into open fields or pastures, with the outcome that the initial stages of secondary succession are often characterized by substantially reduced rates of seed predation on seed shadows. The relevant habitat destruction need not apply to the entire vegetative cover; one may speculate that the removal (by hunting) of the large herds of wild pigs, rhinos, forest elephants and other such animals that once fed heavily on dipterocarp seeds in Malaysia has forever altered the seed predator structure and recruitment regimes of those habitats. In this context, a specimen does not a population make. Many animals may still occur in a wildland forest, but at densities that do not even begin to approximate prior conditons. On the other hand, certain species of seed predators occur at higher density in disturbed forest than in primary forest.

It is tempting to believe that if wildland forest restoration schemes could be set up such that the seed predators (peccaries, parrots, mice, beetles, etc.) were missing (owing to the site having been a relative desert before), then reforestation would occur more rapidly and desired trees might be more abundant in the new forest. Such could happen, but, in general, the novel habitat structure so generated would have much more the appearance of island vegetation (fewer tree species in relatively monospecific stands, each to its own habitat) than of mainland vegetation. In other words, the seed predators are in fact removing large numbers of offspring before they have a chance to express their competitive superiority, and many of these species may well be species that are not desired by the manager. Finally, many seed predators are also seed dispersal agents. Primates, ungulates, and large rodents are the most notorious. They move seeds and they kill seeds. Furthermore, the elimination of a rodent that is, for example, killing a large fraction of a seed crop may simply mean that some other mortality agent takes a novel percentage of the larger number of surviving seeds, seedlings or saplings. Likewise, even if the plant becomes more common or more locally widespread, it may well occur in the pattern desired by the forest manager.

SEED DISPERSAL

What do we know?

The dispersal agents (including wind and secondary dispersal by erosion) generate individual and aggregate 'seed shadows' from the survivors of pre-dispersal seed predation. These seed shadows are then pruned by post-dispersal seed predation, germination, and physical mortality. The resultant 'seedling shadow' is then again pruned by the interdependent action of herbivores, competitors and physical mortality.

Just as is the case with seed predators, animate seed dispersal agents are very sensitive to alterations of the habitat. For some (especially the medium-sized to large vertebrate generalists), a moderately disturbed site may have an elevated density of dispersal agents (though their resultant seed shadows fall on a very different habitat mosaic than before); for others, the increased drying, insolation and windiness, and changed biotic interactions, cause a depression of density and/or activity.

Again, just as with predators, seed dispersers have highly idiosyncratic traits. Their responses to different habitats are different, with the outcome that their activities in moving forest seeds into cleared or different habitats will depend in great part on whether the new site is of interest to them. Monkeys do not venture out into pasture and old fields, for example, and therefore they make minimal contribution to dispersal of forest trees into the early stages of large areas of secondary succession; baboons, on the other hand, may freely cross (and defaecate tree seeds in) abandoned pastures and fields. However, forest monkeys often traverse the edges of small disturbances in wildland forest and may be major dispersers of forest seeds into these microhabitats.

Different forests have very different collective reproductive biologies. For example, a Malaysian dipterocarp forest with a species-poor overcanopy of wind-dispersed mast-seeding species with long inter-mast intervals and a species-rich undercanopy of animal-dispersed species that seed more frequently will invade an old field at very different rates and patterns than will a Costa Rican species-rich dry forest that is a mix of animal- and wind-dispersed species that in aggregate fruit every year.

Likewise, that same Malaysian forest understorey is poor in frugivorous small birds while the Costa Rican one is the opposite; this difference will, in turn, influence the relative rates of flow of forest understorey shrubs from primary into secondary succession.

Implications for management

Seed dispersal agents and processes are essential (though not sufficient) for forest to move onto land that has been cleared and for return of partly perturbed forest to its original state. This places a premium on questions of the distance (in time as well as space) of seed sources from the manipulated forest. The soil seed-bank (see below) contains representatives of only a tiny fraction of the species of trees in a tropical forest (just as is the case in extra-tropical forest), and these seeds are in haphazard proportions having little to do with any particular desired forest structure. They are, however, almost all so-called "pioneer" species. The dispersal agents are an essential link in the establishment of the seed shadows that will generate the seedling–environment interaction that will eventually maintain the multi species pattern of tree species in the forest. Seedling establishment and growth to maturity is highly dependent on the number of 'tries' at a given site, which is in turn dependent on seed dispersal systems. The seeds have to get to a gap or other safe site before they can even try to survive and grow there. Since there is both seed attrition in the soil through death and germination (establishment) attempts, and since aged seedlings have different chances of surviving than do new seedlings if a gap or other favourable growth circumstance is opened up, the pattern of seed input will have a strong inpact on the pattern of both appearance and success of recruitment attempts.

Most tropical forests are mixes of wind and animal-dispersed seeds. As perturbation increases, or as the forest invades an open area, these two processes are differentially affected. In a forest from which the vertebrates have been largely removed, not only do the animal-dispersed species begin to decline in numbers and change their relative abundances, but the wind-dispersed species do the opposite. Hunting does not destroy the wind, and the first seeds of large trees to arrive in a large clearing or pasture adjacent to forest are often wind dispersed; furthermore, wind-dispersed seeds almost always mature in the driest part (windiest part) of the year, with the outcome that their seeds are more resistant to the desiccating conditions of an open site than are those of many other primary forest trees.

While a given species of tree often has a multi species disperser coteries, each disperser tends to make a different contribution to the tree's seed shadow. The outcome is that the selective removal of one species of disperser may not result in a significantly diminished rate of seed removal from the parent tree yet result in the plant generating quite a different seed shadow (which in turn generates a quite different seedling shadow). Different dispersers display different timings of fruit removal (which in turn alters the percentage of the seeds that are killed by seed predators) and different dispersal agents generate seed shadows with different susceptibilities to seed predators.

In contrast to seed predators, certain seed dispersers may be sufficiently controllable that they become management tools. Cattle and horses in the neotropics are in this category if used with care at moderate density; they consume large quantities of certain fruits and disperse the seeds among seasons as well as over large areas. Simultaneously, they may trample and graze herbaceous plants in a manner so as to reduce their competition with tree seedlings and saplings, and reduce the amount of fuel during wildfires. Certain wild animals may also have these effects, and therefore have importance in forest management, quite aside from the question of whether the forest is maintaining them in their own right or whether they are being harvested.

Management of animals in forest regeneration, whether as seed dispersers, pollinators or food dispersers, may also legitimize concern for management of areas that are far removed from the site in question. Many frugivorous birds are highly migratory (as are some of the moths that are important pollinators) both within the tropics and between the tropics and extra-tropical regions.

DORMANCY

What do we know?

While seedling biology is beyond the scope of this essay, seeds are just the first sentence in a long paragraph on the biology of waiting. Dormant seeds are waiting, just as are most seedlings in the forest, for the ‘right’ conditions. ‘Seed’ is a term based on morphology rather than ecology. If we used a term that meant ‘waiting juvenile’ rather than seed, it would draw attention to the numerous traits of seeds and seedlings that have their significance in how a plant stays alive between dispersal and its (spurts of) upward growth. While this essay is focused on seeds, the seedling- and sapling-bank may be much more usable and manipulatable than are seed crops. Relative dormancy is a large part of the biology of the seedling- and sapling-bank.

Tropical perennial plant seeds are extremely variable in their dormancy traits. Inter-specific dormancy ranges from dry forest tree seeds that can remain dormant for 10 years in a bottle or soil (even wet soil) to those that are already growing when they hit the ground. Germination within a few weeks after dispersal is the rule in rain forests but there are many exceptions. These are based both on traits intrinsic to the seed and on the weather at the time of dispersal. Within an individual's crop, great variance is also induced by the individual-specific reception of, and sensitivity to, germination cues. Further, just as with all other traits of organisms, seed germination times can be significantly different among conspecific parents. The greater the capability of a seed to remain dormant in an environment free of germination cues (as reflected in cool and/or dry laboratory storage), the greater will be the realized intra-specific variation in nature in the duration of dormancy for that species. However, owing to response to false cues and repeated occurrence of legitimate cues in nature, realized dormancy is always of shorter duration than is potential dormancy. Furthermore, seed predation (by animals, fungi and bacteria) as well as germination usually exhausts a cohort of potentially dormant seeds in nature well before those seeds reach their physiological limits of dormancy.

The soil seed pool is very different among forests and disturbance regimes. First, the wetter the forest throughout the year, the poorer is its soil in seeds of forest plants. This is usually because, the wetter the forest, the less likely are the seeds to be dormant and waiting for germination cues at the time of dispersal. Second, the soil seed-bank is almost entirely made up of pioneer species with a high turn-over rate; these ruderals are usually professional colonizers of new light gaps or long-term disturbances such as river banks and steep slopes. Measurement of their presence in so-called rain forest soils is severely confounded by the fact that most sample sites in the forest have been within a few kilometres of extensive tracts of secondary succession that generate extremely dense and far-flung seed shadows (produced by animals and by wind) that overlay apparently pristine forest. When tree falls occur in large expanses of truly pristine neotropical flatland forest, many of the so-called gap colonizers (e.g. *Cecropia*, *Trema*, *Ochroma*, large *Piper*) are absent or scarce in the regeneration. Even where they are present, there is some doubt as to whether they are derived from a breeding population supported by the small amount of habitat in natural tree falls, or whether their presence is maintained by seed flow from nearby large areas of repeated disturbance (river banks, road-sides, old fields, etc.).

The dormancy of seeds in wildland soils is not distributed randomly across the species in a site. Water-rich seeds almost always germinate within a few weeks (e.g. Dipterocarpaceae), while harder seeds or seeds with harder seed coats (e.g. some Leguminosae) may remain dormant for months (or longer if the weather is dry). Species that are dispersed by passage through the gut of a mammal are often more likely to display moderate dormancy than are those that are dispersed by regurgitation by a bird. Moraceae also display the same sort of variation as do legumes; in neotropical wet forests, *Brosimum*, *Poulsenia*, *Pseudolmedia*, and *Trophis* occupy the dipterocarp end of the spectrum, *Ficus* and *Chlorophora* have dormancy of a few months, and *Cecropia* (like other so-called gap colonizers) has small and dormant seeds that require an increase in light (and sometimes temperature) to germinate and can remain dormant as long as 1–2 years (if not killed by animals or fungi). In the case of the dipterocarps (with their highly edible large seeds), seed escape from predators was originally through predator satiation (though today the most important vertebrate seed predators have been almost extinguished); rapid germination by the abundant and obvious seeds is the next step of escape. Large tree seeds that remain dormant tend to be relatively poisonous to vertebrates, occur at low density in the forest, be widely scattered in space or time, be buried as part of the dispersal process, or otherwise be relatively unavailable to seed predators.

In the dry habitats, there has been realized selection for the ability of seeds to remain dormant during the dry season if the fruits mature in the dry season (just as is the case in extra-tropical trees with respect to the winter). In more humid forests, a large proportion of species may wait for the beginning of the rainy season to mature fruits or to germinate; in dry forest with a 6-month rain-free period, there is virtually no germination of seeds until the rainy season begins. Once into a dormant mode, even large tree seeds may simply wait (a long time in captivity) until a cue comes along; from this standpoint, seeds that mature and disperse in the rainy season are likely to display a very different dormancy pattern than are seeds that mature in the dry season. Once dormancy has evolved, there is then the obvious opportunity for further selection towards spreading the germination pattern of a seed crop in time as well as space.

Generalizations about a seed trait do not necessarily apply to any particular species. For example, large primary forest tree seeds do not generally remain dormant for long periods, yet some of the 2–6 g seeds of *Hymenaea courbaril* (a 100 to 400-year-old large evergreen tree of the primary dry forest of the Pacific side of Mesoamerica), remain hard and dormant in the soil for at least 1–3 years. Within any tropical forest, intra- and inter-specific variation in seed dormancy coupled with the same variation in seed production produces an aggregate seedling shadow that is highly heterogeneous in time and space.

Seed dormancy is a trait that has its origins in a multitude of selection pressures and is also influenced by the raw material on which selection is operating. These include hard seed coats, time of year of dispersal, brutality of the dispersal agent, fruit traits, pattern of seed predation, seed chemistry, seed perception of the environment, seed volume, regime of appearance of seedling growth opportunities, taxonomic lineage, and a multitude of other traits. All these traits are evolutionarily involved in determining seed mass, specific gravity, seed shape, etc.

The variation in the duration of seed dormancy and the conditions that prolong it has long attracted attention. The term 'orthodox' has even been coined for seeds that can be dried and stored under cool conditions and 'recalcitrant' seeds are those that are killed by desiccation and die soon if not allowed to germinate, no matter what the conditions. Such a terminology is akin to labelling an animal 'recalcitrant' if it cannot be maintained without food. The never-ending pursuit of technology to allow storage and transport of tropical forest seeds (as opposed to the messiness of letting the population bubble along in a conserved wildland habitat, and harvest seed as needed and available) has resulted in detailed study showing that desiccation of non-dormant rain forest seeds leads to irreversible structural and chemical changes in seed cell walls. What seems in general to be needed by rain forest seeds for dormancy is reduction of moisture and temperature to the lowest limits tolerable, good ventilation, and liberal application of fungicides.

Just as with other seeds, tropical forest tree seeds display two-to-ten variation in seed mass, seed dimensions and moisture content within and among conspecific crops. There are two conspicuous correlations between these parameters and seed dormancy. The drier the habitat (or season of seed production), the less water there is in mature seeds; the less water, the more uniform the average water content. Second, if a tree ranges over several habitats that differ strongly in their moistness, the seeds from the wetter end of the spectrum are likely to have a higher water content and remain dormant for a shorter time.

Implications for management

Since large tree seeds are not maintained in the soil seed-bank, more attention must be paid to the relationship of the locations of seed trees, seedling/sapling pools, seed predators and dispersal agents, to recruitment. Equally difficult, even where there are barriers between seed trees and seed predators and seed dispersers, large stores of seeds do not accumulate in the soil beneath maternal parent trees (but management of seedling pools may be in order). Finally, there is the annoying fact that, while the large seeds of many species of large trees do not remain dormant in the soil, a relatively rich flora of ruderal herbs, vines and treelets may be dormant there at the time of forest perturbation.

It is particularly important not to use the potential dormancy of seeds in laboratory storage as a measure of their likelihood of dormancy in wildland soils. Many tropical seeds in the drier end of the habitat spectrum disperse their seeds during the dry season, and, if these dry season conditions are maintained during seed storage, at least several years of dormancy may be achieved. In nature, however, germination at the beginning of the rainy season, coupled with continuous post-dispersal seed predation, soon eliminates the seed reservoir (though a substantial seedling reservoir may persist). This caveat applies even to legumes with extremely hard and dry seeds that can last for 10 years in the herbarium.

If management involves the accumulation, storage and dispersal of seeds, it is critical to realize that there is an obvious decline from dry forest to very wet forest in the proportion of the tree species whose seeds will remain dormant for several years or more. However, even in the driest sites, a substantial fraction of the trees will have seeds that cannot tolerate more than one dry season (if that). Furthermore, if management involves introducing seed stocks from elsewhere so as to replace an extinguished local population, it is likely that the incoming seed genotypes will have dormancy traits that match their home habitat better than the destination habitat (this is, however, not necessarily detrimental to a management programme). On the other hand, it is also possible that there is considerable within-population variation in genetic traits for dormancy, opening up the possibility of rapid ecological or evolutionary selection for stocks that have exceptionally dormant seeds.

DISCUSSION

Seeds are one of the many kinds of software needed to restore or manipulate a forest, be it tropical or elsewhere. How relevant the software is, and how well it performs depends on having it in the right computer and having chosen its traits appropriately. Here is a piece of abandoned tropical pasture. What do we do to turn it back into forest? Given a piece of 20-year-old and highly deciduous secondary succession on an old tropical cornfield, what do we do to hasten its return to a semi-evergreen dry forest with an abnormally high density of valuable hardwoods? Here is a dipterocarp forest. How should it be harvested so as to speed its return to a facsimile dipterocarp forest 100 years hence? Seeds and seed biology obviously play a role in the answers to these kinds of questions.

However, studying seeds *in vacuo* in the hope of having the answers fall out is rather like counting the hairs on a horse so as to determine the outcome of a horse race. An analysis of the outcome of past experiments (races) is generally a better (and cheaper) indicator, once you have some understanding of the natural history of horses, jockeys and horse races. And neither of us knows of a single completed case study (or experiment) in either tropical restoration ecology or forest management where the seed component has been teased out and followed through the overall web of interactions that generated this or that persistent vegetation type. Yes, we do know of many little pieces that should influence this or that process in this or that direction (just as does our readership). The papers listed in the bibliography contain reference to many such bits. We also know some experiments that are in progress that are designed to achieve a more analytical description derived from the naturally occurring syntheses in field trials. We even know enough to lay down some general rules for management. But a plethora of case studies there is not.

Seed biology appears logically to have a great deal to do with forest management and reforestation. However, trees are relatively large and recalcitrant beasts, ones that do not easily allow us to manipulate their seed biology even when we understand the relevant traits. Yes, one can make a *Tabebuia ochracea* tree flower in the dry season by cooling it; the hard part is getting the tree into the refrigerator. This recalcitrance leads to a mining attitude about seed traits. Seeds are harvested happenstance, animals and their idiosyncratic ways of dispersal are tolerated, and by and large the goal seems to be to strip away the environment of the tree(s) and replace it with just those parts needed (usually a human draft animal that doubles as pollinator, disperser, protective chemicals, and/or dormancy physiology). However, we feel that the goal should be to minimize the human labour input and maximize the accuracy of the manipulation of the non-human participants to where forest management and regeneration largely takes care of itself. Tropical labour may be cheap today, but it won't be tomorrow; it is cheap today only because we do not add in the cost of assigning a human to draft animal status.

A second departure from convention is called for in applying seed and seedling biology to management of tropical wildland forests. There may in fact be many cases where recruitment of adult plants is much more certainly and rapidly obtained by removing competing plants growing near healthy, if supressed individuals of the desired species.

In the tropics, there is much confusion about how to apply science to management; the area of seed biology is no exception. Detailed studies of how a rain forest tree seed shrivels when dried to death may appear productive, but it is wasteful of human and temporal resources. Continued pursuit of the details of how particular wildland forest systems function will increase understanding but will only very slowly increase our ability to manage any particular forest site in a wildland state unless the studies are focused on that site. Site-specific experiments are critical.

At least the following management rules seem unambiguous.

(1) No matter how compelling the logic, a supposition about a process in seed biology requires field trial in a particular situation before it is certain that it will apply to that situation; a rough and dirty field experiment is worth a thousand logics, at least at this primordial stage.

(2) A given effect (e.g. increase in density of mahogany seedlings appearing in an abandoned pasture) can be produced by altering many different parts of the overall system, rather than just by changing the relevant seed parameters.

(3) Interactants with seeds, be they animals, fungi, bacteria or other plants, are only partly interchangeable; 'animals' do not prey on and disperse seeds, but rather species and individuals do.

(4) Participants should be viewed as ecologically rather than evolutionarily fit together, and certainly should not be viewed as 'co-evolved' unless demonstrated to be; it is certain that many, if not most, of the organisms that initially selected for the seed traits now observed are no longer interacting with those seeds, and that most mainland habitats are populated largely by animals and plants that arrived by immigration from other sites where they did evolve.

(5) Only the elimination (extinction) of species is irreversible; all habitat structure can be regained if the participants are still present and sufficient time and/or resources can be allocated to the project.

(6) Proximity in time and space to seed, disperser and predator source matters, but not necessarily linearly.

(7) Location of the beginning of a seed experiment matters as much in time and season as in space.

(8) The more detailed experience that a project manager has with the details of other tropical restoration and management projects, the more likely he or she is to be able to identify those processes that will dramatically alter other processes in the focal project.

(9) Seeds (and their products, plants) are only partly interchangeable; seeds are much less monomorphic than their appearance would lead one to suspect.

(10) Small plants ('juveniles') may come from sources other than seeds; likewise, a seed-bank in the soil or a remnant seed-bearing plant does not guarantee either persistence or influence of a plant species.

CONCLUSION

It seems evident to us that if the esoteric world of plant and animal ecology is to offer substance to tropical restoration ecology and forest management, it has to partly abandon its traditional and truly haphazard approach to the choice of topics and focus on particular components of projects under scrutiny. The study of tree natural history has got to be given priority rather than assume that we have the time to study all and then choose among these data. A penetrating exploration of the bruchid beetle that eats *Pterocarpus* seeds in pristine forest will not have the double yield offered by the same study done on the bruchid beetle that eats *Acacia* seeds in that *Acacia* invades old grasslands. Evermore detailed analysis of the physiology of why seeds die in the guts of horses may give substantially lower overall returns than will that same level of analysis applied to the question of why the seedlings from seeds in dung that wait to germinate do better than those that germinate immediately upon leaving the disperser. Experimental selective logging can be set up so as to be doing with a bulldozer and a chainsaw what field ecologists have been doing with shotguns to territorial animals for decades.

There is, of course, an ulterior motive. If the esoteric world can meld its research with the projects of the management world, and thereby come to a substantially more rapid understanding of how to restore and manage tropical wildlands, we just might be in time to hang on to a tiny piece of tropical wildlands, really integrate that piece with the remainder of its agro-ecosystem. The tropics are destined to be a green hell, occupied by people who are well fed and bored out of their minds. Our only hope is to show the world what tropical wildland living classrooms and libraries have to offer in intellectual stimulation as well as material goods. The complexity of the interaction of seeds with the rest of the ecosystem is just one part of those largely unoccupied libraries and classrooms.

ACKNOWLEDGEMENTS

This paper was supported by NSF BSR-84-03531 and the Servicio de Parques Nacionales de Costa Rica (DHJ), and by travel funds from the Universidad Nacional Autonoma de Mexico (C. V.-Y.). The manuscript was constructively criticized by C. Herrera, F. Ng, N. Garwood, W. Hallwachs,

C. Uhl, and G. Maury-Lechon. C. Herrera aided greatly with references and Dr. H. Barner kindly permitted a search for literature at the Danida Forest Seed Centre in Hunlebaek, Denmark.

REFERENCES*

Alexandre, D.Y. (1977). Régénération naturelle d'un arbre caractéristique de la forêt équatoriale de Côte-d'Ivoire: *Turraeanthus africana* Pellegr. *Oecologia Plantarum,* **12,** 241–62

Alexandre, D.Y. (1978). Le rôle disséminateur des éléphants en fôret de Tai, Côte-d'Ivoire. *Revue d'Ecologie (Terre et Vie),* **32,** 47–72

Alexandre, D.Y. (1978). Observations sur l'écologie de *Trema guineensis* en basse Côte-d'Ivoire. *Cahiers ORSTOM, Série Biologie,* **13,** 261–6

Alexandre, D.Y. (1980). Caractère saisonnier de la fructification dans une fôret hygrophile de Côte d'Ivoire. *Revue d' Ecologie (Terre et Vie),* **34,** 47–72

Alexandre, D.Y. (1982). Aspects de la régénération naturelle en fôret dense de Côte-d'Ivoire. *Candollea,* **37,** 579–88

Alvarez-Buylla R. (1986). Demografía y dinamica poblacional de *Cecropia obtusifolia* Bertol. (Moraceae) en la selva de Los Tuxtlas, México. MS thesis, Universidad Nacional Autónoma de México, Mexico.

Augspurger, C.K. (1983). Offspring recruitment around tropical trees: changes in cohort distance with time. *Oikos,* **40,** 189–96

August, P.V. (1981). Fig fruit consumption and seed dispersal by *Artibeus jamaicensis* in the Llanos of Venezuela. *Biotropica Supplement.* Reproductive Botany, pp. 70–6

Becker, P. and Wong, M. (1985). Seed dispersal, seed predation, and juvenile mortality of *Aglaia* sp. (Meliaceae) in lowland dipterocarp rain forest. *Biotropica,* **17,** 230–7

Bonaccorso, F.J., Glanz, W.E. and Sandford, C.M.. (1980). Feeding assemblages of mammals at fruiting *Dipteryx panamensis* (Papilionaceae) trees in Panama: seed predation, dispersal, and parasitism. *Revista de Biología Tropical,* **28,** 61–72

Garwood, N.C. (1983). Seed germination in a seasonal tropical forest in Panama: a community study. *Ecological Monographs,* **53,** 159–81

Cheke, A.S., Nanokorn, W. and Yankoses, C. (1979). Dormancy and dispersal of seeds of secondary forest species under the canopy of primary tropical rain forest in northern Thailand. *Biotropica,* **11,** 88–95

Cole, N.H.A. (1977). Effect of light, temperature, and flooding on seed germination of the neotropical grass *Panicum laxum* Sw. *Biotropica,* **9,** 191–4

Cruz, A. (1981). Bird activity and seed dispersal of a montane forest tree (*Dunalia arborescens*) in Jamaica. *Biotropica Supplement.* Reproductive Botany, pp. 34–44

Denslow, J.S. (1980). Gap partitioning among tropical rain forest trees. *Biotropica,* **12** (Supplement), 47–55

DeSteven, D. and Putz, F.E. (1984). Impact of mammals on early recruitment of a tropical canopy tree, *Dipteryx panamensis*, in Panama. *Oikos,* **43,** 207–16

Enright, N. (1985). Existence of a soil seed bank under rain forest in New Guinea. *Australian Journal of Ecology,* **10,** 67–71

Estrada, A. and Coates-Estrada, R. (1984). Fruit eating and seed dispersal by howling monkeys (*Alouatta palliata*) in the tropical rain forest of Los Tuxtlas, Mexico. *American Journal of Primatology,* **6,** 77–91

Fenner, M. (1985). *Seed Ecology.* (Chapman and Hall: London)

Fleming, T. H. (1981). Fecundity, fruiting pattern, and seed dispersal in *Piper amalago* (Piperaceae), a bat-dispersed tropical shrub. *Oecologia,* **51,** 42–6

Fleming, T.H. and Heithaus, E.R. (1981). Frugivorous bats, seed shadows, and the structure of tropical forests. *Biotropica,* **13,** 45–53

* This bibliography is offered as a reading list on the topics addressed in this chapter.

Fleming, T.H., Williams, C.F., Bonaccorso, F.J. and Herbst, L.H. (1985). Phenology, seed dispersal, and colonization in *Muntingia calabura*, a neotropical pioneer tree. *American Journal of Botany,* **72**, 383–91

Foggie, A. (1960). Natural regeneration in the humid tropical forest. *Caribbean Forester,* **21**, 73–80

Foresta, H. de, Charles-Dominique, P., Erard, C. and Provost, M.F. (1984). Zoochorie et premiers stades de la régénération naturelle après coupe en fôret guyanaise. *Revue d'Ecologie* (*Terre et Vie*), **39**, 369–400

Gautier-Hion, A. (1984). La dissémination des graines par les cercopithecides forestiers africains. *Revue d'Ecologie* (*Terre et Vie*), **39**, 159–65

Gómez-Pompa, A. and del Amo R, S. (eds.) (1985). *Investigaciones sobre la Regeneracion de Selvas altas en Veracruz, Mexico*. (Alhambra Mexicana: Mexico)

Gottsberger, G. (1978). Seed dispersal by fish in the inundated regions of Humaita, Amazonia. *Biotropica,* **10**, 170–83

Guevara, S.S. (1986). Plant species availability and regeneration in Mexican tropical rain forest. Acta Univ. Uppsala, (Comprehensive Summaries of Uppsala Dissertations from the Faculty of Science), **48**, 1–25

Guevara, S. and Gómez-Pompa, A.. (1972). Seeds from surface soils in a tropical region of Veracruz, Mexico. *Journal of the Arnold Arboretum,* **53**, 312–35

Hall, J.B. and Swaine, M.D. (1980). Seed stocks in Ghanaian forest soils. *Biotropica,* **12**, 256–63

Hartshorn, G.S. (1978). Treefalls and tropical forest dynamics. In Tomlinson, P.B., and Zimmermann, M.H. (eds.) *Tropical Trees as Living Systems*, pp. 617–38. (Cambridge University Press: New York)

Holijen, A.M.A. and Boerboom, J.H.A. (1982). The *Cecropia* seed bank in the Surinam lowland rain forest. *Biotropica,* **14**, 62–8

Hopkins, M.S. and Graham, A.W. (1983). The species composition of soil seed banks beneath lowland tropical forests in north Queensland, Australia. *Biotropica,* **15**, 90–9

Howe, H.F. and Richter, W.M. (1982). Effects of seed size on seedling size in *Virola surinamensis*: a within- and between-tree analysis. *Oecologia,* **53**, 347–51

Jackson, J.F. (1981). Seed size as a correlate of temporal and spatial patterns of seed fall in a neotropical forest. *Biotropica,* **13**, 121–30

Janzen, D.H. (1970). Herbivores and the number of tree species in tropical forests. *American Naturalist,* **104**, 502–28

Janzen, D.H. (1974). Tropical blackwater rivers, animals, and mast fruiting by the Dipterocarpaceae. *Biotropica,* **6**, 69–103

Janzen, D.H. (1976). Why bamboos wait so long to flower. *Annual Review of Ecology and Systematics,* **7**, 347–91

Janzen, D.H. (1978). Seedling patterns of tropical trees. In Tomlinson, P.B. and Zimmerman, M.H. (eds.) *Tropical Trees as Living Systems*, pp. 83–128. (Cambridge University Press: New York)

Janzen, D.H. (1982). Cenizero tree (Leguminosae: *Pithecellobium saman*) delayed fruit development in Costa Rican deciduous forests. *American Journal of Botany,* **69**, 1269–76

Janzen, D.H. (1983). No park is an island: increase in interference from outside as park size decreases. *Oikos,* **41**, 402–10.

Janzen, D.H. (1985*a*). *Spondias mombin* is culturally deprived in megafauna-free forest. *Journal of Tropical Ecology,* **1**, 131–55

Janzen, D.H. (1985*b*). On ecological fitting. *Oikos,* **45**, 308–10

Janzen, D.H. (1986*a*). Mice, big mammals, and seeds: it matters who defecates what where. In Estrada, A. and Fleming, T.H. (eds.) *Frugivores and Seed Dispersal*, pp. 251–71, (Dr. W. Junk, Publishers: The Hague)

Janzen, D.H. (1986*b*). The future of tropical ecology. *Annual Review of Ecology and Systematics,* **17**, 305–24.

Janzen, D.H. (1986*c*). *Guanacaste National Park: Tropical Ecological and Cultural Restoration.* (UNED: San José, Costa Rica)

Janzen, D.H. (1988). Management of habitat fragments in a tropical dry forest: growth. *Annals of the Missouri Botanical Garden,* **75**, 105–16

Janzen, D.H. and Martin, P.S. (1982). Neotropical anachronisms: the fruits the gomphotheres ate. *Science,* **215**, 19–27

Jordano, P. (1983). Fig seed predation and dispersal by birds. *Biotropica,* **15**, 38–41

Kellman, M.C. (1974). The viable weed seed content of some tropical agricultural soils. *Journal of Applied Ecology,* **11**, 669–77

King, M.W. and Roberts, E.H. (1979). *The Storage of Recalcitrant Seeds; Achievements and Possible Approaches*. Report for the International Board for Plant Genetic Resources, FAO, Rome.

Leite, A.M.C. and Rankin, J.M. (1981). Ecología de sementes de *Pithecellobium racemosum* Ducke. *Acta Amazónica,* **11**, 309–18

Lieberman, D., Hall, J.B., Swaine, M.D. and Lieberman, M. (1979). Seed dispersal by baboons in the Shai Hills, Ghana. *Ecology,* **60**, 65–75

Liew, T.C. (1973). Occurrence of seeds in virgin forest soil with particular reference to secondary species in Sabah. *Malaysian Forester,* **36**, 185–93

Longman, K.A. (1969). The dormancy and survival of plants in the humid tropics. *Society of Experimental Biology Symposium,* **23**, 471–89

Macedo, M. (1977). Dispercao de plantas lenhosas de uma campina amazonica. *Acta Amazonica,* **7** (Supplement), 1–69

Martinez-Ramos, M. and Alvarez-Buylla, E. (1986). Seed dispersal, gap dynamics and tree recruitment: the case of *Cecropia obtusifolia* at Los Tuxtlas, Mexico. In Estrada, A. and Fleming, T. (eds.) *Frugivores and Seed Dispersal*, pp. 333–46. (Dr. W. Junk Publishers: The Hague)

Ng, F.S.P. (1974). Seeds for reforestation. A strategy for sustained supply of indigenous species. *Malaysian Forester,* **37**, 271–7

Ng, F.S.P. (1976). The problem of forest tree seed production with reference to dipterocarps. In Chin, H.F., Enoch, I.C. and Raja Harum, R.M. (eds.) *Seed Technology in the Tropics*, pp. 181–6. (University Pertanian Malaysia: Serdang)

Ng, F.S.P. (1978). Strategies of establishment in Malaysian forest trees. In Tomlinson, P.B. and Zimmerman, M.H. (eds.) *Tropical Trees as Living Systems*, pp. 129–62. (Cambridge University Press: Cambridge)

Ng, F.S.P. (1980). Germination ecology of Malaysian woody plants. *Malaysian Forester,* **43**, 406–37

Nunez Farfan, J.S. (1985). *Aspectos ecologícos de especies pioneras en una selva húmeda de México.* Tesis Profesional, Universidad Nacional Autónoma de México, Mexico, D.F.

Perez Nasser, N. (1985). *Viabilidad en el suelo de las semillas de once especies de la vegetación de Los Tuxtlas, Veracruz.* Tesis de Licenciatura, Universidad Nacional Autónoma de México, Mexico, D.F.

Prèvost, M.F. (1983). Les fruits et les graines des espèces végétales pionères de Guyane francaise. *Revue d'Ecologie* (*Terre et Vie*), **38**, 121–45

Reis, N.R. and Guillaumet, J.L. (1983). Les chauves–souris frugivores de la région de Manaus et leur rôle dans la dissémination des espèces végétales. *Revue d'Ecologie* (*Terre et Vie*), **38**, 147–69

Richards, P.W. (1952). *The Tropical Rain Forest, an Ecological Study*. (Cambridge University Press: Cambridge)

Sasaki, S. (1980). Storage and germination of some Malaysian legume seeds. *Malaysian Forester,* **43**, 161–5

Schemske, D.W. and Brokaw, N. (1981). Treefalls and the distribution of understory birds in a tropical forest. *Ecology,* **62**, 938–45

Smythe, N. (1970). Relationships between fruiting seasons and seed dispersal methods in a neotropical forest. *American Naturalist,* **104**, 25–35

Stocker, G.C. (1981). Regeneration of a North Queensland rain forest following felling and burning. *Biotropica,* **13**, 86–92

Stocker, G.C. and Irvine, A.K. (1983). Seed dispersal by cassowaries (*Casuarius casuarius*) in north Queensland's rain forests. *Biotropica,* **15**, 170–6

Symington, C.F. (1933). The study of secondary growth on rain forest sites in Malaya. *Malayan Forester*, **2**, 107–17

Uhl, C., Clark, K., Clark. H. and Murphy, P. (1981). Early plant succession after cutting and burning in the Amazon basin. *Journal of Ecology,* **69**, 631–69

Uhl, C. (1983). You can keep a good forest down. *Natural History,* **83(4)**, 71–8

Uhl, C. and Clark, K.. (1983). Seed ecology of selected Amazon basin successional species. *Botanical Gazette,* **144**, 419–25

Uhl, C. and Jordan, C.F. (1984). Successional and nutrient dynamics following forest cutting and burning in Amazonia. *Ecology,* **65**, 1476–90

Uhl, C. and Buschbacher, R.. (1985). A distributing synergism between cattle ranch burning practices and selective tree harvesting in the eastern Amazon. *Biotropica,* **17**, 265–8

Vázquez-Yanes, C. and Orozco Segovia, A. (1984*a*). Ecophysiology of seed germination in the tropical humid forests of the world: a review. In Medina, E., Mooney, H.A. and Vázquez–Yanes, C. (eds.) *Physiological Ecology of Plants of the Wet Tropics*, pp. 37–50. Tasks for Vegetation Science 12. (Dr. W. Junk Publishers: The Hague)

Vázquez-Yanes, C. and Orozco Segovia, A. (1984*b*). Fisiologia ecologia de las semillas de arboles de la selva tropical. *Ciencia,* **35**, 191–201

Whitmore, T.C. (1983). Secondary succession from seed in tropical rain forests. *Forestry Abstracts,* **44**, 767–79

Whitmore, T.C. (1984). *Tropical Rain Forests of the Far East.* Second Edition. (Clarendon Press: Oxford)

Willan, R.L. (1985). *A Guide to Forest Seed Handling with Special Reference to the Tropics.* Danida–FAO Forestry Paper, **20(2)**, (FAO: Rome)

Wilson, M.F., Porter, E.A. and Condit, R.S. (1982). Avian frugivore activity in relation to forest light gaps. *Caribbean Journal of Science,* **18**, 1–6

Yap, S.K. (1981). Collection, germination and storage of dipterocarp seeds. *Malaysian Forester,* **44**, 281–300

CHAPTER 7

NUTRIENT CYCLING PROCESSES AND TROPICAL FOREST MANAGEMENT

C.F. Jordan

PROCESS RATES IN THE HUMID TROPICS

Lowland and lower montane forests of the wet and moist tropics and subtropics are strongly influenced by a warm or hot, and continually or seasonally humid climate (Walter 1971). The tropical climate, and its influence on ecosystem processes, is a major reason why nutrients often become limiting when native rain forests are cut and converted into agriculture or pasture (Jordan 1985). Three processes which are especially important in regulating the cycles of nutrients in ecosystems are primary production, decomposition, and soil respiration.

Because of year-round conditions favourable for photosynthesis, annual rates of litter plus wood production in tropical rain forests are usually higher than in other regions (Whittaker and Likens 1975, Jordan 1983). The conditions favourable for continuous photosynthesis also favour continuous decomposition, and consequently this process should have a high annual rate in tropical forests that do not have prolonged dry seasons. Early studies and comparisons of litter decomposition at various latitudes have suggested that rates in fact are higher in tropical rain forest environments (Jenny *et al.* 1949, Madge 1965, Olson 1963). Anderson *et al.* (1983) have questioned whether decomposition really is any faster in the tropics. However, if litter production is generally higher in the wet tropics, as suggested by the data synthesis of Jordan (1983), then annual decomposition also must be higher. If annual decomposition were not higher, then there would have to be higher levels of soil organic matter in the wet tropics compared to other regions. Zinke *et al.* (1984) show that, in general, there is little build up of soil organic matter in the tropics compared to higher latitudes. Levels of soil organic matter generally decline along gradients from high to low latitude, and from high to low elevation (Post *et al.* 1982). However, there are a few exceptions to this trend, such as tropical forests on nutrient-poor podsols (Jordan 1985).

Soil respiration is a measure of the activity of both decomposers and roots in the soil. Therefore, global patterns of soil respiration should in general follow global patterns of net primary production and decomposition. Where the latter processes are high on an annual basis, soil respiration should also be high. There are few data to support this assertion, but comparisons along a climatic gradient in Venezuela (Medina and Zelwer 1972) indicated that rates were positively correlated with temperature and water availability.

The high annual rates of soil respiration in the humid tropics result in high production of carbonic acid in the soil (Johnson *et al.* 1977). Dissociation of carbonic acid into hydrogen and bicarbonate ions creates a high potenetial for leaching, because these ions replace nutrients exchanged on soil surfaces (Jordan 1985). Nutrients dissolved in the soil solution are transported by percolating soil water, and when annual rainfall totals are high, as in tropical rain forests, leaching losses can be high.

The year-round acidification of the soil due to continuous soil respiration also creates a high potential for fixation of phosphorus, frequently by aluminium, which generally becomes more soluble as pH decreases. Although phosphorus is generally not lost from the ecosystem through acidification of the soil as are other nutrients, phosphorus is lost in the sense that it becomes unavailable for uptake by many crop plants. Low level of phosphorus availability is the most common factor limiting plant growth in tropical humid environments (Sanchez 1976, Fox and Searle 1978, Uehara and Gillman 1981).

The continuously warm and humid climate of rain forests also gives potentially high annual losses through volatilization. Microbial activity governs the rate of nitrogen volatilization through a number of pathways such as ammonium volatilization or denitrificaton (Delwiche 1977), and microbial activity is determined in part by temperature and moisture. Sulphur also may be lost through volatilization (Delmas *et al.* 1980).

NUTRIENT-CONSERVING MECHANISMS

How do naturally occurring tropical rain forests survive in this environment which has such a high potential for nutrient loss? It appears that a number of mechanisms have evolved in rain forest species which enable them to minimize losses. Some of the more important nutrient-conserving mechanisms of tropical rain forest species are listed in the following sections.

Trees

Large root biomass
A high root/shoot ratio may be characteristic of some species genetically adapted to nutrient-poor soils (Chapin 1980), while others may have a high root biomass only on nutrient-poor soils (Gerloff and Gabelman 1983). High root biomass occupies more fully the volume of soil where nutrients are held after release from decomposition. High root biomass also increases the surface area on which nutrients can be strongly adsorbed.

Root concentration near surface
A large concentration of roots on or near the soil surface may be caused in some cases by anaerobic conditions or soil impermeability, but often it appears to be a response to low nutrient availability. For example, in the upper Rio Negro region of the Amazon Basin, where soils are very low in fertility, relatively thick mats of roots occur on top of the soil surface (Stark and Jordan 1978). Trees with roots near the soil surface may be better able to compete with decomposers for nutrients, an important factor in low fertility soils. Surface roots also take up nutrients before they reach the mineral soil, where they are held only by exchange and thus are more susceptible to leaching.

While continuous thick root mats on top of the soil surface as at San Carlos are somewhat unusual, many areas in eastern Amazonia have almost continuous thin layers on the soil surface, with thicker patches in soil depressions, and a very definite high concentration of fine roots in the upper few centimetres of soil. Such rooting habits have been observed in forests of eastern Amazonia near the towns of Paragominas, and Maraba, and on the Jari Plantation near Monte Dourado.

Aerial roots
Mats of living and dead bryophytes, lichens, mosses, bromeliads, ferns, and other epiphytes often occur on the branches and stems of rain forest trees. Adventituous tree roots penetrate the mats, and appear to take up nutients captured by the epiphytes (Nadkarni 1981).

Mycorrhizae
The symbiotic relationship between roots of higher plants and mycorrhizal fungi seems to be important for plant nutrition almost everywhere in the world (Ruehle and Marx 1979). Mycorrhizae undoubtedly play an important role in the mineral nutrition of rain forest trees (Janos 1983, St. John and Coleman 1983), but difficulty in quantifying the functional role of mycorrhizae prohibits the conclusion that they are more important in the humid tropics than elsewhere.

Tolerance of acid soils

Many tropical soils are acid, and have high concentrations of labile aluminium, which can inhibit root development of non-adapted plants. Tropical trees that naturally occur on these soils have a high tolerance for aluminium (Baker 1976). While aluminium tolerance is not in itself a nutrient conserving mechanism, it does help explain how rain forests survive on acid soils.

Nutrient uptake kinetics

Where soil nutrient availability is low, species with low requirements will survive and grow, whereas other species with higher requirements may not. Most native species of primary tropical rain forests appear to have low nutrient requirements compared to crop or pasture species. Many crop species have been bred for high growth rates under conditions of high nutrient availability, and consequently are at a disadvantage on low fertility soils when competing with native species having low requirements for nutrients (Olson *et al.* 1981, Chapin 1983).

Long life span

A long life span, a property of many native rain forest trees, enables individuals to take up nutrients beyond their immediate need during seasons of nutrient abundance and store them for later use during seasons of nutrient scarcity (Chapin 1980). However, large above-ground biomass does not in itself seem to be a particular nutrient conserving adaptation of rain forest trees, since trees in many other regions have equally large, or larger, biomass (Jordan 1985).

Leaf morphology and physiology

Many late successional and primary evergreen forest trees of species on nutrient-poor soils of the tropics have leaves which are relatively thick and tough (scleromorphic or sclerophyllic, Klinge and Medina 1979). Leaves that are long lived, tough, and insect resistant may be advantageous to plants in a nutrient-poor area where leaf replacement is energetically expensive (Chapin 1980, Coley 1982, 1983). Long-lived leaves reduce the necessity for nutrient uptake to replace leaves which are shed or eaten.

Allelopathy

Many higher plants synthesize substantial quantities of secondary compounds that can repel or inhibit other plants or herbivores (Whittaker and Feeny 1971). Like scleromorphy, allelopathy is especially beneficial in low nutrient environments, because replacement of leaves damaged by herbivores can be difficult when nutrient reserves are low.

Nutrient translocation
Phosphorus, nitrogen, and potassium are readily retranslocated from leaf to twig before leaf abscission (Charley and Richards 1983). This mechanism probably is more common in species with less sclerophyllous leaves.

Efficiency of nutrient use
The relatively low nutrient concentrations in tissues of many rain forest species suggests that they are able to produce a unit of biomass with a low amount of nutrients (Vitousek 1982, 1984, Klinge 1985).

Reproduction
Many tree species in infertile habitats do not produce a large seed crop every year. Often several years elapse between large seed crops. This reproductive pattern not only reduces nutrient use, it also may keep populations of seed predators at a relatively low level (Janzen 1974).

Silica concentration
The high concentration of silica in many rain forest species (Koeppen 1978) may be related to the low phosphorus availability in many tropical soils. Silicates have the ability to replace phosphates bound with iron and aluminium in the soil, thereby releasing phosphates in soluble form (Taylor 1961, D'hoore and Coulter 1972). Because upper horizons of many tropical soils are low in silica, storage and retranslocation of silica to the soil surface could be an important mechanism for ensuring a supply of phosphate for surface roots (Jordan 1985). Such retranslocation could occur if fine surficial roots are high in silica, and if such roots are commonly shed. Although this seems possible, there is as yet no evidence, but studies that could test the idea have not been carried out.

Bark
The bark of some trees species in nutrient-poor regions is relatively thick. In the upper Rio Negro region of the Amazon Basin, bark constitutes about 10% of the trunk weight of the average tree (Jordan and Uhl 1978), and, in a lower montane forest in New Guinea, it was 7% (Grubb and Edwards 1982). Thick bark may help protect trees from physical injury and subsequent attack by bacteria, fungi, and insects.

Epiphylls
Leaves in the humid tropics are often covered with epiphylls such as algae, lichens, and mosses. Some of the lichens and algae fix nitrogen (e.g. Forman 1975). Almost all epiphylls appear to be effective in scavenging nutrients from rain water moving across the leaves. They can be as effective in nutrient removal as a laboratory ion exchange column

(Witkamp 1970). Studies of radio-active fallout have shown that ions rapidly adsorbed by lower plants during rain storms are later gradually released (Whicker and Schultz 1982). Thus, nutrients captured by epiphylls are slowly released, and probably become available to the host plant through foliar uptake. The relationship between leaves and epiphylls is mutualistic, in the sense that leaves provide physical support for the epiphylls, while the epiphylls increase the nutrients available to the leaves.

Drip tips

Drip tips are often cited as being important for nutrient conservation, because they may reduce the amount of water on the leaf, and consequently the potential leaching (Dean and Smith 1978). However, drip tips may simply be the result of a trend toward sclerophylly, in that the area of the leaf blade is reduced, but the petiole retains its original length (Jordan 1985).

The below-ground ecosystem

The preceding nutrient-conserving mechanisms appear to have evolved in tree species as a result of selection pressures in the nutrient-poor environment. The mechanisms seem to enable individuals to overcome, at least to some extent, limitations imposed by low soil fertility and low pH. There is another mechanism in naturally occurring forests which also conserves nutrients. In contrast to the mechanisms associated with tree species, this mechanism may or may not have evolved as a result of selective pressures in a low nutrient environment. Regardless, it does serve to reduce nutrient losses from the entire ecosystem, and it appears to be more important in nutrient-poor soils than in nutrient-rich soils. That mechanism is the community of lower organisms which live on the soil surface and within the mineral soil environment.

In the undisturbed forest, nutrients released from dead plants and animals usually do not move directly to the mycorrhizae and roots of trees, but instead pass through a whole series of small-scale cycles or “spirals” within the organic matter portion of the soil, similar to spirals of nutrients in streams (Newbold *et al.* 1982). Cycles sometimes begin with soil arthropods. As particles pass through their digestive systems, the complex organic compounds are changed, often by symbionts, to simpler compounds which are more readily utilized by other soil organisms. Decomposition can also begin with invasion of the tissue by bacteria and fungi. If nutrient concentrations in tissues are low, fungi may be the first invaders. As exoenzymes excreted from fungal hyphae break down complex organic compounds, bacterial colonization may be favoured. Further complicating the spirals are fungivores, bacterivores, and carnivores. The inter-relationships between organisms during decomposition may be very complex, and almost every situation is unique (Swift *et al.* 1979).

Nutrients in the soil are relatively susceptible to loss from the soil when they are in the soil solution, or when they are adsorped on surfaces of clay minerals. In contrast, nutrients incorporated in the tissues of organisms of the below-ground community cannot easily be lost through leaching, volatilization, or reaction with iron and aluminium in the case of phosphorus. For this reason, the entire living below-ground community in a natural occurring rain forest can be considered a nutrient-conserving mechanism.

IMPLICATIONS FOR FOREST MANAGEMENT

An important aspect of all the nutrient-conserving mechanisms is that they are all an integral part of the undisturbed functioning rain forest ecosystem. Many naturally occurring rain forests have some or all of the adaptations listed. Although there is little direct evidence for the effectiveness of the mechanisms, changes in nutrient dynamics when rain forests are cut down and the retention mechanisms destroyed suggests that they are important in maintaining the stocks of nutrients in most undisturbed rain forests (Jordan 1985).

What happens when the rain forest is cut down so that the site can be used for agriculture or plantation forests? Since the nutrient conserving mechanisms are part of the naturally occurring rain forest itself, cutting and burning the forest destroys the mechanisms and increases the potential for nutrient loss. A review of 10 case studies of nutrient dynamics in forests of the Amazon basin following deforestation showed that, in most cases, cutting and burning resulted in an increased rate of nutrient loss, diminution of nutrient stocks in the ecosystem, and reduction in primary production (Jordan 1987). Case studies included deforestation for shifting cultivation, permanent plot agriculture with continuous cultivation, lightly grazed pasture, heavily grazed pasture, sustained-yield forestry, and plantation forestry.

Where economic conditions permit, fertilizers are often used to remedy nutrient depletion in soils. However, conventional inorganic fertilizers are inefficient in the humid tropics because nutrients are lost before plants can take them up. Recognition of this problem has led to the use of slow-release fertilizers. Slow-release fertilizers are "solubilized" in the soil at about the same rate as the plants take up nutrients. Organic mulch, like slow-release inorganic fertilizers, releases nutrients at a rate close to that which is optimal for plant uptake, but organic mulch is often not used because it can be difficult to handle and apply. What rarely has been recognized is an important difference between slow-release inorganic fertilizers and organic mulch fertilizers. Organic litter or mulch is a source of energy as well as of nutrients. Decomposing organic matter provides energy for maintaining the below-ground communities that play a critical role in conserving nutrients and sustaining productivity of higher plants.

The importance of naturally occurring forest structure and function for the maintenance of nutrient-conserving mechanisms suggests that the most effective approach to tropical forest management would be to adopt systems that maintain, as much as possible, a forest structure, especially the living below-ground community. Examples of forest management and agriculture which exemplify this approach are discussed in the following sections.

FORESTRY SYSTEMS

There are various approaches to forest management. Native forests can be removed, and replaced with plantation species. Forests also can be managed for production of native species. Management for native trees can be carried out in monocyclic systems or polycyclic systems (Whitmore 1984). The important question here is the degree to which each approach results in significant nutrient loss from the ecosystem.

At the Jari plantation in Pará, Brazil, Russell (1983) carried out a study to determine the loss of nutrients when native forest was selectively logged, the slash and remaining forest burned, and plantation species established. He found that, of the total stock of nutrients in the ecosystem, including the soil, conversion removed 47% of the calcium, 26% of the magnesium, 71% of the potassium, and almost half the nitrogen in the ecosystem. The losses were due to removal of trees, and to volatilization, leaching, and denitrification during the disturbance. Only a small proportion of the phosphorus in the ecosystem was lost because of the large stocks held in insoluble form in the soil. The plantation forests appeared to be able to mobilize some of this insoluble phosphorus during the approximately 10 year rotation, but, nevertheless, nutrient loss is a problem. Fertilization is now practised at Jari (Hornick *et al.* 1984).

In contrast to clear-cutting for plantation forestry, nutrient losses are much less during polycyclic management systems for native species. In these systems, trees are harvested periodically, but only a portion of those on a site are taken. Trees may be selected on the basis of size, species, form, or other criteria. Remaining trees may be left untouched, they may be killed, or they may be encouraged by removing competing trees.

This kind of polycyclic selective harvesting is frequently called sustained yield forestry, probably because a site that is selectively harvested is expected to be able to produce more or less indefinitely with little or no inputs such as fertilizers. During the first half of the 20th century, selective harvest management systems for naturally occurring tropical forests of Africa and South-east Asia were developed (Baur 1968). In Africa, the systems were often referred to as the "tropical shelterwood system".

In Suriname, Jonkers and Schmidt (1984) found that a selective harvest of 20 m^3ha^{-1} from a native forest resulted in a reduction of only 3.1% of the calcium, 2.5% of the potassium, 2.5% of the magnesium, 1.9% of the phosphorus, and 2.1 % of the nitrogen. Repeated selective harvests would of course result in greater losses, but the important aspect is that selective harvesting does not destroy the basic structure of the forest. Therefore, the nutrient-conserving mechanisms are left for the most part in a functional state, and there is little important nutrient loss due to disturbance. Maintenance of both above- and below-ground diversity also is a nutrient-conserving advantage of this type of management.

In monocyclic systems for native species, all larger timber is cut or poisoned at the same time, but natural regeneration of desirable species is left. In South-east Asia, this type of system is known as the "Malayan Uniform System" (Fox 1976, Appanah and Salleh, this volume). This system is part-way between monoculture plantations and polycyclic systems for native systems, in severity of nutrient loss. Most of the above-ground, nutrient-conserving mechanisms are destroyed, but the below-ground mechanisms may remain partially intact, depending on the amount of logging damage.

Management of tropical forest for maintenance of structure not only prevents nutrient loss, but increases gain, compared to systems where the nutrient-conserving mechanisms described above have been destroyed through clear-cutting. In addition, management systems which encourage natural regeneration usually are not seriously affected by disease because the high species diversity common in tropical forests inhibits spread of disease organisms (Orians *et al.* 1974). Erosion too is not usually a problem, since bare soil is seldom exposed on a large scale (Poore 1976).

Although silvicultural systems for native species are very desirable from an ecological aspect, they frequently seem impractical in the face of the intensive pressure from expanding human populations. Many of the systems have been abandoned because they could not compete economically with agricultural use of the land (Lowe 1977).

AGRO-ECOLOGY AT TOME-ASSU

There are a variety of approaches to agricultural management in the tropics. Systems which utilize tree crops, or other perennials, in combination with annual food crops, and sometimes with animals, are often called agro-ecosystems, or agro-forestry systems. Weaver (1979), among others, has described and classified the various systems of agro-forestry. The aspect of agro-forestry systems that apparently makes them relatively sustainable, despite the high potential for nutrient loss in the humid tropics, is that a large proportion of the cropping system, either in time or in space, is occupied by trees.

In addition to maintenance of a forest structure, there are other ways to improve efficiency of nutrient-use in the humid tropics. Near Tome-Assu, in Pará State of Brazil, a group of farms was established by the Japanese early this century (de Oliveira 1983). The farms have banded together to form a co-operative that grows agricultural crops using methods that are very conservative of soil and nutrients. Many of the farms could be described as agro-ecosystems. Efficient use of nutrients comes about not only through agro-forestry within individual farms, but also through co-operation and interaction between farms. A description of some of the farms follows (from Jordan 1987).

A cycle of production on a particular field may start with the clearing of forest. Some trees may be used for lumber, and others converted to charcoal. After the plot is cleared, the remaining slash is burned, and seedlings of a variety of rubber trees tolerant of the low fertility soils may be planted at intervals of several metres. For several years, before the canopies of the rubber trees close, other crops are planted to take advantage of the freshly cleared soil, which is relatively nutrient-rich due to decomposing roots and surface organic matter. The crops also serve to maintain soil cover. First, corn is planted between the rubber seedlings. When the corn is half a metre tall, ginger is planted. When the corn is harvested, the stalks and leaves may be left in the field as mulch, or they may be collected and spread around the base of fruit tree seedlings in neighbouring plantations to conserve moisture and improve soil quality.

Following the corn harvest, cotton, winged beans and peanuts are planted in the same plot. These crops do well due in part to the fertilizing effect of decaying corn roots. If a second rotation of corn–beans–cotton is carried out, some fertilization is required. After 2 years, the canopy of the rubber trees begins to close. A high-yield variety stem of rubber is grafted to the root stock, and later, a second, fungal-resistant graft may be added to form the crown. Non-resistant leaves in Amazon rubber plantations will usually be killed by attacks of the leaf fungus *Dothidella ulei* (Sioli 1973).

Pepper is an important crop at Tome-Assu. When the pepper is harvested, the fruit shells, which the farmers claim repel nematodes, are spread on nearby beds of onions susceptible to the nematodes. Pepper is a vine, and occasionally living trees of nitrogen-fixing species are used to support the vines. Pepper can be cultivated for 5–10 years before problems with rot force abandonment. Several years before a pepper plantation is abandoned, tree species with high commercial value for wood may be planted amongst the pepper vines. Thus, when the pepper plants are abandoned, the plot is already supporting a vigorously growing forest of valuable species.

Cacao is also an important crop, now that techniques to control disease have been developed. Cacao is naturally an understorey tree, and a variety of species such as valuable timber species, nitrogen-fixing trees especially of the genus *Erythrina*, rubber, and coconut palms are used as an overstorey. To control fungal disease of cacao, the fruits are transported out of the plantation before the seeds are extracted. Uninfected pods are composted, and then used as organic fertilizer in the cacao plantation, or in other fields. Infected pods are burned, and the ash is used as fertilizer. Commercial fertilizer is also sometimes used. To gauge the amount of fertilizer that will be required during a season, the farmer will inspect the plantation carefully during the flowering season. If a large number of flowers are observed, a large crop is anticipated and heavy applications of fertilizer are made.

"Witches broom" growths on cacao trees caused by the fungus *Crinipellis perniciosa* (Stahel) is controlled by pruning branches every year, which increases wind flow, and decreases humidity. Pruning also stimulates fruit production, and makes harvesting easier. In some plantations, vanilla, a vine, is planted at the base of trees of cacao which provide support. Vanilla is an extremely high value product, but pollination, which is done by hand because natural pollinators are lacking, is a very tedious process.

Besides cacao, a variety of other fruit trees are grown at Tome-Assu. Seedlings are raised in a nursery before they are planted in the field. Enriched and sterile soil from beneath incinerators is used as a medium for establishing seedlings in the nursery. Palm species valuable for heart of palm, and for fruit used for juice and as pig food in neighbouring farms, are sometimes seeded in along margins of fields by parrots and macaws, and these palms may be left to grow.

Several of the farms belonging to the co-operative at Tome-Assu concentrate on animal production. In one, chicken manure is converted into organic fertilizer. Husks from rice grown on another farm are spread on the floor of the chicken houses. Every few months, the husk-manure is bagged up and used to fertilize plantations of fruit trees. At another farm, pigs are fed corn, rice, and manioc grown on neighbouring farms, and supplemented with minerals. Organic waste from the pigs is used to fertilize pepper plantations.

These are just a few of the combinations of crops at Tome-Assu. Others include pineapple, oil palm, taro, several types of citrus, papaya, and platano.

The farming practices can be summarized in a set of management principles:

(1) Utilize, as much as possible, tree species, in which the biomass harvested such as fruit or latex is only a small proportion of the total plant biomass, thereby minimizing nutrient loss from the system during harvest, and minimizing disturbance to the soil.

(2) For annual crops, use only species of high economic value, and plant them only once, or twice at the most. Immediately follow with species of lower economic value, but which are less demanding of soil nutrients. When possible, use species that can both enrich the soil and provide economic benefit.

(3) Maintain complete ground cover as much as possible, to minimize erosion and deterioration of soil physical properties.

(4) Maintain as high a diversity of crops in an area as possible, to fully exploit soil nutrients and sunlight, and to inhibit problems of diseases and insects. Diversity of species also maintains a more stable economic income.

(5) Recycle, as much as possible, both animal and vegetal organic matter into the soil.

Although the agro-ecology project at Tome-Assu is ecologically desirable, management is closely tied to a strong traditional culture, which contributes to the project's success. If such a project were to be attempted outside that cultural setting, success might be more difficult.

DIRECTIONS FOR FUTURE WORK

The sustained-yield forestry project in Suriname, and the agro-ecology project at Tome-Assu, suggest that a lot is already known about good management techniques in the humid tropics. Knowledge gained from other recent approaches to development in the Amazon have been reviewed by Jordan (1987). However, the knowledge is empirical. Various mangement systems were tried, and from them, the best ones were selected and maintained. Why do they work? We know in a general way. They incorporate to varying degrees the nutrient-conserving mechanisms of the native forest. However, there is still much that can be done scientifically to increase our understanding about why sustainable systems are sustainable, and thereby perhaps increase our ability to improve future systems.

In November 1985, the US National Science Foundation sponsored an international workshop that brought together experts from throughout the world to advise on scientific priorities for promoting research to devise management practices for stable and sustainable production systems in the tropics. The workshop participants were able to translate their perceptions of research priorities into a major research objective, as follows (Crossley and Freckman 1986):

"A major research objective of future tropical ecological research should be to quantify spatial/temporal variation in soil biological processes controlling nutrient conservation in natural and agricultural systems in the tropics. Specifically, future research should strive to: first, identify the key processes which couple mineral nutrient availability and plant uptake, thereby contributing to sustained production; and second, quantify the effects of the frequency and intensity of perturbation on key processes in natural and managed systems in relation to major climate, soil and vegetation types and in relation to the asynchrony of decomposition and nutrient uptake in the habitat mosaic". A list of the specific research questions from Crossley and Freckman (1986) is given in Appendix A.

IMPLEMENTATION

Neither empirical agricultural trials nor process level studies alone will lead toward significant advances in development of more effective production systems. Empirical trials may be guided by experience or intuition, but they are basically a trial and error approach. By themselves, they usually result in slow progress, in part because they often do not explain the mechanisms which contribute to the success or failure of the system. They show which of the systems tried work best, but usually not why they work.

In contrast, process level studies can produce results which predict which systems *should* work best, but predictions are sometimes wrong, due to unforeseen physical conditions such as climatic variation, or biological conditions such as unexpected species interactions. It is extremely difficult to integrate all processes into a model that will make an accurate prediction. Usually, it takes longer to make, and then validate, such a model, than it does to try a series of systems which intuitively seem good, and make the model after the fact. Like empirical studies taken alone, process level studies alone rarely solve practical problems. The best way to make progress may be to alternate between process level studies, which give the understanding of the mechanisms of nutrient cycling and conservation, and landscape-level or ecosystem-level experiments, which are a practical way to evaluate how well all the process-level phenomena are working under given conditions of climate, soil, and species interactions.

Example of implementation

The following paragraphs give a specific example of how studies at both the ecosystem level and process level have been combined to increase our understanding of how to manage ecosystems in the humid tropics.

Ecosystem study

Between 1974 and 1984, an ecosystem-level study was carried out in the upper Rio Negro region of Venezuela's Amazon Territory (Jordan 1987, 1989). The objective of the study was to quantify nutrient cycles in an undisturbed rain forest, and the change in the dynamics due to slash-and-burn agriculture. A hypothesis tested by the project was that cutting and burning of the forest causes a pulse of nutrients from the ash, and this results in good crop productivity for 1 or 2 years: but, as the ash is leached out of the soil, the reduced soil fertility results in decreased crop production.

The study did find a pulse of ash, but much unburned organic matter remained on top of the soil after the burn. The decline in crop growth during the second and third years of cultivation followed more closely the decline in undecomposed organic matter on the soil surface than stocks of nutrients in the soil. At the time the cultivation site was abandoned, stocks of exchangeable calcium, potassium, magnesium, and total nitrogen and phosphorus were still considerably higher than in the soils of the pre-disturbance forest. There was no clear implication of factors other than decline of surface organic matter in the decrease of crop productivity. A study of herbivory showed that this factor was not important in crop decline, and observations of soil erosion suggested only very localized effects (Jordan 1987).

An observation of practices used by local people to sustain crop growth in shifting cultivation sites also suggested a relationship between surface organic matter and crop productivity. During the third year, when residual unburned slash on the soil surface is almost completely decomposed, these cultivators rake together litter, or gather it from the edge of nearby forest, and pile it around the base of crop plants. An informal experiment was carried out on some of the plants in the project's shifting cultivation site to determine the effect of this treatment. Coarse litter was piled around the base of several plants where mineral soil was almost completely exposed, and growth was compared to plants without the litter treatment. Growth of the litter treated plants was substantially greater.

The reason for the increased growth following litter application was not clear. Decomposing litter is a source of nutrients, but the analysis of the soil before the litter application had shown that stocks of exchangeable calcium, potassium, and magnesium, and total stocks of nitrogen and phosphorus were still relatively high, and unlikely to be limiting to growth of crops. One possible explanation seemed to be that phosphorus might occur in variously available forms not distinguished during the shifting cultivation experiment. The dynamics between the different forms might be responsible for the influence on growth, and organic matter, or products from its decomposition, could affect these dynamics.

Process-level study

The results of the slash-and-burn ecosystem study pointed strongly toward the organic matter as being an important regulator that controls phosphorus availability and thus primary productivity in highly weathered tropical soils. The next step was to implement a process-level study which tested this idea. Consequently, in 1987, a study was initiated to test the hypothesis that organic material on the surface of highly weathered tropical soils can increase the fraction of labile phosphorus in the soil, over and above the amount which can be attributed to mineralization of phosphorus in the organic matter itself.

Ideas on how organic matter or its products could solubilize bound phosphorus have been discussed for quite a few years. Following an early report by Gerretsen (1948), many researchers examined the relationships between organic acids in the rhizosphere and plant uptake of phosphorus from mineral sources. Swenson *et al.* (1949), Struthers and Seiling (1950), and Dalton *et al.* (1952) suggested that organic acids liberated during decomposition of organic matter were responsible for solubilizing phosphate complexed by iron and aluminium. Powell *et al.* (1980) and Cline *et al.* (1982) showed that naturally occurring organic compounds in the soil were important in chelating iron. Such chelation would decrease the iron available to react with phosphorus. Graustein *et al.* (1977) showed that certain oxalates keep phosphorus available to plant roots through chelation of iron and aluminium.

Another way that decomposing litter could result in higher levels of available phosphorus is that nutrients and carbon leached from decomposing organic matter stimulate microbial growth. Changes in microbial activity and growth also could occur indirectly from changes in the soil microclimate brought about by the surface mulch. Microbial activity, or products of microbial metabolism may play a role in solubilizing phosphate, either through production of phosphatase and breakdown of recalcitrant humus within the mineral soil, or through organic acids which solubilize inorganically bound phosphorus.

Implications for land management

If the process-level experiments confirm that organic matter on or in the soil plays an important role in "solubilizing" phosphorus bound by aluminium or by clay minerals, these results will have important implications for land management in the lowland humid tropics. One of the most important uses of the knowledge will be in rehabilitating degraded and abandoned tropical pastures.

In many parts of the eastern Amazon basin, tropical forest has been converted into pasture. Because of the low soil fertility, grazing without fertilization of pastures can continue for only 10 or 12 years, or even less. Because fertilization is usually uneconomical, the pastures are abandoned, and within a few years are choked with secondary vegetation that has little or no value. In order to bring the land back into useful production, the weedy successional species must be removed, but this has proven to be an extremely difficult task. Cutting and burning the weeds only results in more vigorous resprouting. When economic conditions permit, bulldozers are used to clear the land. However, growth of economically important plants following bulldozing is usually very poor.

If the process-level studies show that organic matter is important in maintaining phosphorus availability in tropical soils, these studies can supply at least a partial explanation of why productivity following bulldozing usually is very low. Bulldozing removes the litter and slash on top of the soil, and the upper few centimetres of soil where soil organic matter is concentrated. Thus, bulldozing destroys the mechanism of phosphorus mobilization. Agronomists generally regard phosphorus deficiency, along with soil acidity and aluminium toxicity, as the most serious agronomic problems in the humid tropics (Olson and Engelstand 1972, Fox and Searle 1978, Stevenson 1986).

Realization that conservation of soil organic matter and surface litter may be extremely important for tropical soil fertility has led to experiments with land management techniques that are less destructive of organic matter. Near Maraba, Pará, in the eastern region of the Amazon Basin, experiments have begun with herbicides as an alternative to bulldozing as a land-preparation technique. There have been major improvements in formulation of herbicides within the past 15 years, and many of the new products appear to be environmentally more acceptable. Most now decompose relatively quickly, and the type used in the Maraba experiment becomes inactivated immediately upon contact with the soil. Herbicides control the weedy vegetation which makes cultivation impossible, while, at the same time, they preserve the surface litter and soil organic matter. Such preservation not only may increase phosphorus availability, it will also increase the energy-supplying carbon compounds to below-ground fauna which improve the structure of the soil.

CONCLUSION

The type of process-level study illustrated by the organic matter-phosphorus solubility work gives us a better understanding of why some types of management systems work better than others in the humid tropics, and why those that work best are often quite different than those common in temperate zones. High organic matter input seems to be important for sustainability of crop production in the humid tropics, because of its role in maintaining the below-ground ecosystem. Thus, systems that maintain organic matter input, such as agro-ecosystems and polycyclic forest management, are more sustainable than systems which do not, such as monocultures of annual crops, and clear-cut plantation forestry.

Maintaining phosphorus solubility is only one of many functions that may result from organic matter input and activity of the below-ground community. Appendix A lists many more research questions related to organic matter dynamics in the soil, its influence on the nutrient cycle, and the interaction with plant roots.

REFERENCES

Anderson, J.M., Proctor, J. and Vallack, H.W. (1983). Ecological studies in four contrasting lowland rain forests in Gunung Mulu National Park, Sarawak. III. Decomposition processes and nutrient losses from leaf litter. *Journal of Ecology*, **71**, 503–27

Baker, D.E. (1976). Soil chemical constraints in tailoring plants to fit problem soils. 1. Acid soils. In Wright, M.J. (ed.) *Plant Adaptation to Mineral Stress in Problem Soils*, pp. 127–40. Proceedings of a workshop, Beltsville, Maryland, Nov. 1976. (Agency for International Development: Washington, D.C.)

Baur, G.N. (1968). The ecological basis of rain forest management. (Forestry Commission: New South Wales)

Chapin, F.S. (1980). The mineral nutrition of wild plants. *Annual Review of Ecology and Systematics*, **11**, 233–60

Chapin, F.S. (1983). Patterns of nutrient absorption and use by plants from natural and man-modified environments. In Mooney, H.A. and Godron, M. (eds.) *Disturbance and Ecosystems: Components of Response*, pp. 175–87. (Springer-Verlag: Berlin)

Charley, J.L. and Richards, B.N. (1983). Nutrient allocation in plant communities: mineral cycling in terrestrial ecosystems. In Lange, O.L., Nobel, P.S., Osmond, C.B. and Zeigler, H. (eds.) *Physiological Plant Ecology*, Vol. **IV**. *Ecosystem Processes: Mineral Cycling, Productivity and Man's Influence*, pp. 5–45. (Springer-Verlag: Berlin)

Cline, G.R., Powell, P.E. Szaniszlo, P.J. and Reid, C.P.P. (1982). Comparison of the abilities of hydroxamic, synthetic, and other natural organic acids to chelate iron and other ions in nutrient solution. *Soil Science Society of America Journal*, **46**, 1158–64

Coley, P.D. (1982). Rates of herbivory on different tropical trees. In Leigh, Jr., E.G., Rand, A.S. and Windsor, D.M. (eds.) *The Ecology of a Tropical Forest: Seasonal Rhythms and Long-term Changes*, pp. 123–32. (Smithsonian Institution Press: Washington, D.C.)

Coley, P.D. (1983). Herbivory and defensive characteristics of tree species in a lowland tropical forest. *Ecological Monographs*, **53**, 209–33

Crossley, D.A. and Freckman, D. (1986). Workshop on below-ground dynamics in tropical ecosystems. Executive summary. (Institute of Ecology, University of Georgia: Athens, GA.)

Dalton, J.D., Russell, G.C. and Sieling, D.H. (1952). Effect of organic matter on phosphate availability. *Soil Science*, **73**, 173–81

Dean, J.M. and Smith, A.P. (1978). Behavioral and morphological adaptations of a tropical plant to high rainfall. *Biotropica,* **10**, 152–4

Delmas, R., Baudet, J., Servant, J. and Baziard, Y. (1980). Emissions and concentrations of hydrogen sulfide in the air of the tropical forest of the Ivory Coast and of temperate regions of France. *Journal of Geophysical Research,* **85 (C8)**, 4468–74

Delwiche, C.C. (1977). Energy relations in the global nitrogen cycle. *Ambio,* **6**, 106–11

D'hoore, J. and Coulter, J.K. (1972). Soil silicon and plant nutrition. In *Soils of the Humid Tropics*, pp. 163–73. (National Academy of Sciences: Washington, D.C.)

Forman, R.T. (1975). Canopy lichens with blue-green algae: a nitrogen source in a Columbian rain forest. *Ecology,* **56**, 1176–84

Fox, J.E.D. (1976). Constraints on the natural regeneration of tropical moist forest. *Forest Ecology and Management,* **1**, 37–65

Fox, R.L. and Searle, P.G.E. (1978). Phosphate adsorption by soils of the tropics. In Stelly, M. (ed.) *Diversity of Soils in the Tropics*, pp. 97–119. ASA Special Publication No. 34. (American Society of Agronomy: Madison)

Gerloff, G.C. and Gabelman, W.H. (1983). Genetic basis of inorganic plant nutrition. In Lauchi, A. and Bielski, R.L. (eds.) *Encyclopedia of Plant Physiology*, pp. 453–80. New Series, Vol. **15B**, Inorganic Plant Nutrition. (Springer-Verlag: Berlin)

Gerretsen, F.C. (1948). The influence of microorganisms on the phosphate intake by the plant. *Plant and Soil,* **1**, 51–81

Graustein, W.C., Cromack, K., and Sollins, P. (1977). Calcium oxalate: occurrence in soils and effect on nutrient and geochemical cycles. *Science,* **198**, 1252–4

Grubb, P.J. and Edwards, P.J. (1982). Studies of mineral cycling in a montane rain forest in New Guinea. III. The distribution of mineral elements in the above-ground material. *Ecology,* **70**, 623–48

Hornick, J.R., Zerbe, J.I. and Whitmore, J.L. (1984). Jari's successes. *Journal of Forestry,* **82** (**11**), 663–70

Janos, D.P. (1983). Tropical mycorrhizas, nutrient cycles, and plant growth. In Sutton, S.L., Whitmore, T.C. and Chadwick, A.C. (eds.) *Tropical Rain Forest Ecology and Management*, pp. 327–45. (Blackwell: Oxford)

Janzen, D.H. (1974). Tropical blackwater rivers, animals, and mast fruiting by the Dipterocarpaceae. *Biotropica,* **6**, 69–103

Jenny, H., Gessel, S.P. and Bingham, F.T. (1949). Comparative study of decomposition rates of organic matter in temperate and tropical regions. *Soil Science,* **68**, 419–32

Johnson, D.W., Cole, D.W., Gessel, S.P., Singer, M.J. and Minden, R.V. (1977). Carbonic acid leaching in a tropical, temperate, subalpine, and northern forest soil. *Arctic and Alpine Research,* **9**, 329–43

Jonkers, W.B.J. and Schmidt, P. (1984). Ecology and timber production in tropical rain forest in Suriname. *Interciencia,* **9**, 290–7

Jordan, C.F. (1983). Productivity of tropical rain forest ecosystems and the implications for their use as future wood and energy sources. In Golley, F.B. (ed.) *Tropical Rain Forest Ecosystems: Structure and Function*, 117–35. Ecosystems of the World, Vol 14A. (Elsevier: Amsterdam)

Jordan, C.F. (1985). *Nutrient Cycling in Tropical Forest Ecosystems*. (Wiley: Chichester)

Jordan, C.F. (1987). *Amazonian Rain Forests: Disturbance and Recovery*. (Springer-Verlag: New York)

Jordan, C.F. (ed.) (1989). *An Amazonian Rain Forest. The Structure and Function of a Nutrient Stressed Ecosystem and the Impact of Slash-and-Burn Agriculture*. Man and the Biosphere Series 2. (Unesco: Paris, and Parthenon Publishing: Carnforth)

Jordan, C.F. and Uhl, C. (1978). Biomass of a "tierra firme" forest of the Amazon Basin. *Oecologia Plantarum,* **13**, 387–400

Klinge, H. (1985). Foliar nutrient levels of native tree species from Central Amazonia. 2. Campina. *Amazoniana* (*Kiel*), **9**, 281–95

Klinge, H. and Medina, E. (1979). Rio Negro caatingas and campinas, Amazonas states of Venezuela and Brazil. In Specht, R.L. (ed.) *Heathlands and Related Shrublands*, pp. 483–8. Ecosystems of the World Vol 9A. (Elsevier: Amsterdam)

Koeppen, R.C. (1978). *Some anatomical characteristics of tropical woods.* In Proceedings of a Conference on Improved Utilization of Tropical Forests, pp. 69–82. Forest Products Laboratory, Forest Service, (US Dept. of Agriculture)

Lowe, R.G. (1977). Experience with the tropical shelterwood system of regeneration in natural forest in Nigeria. *Forest Ecology and Management,* **1**, 193–212

Madge, D.S. (1965). Leaf fall and litter disappearance in a tropical forest. *Pedobiologia,* **5**, 273–88

Medina, E. and Zelwer, M.. (1972). Soil respiration in tropical plant communities. In Golley, P.M. and Golley, F.B. (eds.) *Tropical Ecology with an Emphasis on Organic Productivity*, pp. 245–67. (Institute of Ecology, University of Georgia: Athens)

Nadkarni, N. (1981). Canopy roots: convergent evolution in rain forest nutrient cycles. *Science,* **214**, 1023–4

Newbold, J.D., O'Neill, R.V., Elwood, J.W. and VanWinkle, W. (1982). Nutrient spiralling in streams: implications for nutrient limitation and invertebrate activity. *American Naturalist,* **120**, 628–52

Oliveira, A.E. de. (1983). Ocupacao Humana. In Salati, E., Junk, W.J., Shubart, H.O.R., Oliveira, A.E. de (eds.) *Amazonia: Desenvolvimento, Integracao, e Ecologia*. Conselho Nacional de Desenvolvimento Clientifico e Tecnologico. (Editora Brasiliense: São Paulo)

Olson, J.S. (1963). Energy storage and the balance of producers and decomposers in ecological systems. *Ecology,* **44**, 322–32

Olson, R.A. and Engelstad, O.P. (1972). Soil phosphorus and sulphur. In National Research Council Committee on Tropical Soils (eds.) *Soils of the Humid Tropics*, pp. 82–101. (National Academy of Sciences: Washington, D.C.)

Olson, R.A., Clark, R.B. and Bennett, J.H. (1981). The enhancement of soil fertility by plant roots. *American Scientist,* **69**, 378–84

Orians, G., and ten others. (1974). Tropical population ecology. In Farnworth, E. and Golley, F. (eds.) *Fragile Ecosystems*, pp. 5–65. (Springer-Verlag: New York)

Poore, D. (1976). The values of tropical moist forest ecosystems. *Unasylva,* **28** (**112–113**), 127–43

Post, W.M., Emanual, W.R., Zinke, P.J. and Stangenberger, A. (1982). Soil carbon pools and world life zones. *Nature,* **298**, 156–9

Powell, P.E., Cline, G.R., Reid, C.P.P. and Szaniszlo, P.J. (1980). Occurrence of hydroxamate siderophore iron chelators in soils. *Nature,* **287**, 833–4

Ruehle, J.L. and Marx, D.H. (1979). Fibre, food, fuel, and fungal symbionts. *Science,* **206**, 419–22

Russell, C.E. (1983). *Nutrient cycling and productivity in native and plantation forests at Jari Florestal, Pará, Brazil.* Ph.D. dissertation. (Univ. of Georgia: Athens)

Sanchez, P.A. (1976). *Properties and Management of Soils in the Tropics*. (Wiley: New York)

Sioli, H. (1973). Recent human activities in the Brazilian Amazon region, and their ecological effects. In Meggars, B.J., Ayensu, E.S., and Duckworth, W.D. (eds.) *Tropical Forest Ecosystems in Africa and South America: A Comparative Review*, pp. 321–34. (Smithsonian Institution Press: Washington, D.C.)

Stark, N.M. and Jordan, C.F. (1978). Nutrient retention by the root mat of an Amazonian rain forest. *Ecology,* **59**, 434–7

St. John, T.V. and Coleman, D.C. (1983). The role of mycorrhizae in plant ecology. *Canadian Journal of Botany,* **61**, 1005–14

Stevenson, F.J. 1986. *Cycles of Soil Carbon, Nitrogen, Phosphorus, Micronutrients*. (Wiley: New York)

Struthers, P.H. and Sieling, D.H.. (1950). Effect of organic anions on phosphate precipitation by iron and aluminium as influenced by pH. *Soil Science,* **69**, 205–13

Swenson, R.M., Cole, C.V. and Sieling, D.H. (1949). Fixation of phosphate by iron and aluminium and replacement by organic and inorganic ions. *Soil Science,* **67**, 3–22

Swift, M.J., Heal, O.W. and Anderson, J.M. (1979). *Decomposition in Terrestrial Ecosystems*. Studies in Ecology, Vol **5**. (University of California Press: Berkeley)

Taylor, A.W. (1961). Review of the effects of siliceous dressings on the nutrient status of soils. *Agricultural and Food Chemistry,* **9**, 163–5

Uehara, G. and Gilman, G. (1981). *The Mineralogy, Chemistry, and Physics of Tropical Soils with Variable Charge Clays*. (Westview Press: Boulder)

Vitousek, P. (1982). Nutrient cycling and nutrient use efficiency. *American Naturalist,* **119**, 553–72

Vitousek, P. (1984). Litterfall, nutrient cycling, and nutrient limitation in tropical forests. *Ecology,* **65**, 285–98

Walter, H. (1971). *Ecology of Tropical and Subtropical Vegetation*. (Oliver and Boyd: Edinburgh)

Weaver, P. (1979). Agri–silviculture in tropical America. *Unasylva,* **31** (**126**), 2–12

Whicker, F.W. and Schultz, V. (1982). *Radioecology: Nuclear Energy and the Environment. Vol. I.* (CRC Press: Boca Raton, Florida)

Whitmore, T.C. (1984). *Tropical Rain Forests of the Far East.* (Clarendon Press: Oxford)

Whittaker, R.H. and Likens, G.E. (1975). The biosphere and man. In Lieth, H. and Whittaker, R.H. (eds.) *Primary Productivity of the Biosphere*, pp. 305–28. Ecological Studies 14. (Springer–Verlag: New York)

Whittaker, R.H. and Feeny, P.P. (1971). Allelochemics: chemical interactions between species. *Science,* **171**, 757–70

Witkamp, M. (1970). Mineral retention by epiphyllic organisms. In Odum, H.T. and Pigeon, R.F. (eds.) *A Tropical Rain Forest.* H177-H179. (Division of Technical Information, US Atomic Energy Commission: Washington, D.C.)

Zinke, P.J., Stangenberger, A.G., Post, W.M., Emanuel, W.R. and Olson, J.S (1984). *Worldwide Organic Soil Carbon and Nitrogen Data.* Oak Ridge National Laboratory, Environmental Sciences Division Publication No. 2212. US Dept. of Energy.

APPENDIX A

Summary of research recommendations from Crossley and Freckman (1986), Workshop on Below-ground Dynamics in Tropical Ecosystems.

NUTRIENT AVAILABILITY

Organic matter dynamics

(1) How can soil organic matter be divided into functional entities for study, and what biological, physical and chemical factors regulate their stability?

(2) What is the relative importance of the soil organic matter fractions to nutrient release?

(3) What is the particular significance of the microbial biomass as a fraction of the soil organic matter and what factors regulate the dynamics of this fraction?

(4) What are the controls for the dynamics of microbial biomass?

(5) What are the contributions of wet and dry atmospheric deposition on soil nutrients in relation to nutrient cycling in the plant–soil system and nutrient outputs?

(6) What is the influence of the quality and quantity of above- and below-ground litter input on immobilization and mineralization rates of nutrients, and on the temporal synchronization of nutrient availability and plant demand?

(7) What is the significance of soil fauna to the release of nutrients from litter and soil organic matter?

Nutrient retention

(1) What is the relative importance of leaching, gaseous losses and erosion to nutrient losses?

(2) How does soil structure control the partitioning of nutrient loss pathways?

(3) How do the soil biota influence soil structure?

(4) Are soil charge properties under the direct control of biological processess and how is nutrient retention affected?

(5) What is the importance of below-ground biota as a source and a sink of nutrients?

UPTAKE PROCESSES

Shoots

(1) How is photosynthate partitioned among above-ground plant parts, below-ground parts, and associated microorganisms?

(2) How does shoot growth/turnover moderate long-term nutrient uptake and retention?

Associated organisms

(1) How do different types of mycorrhizal root systems moderate nutrient uptake/retention in different systems?

(2) Do the dynamics of the rhizosphere biota stabilize nutrient availability to the plant by its function as a sink or source of nutrients?

(3) How do components of the soil environment affect the ecology of symbiotic nitrogen-fixing bacteria?

Roots

(1) What are the plant production costs of maintaining the mycorrhizal/fine root absorbing system and the symbiotic N-fixers?

(2) What is the relationship between fine root turnover and the quality and quantity of soil organic matter?

(3) How much of root turnover is due to consumption by herbivores, and what significance does this have on nutrient acquisition by plants?

(4) What is the importance of root exudates in the rhizosphere food chain and how does this affect availability of nutrients?

(5) How is the temporal/spatial development of root architecture and production related to root uptake and to environmental influences?

(6) How can benefits of rhizosphere organisms be maximized in tropical soils to maintain long-term productivity?

(7) What are the physiological and architectural properties of roots of secondary successional plants which enable them to absorb nutrients from soils when crop plants have failed due to nutrient depletion?

CHAPTER 8

TROPICAL RAIN FOREST MANAGEMENT: A STATUS REPORT

R.C. Schmidt

INTRODUCTION

The potential for sustained management of natural forests in the humid tropics continues to be a subject of concern and uncertainty among tropical foresters. The destruction of roughly 7.5 x 10^6 ha of tropical rain forest annually, in addition to about 4 x 10^6 ha of open and savanna woodlands, has heightened the importance of managing tropical rain forests (FAO/UNEP 1981*a*, 1981*b*, 1981*c*, Lanly 1982). The virtual disappearance of commercially productive tropical rain forests is imminent in some countries; in others, this is happening at a slower pace, usually because of inaccessibility.

Amongst the principal causes of deforestation are rural poverty, population growth, and poor organization and funding of forestry institutions. To manage tropical forests for economic production is a key element in their conservation.

Despite many writings on this topic, some of them based on significant and well conceived field programmes, only a very small proportion of existing tropical rain forests are currrently managed in any real sense of the term. Also there is a tendency amongst many, but by no means all, experienced tropical foresters, to dismiss natural forest management as unrealistic, unworkable or impractical, and thus detrimental to other forestry activities more easily implemented. This scepticism becomes indefensible when one considers the alternatives. The absence of management for wood and possibly other products in natural tropical forests reduces the options for tropical land-use to only four: protected natural forests, tree plantations, degraded and depleted forests, or, finally, agricultural, urban or industrial use.

Within this scenario, plantations would eventually supply all timber produced in the tropics. Economic and ecological factors involved make this a very risky alternative. Plantation establishment costs are high (Chapman and Allan 1978, Mengin-Lecreulx and Maître 1986) and although tropical tree plantation production can be spectacular, the technology for equatorial lowland forests is not securely established. Plantations supply great benefits but they cannot replace the functions of tropical rain forest. The essential point is that plantations and natural forest management are not competitive but complementary. They provide different types of products and are suited to different terrains. The combination of plantation and natural forest silviculture makes the most efficient use of tropical forest land resources.

In the absence of economic productivity, the risk is that forests will be regarded as reserves of unutilized land, and restriction of activities and access will be the only way to protect them. Protecting them is clearly appropriate to many areas, but difficult, expensive, and perhaps repressive to sustain where rural poor have no alternatives for food production.

Most natural forests are presently logged for a few commercial species and either converted to other uses or left alone to develop as they will until logging is again economically profitable (if ever). Repeated creaming without management can lead to structurally and genetically degraded forests which are extremely difficult and expensive to rehabilitate (Roche and Dourojeanni 1984). In many cases, logging roads facilitate agricultural use, and this leads to complete clearing. Agricultural productivity of cleared tropical rain forest lands is usually very short-lived. The degradation of the productivity of renewable resources is well advanced in many countries, with consequent ever worsening problems of fuelwood supply and conservation of water, soil and genetic resources. At present, hundreds of millions of people are caught in this vicious circle where meeting their survival needs imperils the support systems they depend on.

This contribution reviews selected attempts at natural management of tropical rain forests in Asia, Latin America and Africa, drawing heavily on FAO experience and that of collaborating national institutes in tropical countries. Emphasis is given to management for timber and wood products. Management is planning (and executing the plan), and planning is perceiving where the expected future differs from a desirable future and developing objectives and activities to change that expected future to a more desirable one (Christakis and McDonald 1982). At the moment, the expected future in the majority of tropical forest areas is a lack of natural management resulting in the land-use options enumerated above. In light of this, specific constraints and opportunities for future actions will be considered.

NATURAL FOREST MANAGEMENT IN THE HUMID TROPICAL REGIONS

Natural forest management in humid tropical Asia[1]

Malaysia

Significant efforts to manage natural tropical forests have been made in Malaysia. It was found that, when the marketable trees were cut, either commercial species regenerated, or existing regeneration developed rapidly. While this did not always happen, it did occur under sufficiently different circumstances to be encouraging. Although few studies on humid tropical forests outside Malaysia have clearly confirmed this finding (Leslie 1985), there is little to indicate that it is not true in other areas with appropriate harvesting, and it would seem that silvicultural knowledge is adequate to commence management operations in many humid forests throughout the tropics.

The Malayan Uniform System (MUS), developed after the Second World War, converts the virgin tropical lowland rain forest (a rich, complex, multi-species, multi-aged forest) to a more or less even-aged forest containing a greater proportion of commercial species. This transformation is achieved by a clear-felling release of selected natural regeneration of varying age, aided by systematic poisoning of unwanted species (Wyatt-Smith 1963).

There are five important factors in this system relevant to any humid tropical forest natural management operation.

(1) The stocking of regeneration must be adequate.

(2) The original, partially harvested canopy must be removed.

(3) There must be no tending until regrowth has passed the ephemeral climber stage.

(4) An adequate new canopy must be maintained to prevent the redevelopment of climbers.

(5) Linear sampling must assess regeneration status.

[1] Important management methodologies have developed in dry deciduous forests in which *Shorea robusta* and *Tectona grandis* are significant species (especially in India); these forests will not be considered in this paper as the management and ecology of them are quite different from those of wet or moist tropical forests.

Table 8.1 Sequence of operations of the modified Malayan Uniform System, Sabah (Based on Fox and Hepburn (1972) and Chai and Udarbe (1977))

Year	*Operation*
n–2 to	Allocation of coupe
n–1	First silvicultural treatment – protective tree marking and climber cutting
n	Felling operation
n + 0–1 month	Clearance inspection
n + 0–2 month	Assessment of regeneration through Linear Sampling Milliacre (LSM) (2 m x 2 m plots)
n + 3–6 month	Second silvicultural treatment – First poison girdling of unwanted and defective trees, climber cutting if necessary
n + 10 to	Assessment of regeneration through Linear Sampling
n + 15	Half-Chain Survey (LS 1/2) (10 m x 10 m) plots Third silvicultural treatment – Liberation treatment where necessary

Although the MUS was highly successful, the lowland forest for which it was developed has been largely converted to other land-uses. The MUS was judged less successful in the hill dipterocarp forests. Contributing factors included rough and variable terrain, consequent uneven stocking and variable regeneration, heavier damage to residual stands during steep slope logging, and irregular seeding of principal commercial species.

The modified MUS (Table 8.1) as practised in Peninsular Malaysia and Sabah introduced certain refinements for specific sites and circumstances, and emphasized post-harvest sampling to determine appropriate silvicultural treatment.

Recently, the so-called Selective Management System (SMS) has been developed in Peninsular Malaysia (Salleh and Baharudin 1985). This system advocates a choice between different management options, based on a pre-harvesting inventory of stocking levels to determine diameter limits and species selection for harvesting. Options include: intermediate harvests within a polycyclic system where part of the growing stock is left for harvest after about 25 years; clear-felling release of selected natural regeneration (MUS); and enrichment planting or conversion to plantations (Lee 1982, Mok 1979, in Whitmore 1984). Climber cutting prior to harvest and marking for directional harvest attempt to minimize logging damage to residual stands. The key step of post-harvest treatment is left undefined.

SMS is not a true selection system according to standard silvicultural terminology, where single stems or very small groups of trees are removed as they reach maturity on a more or less continual (polycyclic) basis. True selection systems have not functioned successfully on a wide scale anywhere in the humid tropics. It is therefore, perhaps more accurate to describe this as a system which leaves the manager with wide discretionary power to determine where silvicultural treatment will be most advantageous from a cost/benefit standpoint. Silvicultural treatment does continue in Peninsular Malaysia, and poison girdling of non-commercial species was applied on more than 62 000 ha in 1982 (Salleh and Baharudin 1985). However, the silviculturally treated area is far less than the total area being harvested.

In Sabah, a long history of exploitative logging was countered after 1971 by application of the Modified Malayan Uniform System to 141 000 ha of logged forests (Table 8.1). In 1978, poison girdling was halted because increased logging intensity had opened up the stands to such an extent that additional opening was considered counterproductive. The cost effectiveness of such treatment was uncertain, and market acceptablity was changing so quickly that poisoning undesirables became questionable (Chai and Udarbe 1977).

An alternative silvicultural system was developed in Sarawak from 1974 to 1980, when the Sarawak Forest Department in collaboration with FAO developed and instituted silvicultural practices for the hill dipterocarp forest (UNDP/FAO 1982*a*, Hutchinson 1981, 1987*a*, 1987*b*). The conceptual and operational point of departure for this system is not undisturbed forest, but the increasingly large areas of forest which have been selectively logged for valuable species, where liberation thinnings are performed. The concept is not new or even exclusively tropical: liberation thinnings are logically indicated in any situation where a young crop of potentially good trees is overtopped by older, distinctly less desirable trees. If the overtopped trees will respond vigorously and speedily to form a new stand of good quality, quick and cheap silvicultural transformation to a productive stand is possible (Smith 1962). To recover the cost of rain forest treatment, the maximum rate of increment must be concentrated on what will prove to be the final crop trees (Baur 1968).

Where there is a low proportion of commercially valuable species (a frequently cited constraint to tropical forest management), selective logging is often light. From 1974 to 1980, selective logging in Sarawak extracted 5–15 trees ha^{-1}, representing a volume of 10–50 m^3, compared to total commercial volumes of 150–250 m^3. At these levels of extraction, 60% of Sarawak's logged tracts retained residual forest consisting of 20% undisturbed and 40% disturbed in some way by extraction. Both overstorey removal and liberation thinning were applied. Overstorey removal is cheap, but the increase in diameter at breast height (dbh) increment is correspondingly modest. The distinguishing characteristic of liberation thinning is that it opens wells around trees selected individually to be potential final crop trees. Thus, trees are poisoned in localized patches throughout the forest, their number being inversely related to the minimum dbh specified for the selection of final crop trees (Hutchinson 1987*b*).

The detailed procedures of the system have been described by Wadsworth (1969). The essential principle is that trees are not systematically eliminated according to species and size across forest stands, but according to simple, set rules. Training is necessary, but the system economizes the time needed to treat the forest and minimizes the number of trees eliminated. Proliferation of vines or non-commercial pioneers following silvicultural treatment has often been a serious problem, but can be minimized by the more limited opening of liberation thinnings. Also, in most countries, more and more species are "discovered" as commercially valuable, although relative values have seldom reversed. Therefore, a tree should be eliminated only if it is directly competing with a more valuable one.

Philippines

The Philippines have extensive areas of productive tropical forests, and forestry institutions and educational systems are well established. The approach to management has been that properly implemented selective logging will leave a residual stand which will develop in such a way that another commercial cut will be possible in 30 to 45 years (UNDP/FAO 1970*a*). An extensive and comprehensive inventory system has been developed which is quite similar in scope and approach to the national forest inventory in the USA. Permanently marked plots are randomly selected within forested areas and periodically remeasured. Regeneration, growth and volume are followed, and aerial photos are thoroughly evaluated to contribute data (Nilsson *et al.* 1978). Table 8.2 (Leslie 1985) shows areas harvested and silviculturally treated.

From 1975 to 1981, a Philippines-German Timber Stand Improvement Project experimented and developed methods of post-logging treatment. The treatment prescribed was basically a selection of the best potential crop trees (Leslie 1985). This is consistent with Hutchinson's recommendations in Sarawak.

Changes in government administrations may have significant effects on forest management. With recent changes in the Philippines, many forest logging concessions have been cancelled and forest policy is being re-evaluated. The Philippines now have an unparalleled opportunity to develop optimally their forest and human resources.

Natural forest management in humid tropical America

There are currently no large scale sustained yield management programmes being implemented in the vast closed broadleaved forests of tropical

Table 8.2 Harvesting and silvicultural treatment in the Philippines

Year	*Area logged* (ha)	*Destruction** (%)	*Adequately stocked residual forests* (ha)	*Silviculturally treated* (ha)
1975	62 660	28	45 100	14 000
1976	73 840	28	53 165	16 000
1977	67 260	25	50 445	33 000
1978	64 090	24	48 700	47 000
1979	57 275	20	45 820	44 000
1980	66 932	19	54 215	53 000
1981	58 416	19	47 310	49 200
1982	52 596	18	43 130	56 300
1983	51 993	16	43 670	33 900

*Proportion of forest area inadequately stocked due to logging damage to residual stand

America. This is certainly not for lack of resources that could supply enormous benefits if brought under adequate management: tropical America, as estimated to 1985, has 491.8 x 10^6 ha of productive closed broadleaved forests, of which 54.7 x 10^6 ha have been logged over. Nor is the current absence of this type of management due to lack of experiments, recommendations, and attempts at pilot demonstration programmes, although more effective efforts must be conceived and designed.

Brazil

In 1978, the Brazilian government and the FAO Forestry Department initiated a long-term, large scale pilot project to manage the Tapajos National Forest in the Amazon basin for multiple uses including timber production through natural silvicultural systems (UNDP/FAO 1983b). Earlier, a long series of inventories and experimental treatments had been conducted co-operatively by the Brazilian government and FAO through the 1950s and 1960s. A comprehensive management plan was prepared. The Forest Inventory in 1978 had found the forest contained 54 m^3 ha^{-1} of roundwood in stems > 45 cm dbh. The 28 commercial species made up to 36 m^3 of this. The experimental extraction of 64 ha in 1979 produced a total volume of 72 m^3 ha^{-1} roundwood of which 64 m^3 was commercial. The gross volume of all trees > 55 cm dbh was 132 m^3 ha^{-1} (UNDP/FAO 1980, 1983*a*).

The Project (BRA/78/003) Terminal Report reflects the comprehensive and intensive study that took place on a great variety of technical aspects and background information necessary to begin the development of the Tapajos National Forest as a model in tropical forest management. Fifteen international specialists and 30 Brazilian scientists participated in a study whose results indicated that it would be technically possible and economically viable to develop forest-based industries in selected areas of the Amazon similar to Tapajos National Forest while maintaining permanent tree cover.

A 1976 review mission concluded that commercial operations might commence in 1978. The pre-feasibility study, including a sensitivity analysis, had shown a 53% internal rate of return which could be maintained at 20% even with increased operating costs and sharply reduced prices. Subsequent studies included substantial operational components, as 27 000 m^3 were felled in 1980, 25 000 m^3 in 1981 and 17 000 m^3 in 1982 (UNDP/FAO 1983*c*).

The 1982 UNDP/FAO review mission noted the optimism and entrepreneurial opportunity that existed, but observed that no real forest management had yet commenced. In order to process Tapajos material, controlling outside procurement of logs was indispensable (UNDP/FAO 1982*b*). Existing mills in the area were supplied at operational capacity, typically from land clearing operations sponsored by the Instituto Nacional de Colonizacao e Reforma Agraria in which logs were obtained virtually free at the roadside in return for clearing the land. The review mission still concluded that it should have been possible to manage Tapajos profitably, and that the failure to initiate commercial harvesting by the middle of 1983 should be viewed with concern.

At the request of the Government of Brazil, the UNDP/FAO project had shifted emphasis in 1985, and began to concentrate its efforts in the dry northeastern part of Brazil. The Brazilian government continues to maintain a study project in Tapajos, and various institutes are conducting research on rain forest regeneration and management in several sites, with some international co-operation. However, significant management programmes, producing saw-timber and based on sustained yield principles, are not occurring anywhere in the humid tropical forests of the country (though see contribution by Dubois in this volume).

Peru

In 1971, the Peruvian government requested UNDP funding for a demonstration project for forest management in the Alexander von Humbolt National Forest in the lowland Amazon basin. The project operated from 1974–1978. The long term goals were to:

(1) increase forest production in the Amazon lowland basin of Peru through demonstration of management, improvement, protection and utilization for sustained yield of national forests;

(2) contribute to the development of organized timber industry to obtain the maximum social and economic benefits.

The short term objectives were to:

(1) determine the technical and economic feasibility of developing a wood processing complex,

(2) research the regeneration of commercially valuable species,

(3) formulate a management plan for the von Humbolt National Forest,

(4) interpret all study results in feasibility studies, and

(5) train people at all levels (UNDP/FAO 1979).

An inventory of 200 000 ha showed the forest to be very heterogeneous in floristic composition but quite homogeneous in volume actually commercialized (VAC). Although 300 tree species were identified in the inventory, 28 of these made up 70% of the total. Three quarters of the commercial volume was in 21 species, and the regeneration of 15 species comprised 85% of the trees. The VAC rose from 15 to 30 m^3 ha^{-1} during the term of the project since 20 new species became commercially valuable. The economic analysis indicated that a total investment of US $26 million was necessary to operate the project, the internal rate of return was calculated as 12–17% and cash flow was always positive. A comprehensive management plan allocated different areas for agro-silviculture, plantations, natural production forest and protection forest. Silviculturally, the production forest was to be managed on a 60-year rotation with a 30-year cutting cycle.

With the termination of the project, momentum was lost, and management activities were not implemented as foreseen. Landless poor occupied many cut-over sites, including some research plots, and initiated largely unsustainable agricultural practices. At present in the Peruvian Amazon, the Japanese government is involved in a co-operative project of plantations and natural forest regeneration and the Universidad Nacional Agraria La Molina is carrying out research in tropical silviculture (Romero Mejia 1983). Meanwhile, official requirements for post-logging management activities are not currently enforced and reported timber extractions in Peru have risen as follows:

1977 – 474 205 m^3
1978 – 476 016 m^3
1979 – 526 077 m^3
1980 – 606 594 m^3
1981 – 643 343 m^3

Colombia

In 1965–1970, a similar UNDP/FAO project was carried out in Colombia in the Serrania San Lucas, a moist tropical lowland forest between the Magdalena and Cauca Rivers. This area contained 1.2 x 10^6 ha of forest in 1965 and 1.0 x 10^6 ha in 1970. Inventories indicated 114 m^3 ha^{-1} of total stem volume and 33 m^3 ha^{-1} of commercial volume. Recognizing that long-term studies were a necessary basis for sound silvicultural practice, the project staff were nevertheless convinced that judicious cutting would produce secondary forests that could be managed for equal or greater commercial volumes than the undisturbed forest. Stocking and regeneration were adequate and the forest had begun to develop satisfactorily after sample cuts (UNDP/FAO 1970b). This project site was abandoned in 1970 for security reasons.

On the extremely wet Pacific coast of Colombia, silviculture and management of the "Bosque de Guandal" were analysed during a UNDP/FAO project (Neyra Roman 1979). Some 0.5 x 10^6 ha of this forest, dominated by two species (*Dialyanthera gracilipes* and *Campnosperma panamensis*), were reported in Nariño. A sound management programme has not developed in these forests, although the species structure would make this relatively easy technically.

In Colombia, Carton de Colombia utilizes the wet forests at Bajo Calima on the west coast for pulp. Regeneration is vigorous, approaching the original biomass within 15 years, but, because local farmers utilize the developing secondary forests for a number of products, Carton de Columbia has not attempted subsequent management.

Ecuador

The impact of "colonizacíon", the occupation and clearing of forest lands for family farms, has been a determining factor for the state of forest management in Ecuador. In the early 1970s, there were an estimated 30 x 10^6 ha of productive and well stocked natural tropical forests in Ecuador (Dixon 1971). Some 3 x 10^6 ha of undisturbed easily accessible forests of the Ecuadorian northwest could have been the basis of an increasingly integrated forest industry. A project (UNDP/FAO 1977) for strengthening the Ecuadorian Forest Service observed that 418 600 ha of forest had been occupied as farmland in the Amazon basin from 1971–75, with only about 5% receiving legal title. The farmers had gained access through the petroleum industry roads. The wood extracted (42 000–56 000 m^3 in 1976) supplied 32 sawmills, but represented no more than 2% of timber cut for land clearing. UNDP/FAO (1982*c*) concluded that the problem of colonization had put the entire forest industry into danger, and recognized a wide range of additional problems including the necessity of reforestation in zones where commercial timber could not be grown. Now, government requests for international forestry expertise have shifted to watershed management to extend the life of reservoirs.

Suriname

Natural silviculture in tropical rain forests in Suriname is discussed by Jonkers and Schmidt (1984), Boxman *et al.* (1985) and de Graaf (1986, this volume). Some 90% of this country is still forested, and 9000 ha of *Pinus caribaea* plantations have not fulfilled expectations for profitable wood production. It was found that current felling and extraction of rain forest was haphazard, and that rationally planned skid trails and felling techniques could reduce damage to remaining trees as well as reducing costs of extraction. This is consistent with standard forestry practice in any well managed forest. Poisoning non-commercial trees > 20 cm dbh and cutting lianas increased annual dbh increment from 0.4 cm to 1.0 cm. Refinement of 200 ha required 2.8 worker-days ha^{-1} and 17 litres of 2.5%, 2,4,5-T in diesel oil. It was predicted that this treatment would result in production of 40 m^3 ha^{-1} of harvestable volume in 20 years with 13.5 trees ha^{-1} becoming commercial.

The experimental work in Suriname is well conceived in that it was carried out on a fairly large scale in conjunction with the forest industry. The basic conclusion, which has been reached in many countries, is that the silvicultural problems are manageable. De Graaf (1986) thoroughly describes 20 years of applied silvicultural research in these forests, recommending: a polycyclic system of harvesting, perhaps every 20 to 30 years; controlled harvesting procedures to preserve a residual sapling stand; and periodic and relatively light refinements to eliminate larger non-commercial trees. It is important to note that no established forest estate is supervised by the Suriname Forest Service, which has only a minimal influence over forest concessions (Wood 1982).

Costa Rica

Costa Rica confronts the possible loss of all legally and physically accessible productive forest by the twenty first century. A recent project (UNDP/FAO 1985*b*) identified a study area of 14 000 ha containing 8000 ha of undisturbed forest. A sawmill with 5000 m^3 annual capacity will be installed. In 1985, in a 70 ha pilot operations area where inventories before and after harvesting were conducted, 10–12 ha produced 828 m^3 in 34 species before rain closed down operations. Latest information indicates that follow-up operations did not take place in 1986, but it is hoped that the study will be continued.

French Guyana

In French Guyana, a silvicultural research project was carried forth in the 50–100 km wide strip of coastal forest (Maître 1982). Commercial volumes and harvesting costs were calculated and the economic implications of scenarios with varying levels of harvesting for timber or fuelwood were presented. The highest internal rate of return (7.78%) resulted from a combined strategy removing 1.3 m^3 ha^{-1} yr^{-1} in timber and 3.0 m^3 ha^{-1} yr^{-1} in fuelwood. This forest does not presently experience population pressure. Timber extraction is increasing, but was a modest 120 000 m^3 in 1980. Sustained yield management could preclude the necessity and cost of building new roads into the interior, undisturbed forests.

Other countries
No other large scale commercial sustained yield management programmes are known in tropical America. Without mentioning all countries, Venezuela has only implemented such programmes on an experimental scale, and true programmes of forest utilization ("aprovechamiento") in which natural regeneration is implicit have never existed in Mexico (Gómez-Pompa 1985).

Natural forest management in humid tropical Africa

Silviculturalists in Nigeria and Côte d'Ivoire experimented with natural regeneration and line planting during most of the first half of the twentieth century. Many other African forestry departments tried to take up the challenge of silviculture in moist forests, beginning in the 1950s. Some of the methods relied on natural regeneration, others utilized techniques for improving the dynamics of the stands, and others used artificial regeneration. With or without management, forest harvesting has continued in many of Africa's humid tropical forests.

Natural regeneration
The three main methods based on natural regeneration were the Tropical Shelterwood System (TSS) in Nigeria; the Amelioration des Peuplements Naturels (APN), or improving natural populations, in the Côte d'Ivoire; and the Selection System in Ghana. The TSS was designed in Nigeria in 1944 on the basis of tests which had been carried out for 20 years. The objective was to enhance the natural regeneration of valuable species before harvesting by gradually opening up the canopy (poisoning of undesirable trees, cutting of climbers) to obtain at least 100 one-metre high seedlings per ha over 5 years. The forest thus worked was logged over in the 6th year, then cleaning and thinning operations were carried out over 15 years. Two hundred thousand ha of forest were treated this way by the Nigerian Forestry Department between 1944 and 1966, after which the method was dropped. The main problems encountered were the exuberant spread of

climbers following opening up of the canopy and the failure of the seedlings of valuable species to grow adequately. Besides, poisoning eliminated trees which later turned out to be commercially valuable, e.g. *Pycnanthus angolense* (Myristicaceae).

In 1950, the Forestry Department of Côte d'Ivoire found initial results in Nigeria appealing and gave up line planting in favour of APN, a technique similar to the TSS. Besides technical considerations, an economic motive also accounted for this drastic change: the productive capacity of sawmills for domestic consumption was increasing. The geographical dispersal, as well as the widening of the range of species used, called for scattering operations and lengthening the list of species to be regenerated. The APN method was applied to forests which had been logged over and were well-stocked in valuable trees of average size. The aim was to favour the growth of these average stems and also to ensure regeneration through natural seeding of the valuable species by removing climbers and opening up the canopy. The method was applied by the Forestry Department from 1950 to 1960 on large areas, but was abandoned when results were disappointing.

The Selection System has been applied in Ghana since 1960. Its objective is to assure regeneration of forests well stocked with valuable species. Harvesting occurs about every 15 years, after the Forestry Department has marked the stand to retain some well-distributed seed trees. The method has been found to cause considerable felling damage because of the relatively short rotation. Regeneration has been poor, and less valuable shade-tolerant species sometimes dominate because of insufficient opening up of the canopy.

Improvement of stand dynamics

A technique for the improvement of stand dynamics was utilized in the 1950s in Gabon's *Aucoumea klaineana* (Burseraceae) forest to accelerate the growth of all-sized stems of valuable species in naturally well-stocked stands, without particularly trying to provoke regeneration through natural seeding. The species grows in patches or clumps presumably due to natural seeding of forest trees in the clearings or gaps. The objective was to let these stands attain commercial diameters as quickly as possible through thinning operations, but the production gain was never measured. After thus treating about 100 000 ha, the Forestry Department gave up the technique in 1962 to switch to *Aucoumea klaineana* plantations.

Belgian foresters in the Congo also tried stand improvement techniques in the 1950s: "uniformisation par le haut" and "normalization". Several thousands of hectares were managed through these techniques, mainly in the Mayumbe region (Low Zaire). After independence in 1960, these trials were not continued by the national Forest Department.

Thus, natural regeneration techniques are not practised on a commercial scale in the francophone countries, nor in Nigeria. However, Ghana and Uganda have, in principle, continued moist forest management.

Forest harvesting

Forest harvesting in Africa has evolved considerably over the past few decades. The forest lands harvested before 1980 are estimated to cover about 42 x 10^6 ha out of the 162 x 10^6 ha of productive closed broadleaved forest. The most accessible forests, such as those remaining in Côte d'Ivoire, Ghana and Nigeria, have already been logged-over several times. Approximately 90% of the undisturbed forests are found in Zaire, Gabon, Congo and Cameroon.

Harvesting is still very selective, as it was originally. The net volume of logs extracted from the forest averages from 5 to 35 m^3 ha^{-1}. The number of species utilized is still quite limited.

In densely populated countries such as Ghana (55 inhabitants km^{-2}) and particularly Nigeria (more than 100 inhabitants km^{-2}), domestic markets have overtaken exports, which would normally lead to a diversification of utilized species. On the other hand, in other major wood producing countries such as Côte d'Ivoire, Cameroon, Gabon and Congo, 75% of the timber trade volume consists of exports, and the domestic market is not large enough to absorb the lesser utilized species.

Harvesting rules which existed before 1960 in English-speaking countries (particularly in Ghana, Nigeria, Uganda) and in Zaire, have been gradually abandoned in most countries because of insufficient staff and funds. When the efficiency of silvicultural techniques was questioned, more spectacular planting operations backed by external funding were chosen.

In the 1960s, a system was set up in the Central African Republic whereby silvicultural operations were carried out within concessions, financed mostly by concessionaires. They consisted of marking saplings for protection during harvesting and post-harvesting thinning operations. Unfortunately, this procedure was discontinued after 10 years because of institutional problems.

Forest management in francophone Africa has sometimes consisted of harvesting regulations which fix the terms and areas for concessions and a minimum exploitable diameter for economic species. Stumpage rates paid in harvesting contracts are theoretically earmarked in order to finance management and regeneration costs, but, in practice, the funds often revert to the general State budget.

Although harvesting regulations are not complete management systems, they are a positive start. Les Unites Forestieres d'Amenagement (UFAs) were developed in Congo in the 1960s with the assistance and advice of FAO. Areas for concessions were delineated, and inventories are required to provide information on stocking levels. When an area is judged to be depleted, it is closed to harvesting for recuperation. Progress is being made toward control of product flow and the development of the institutional structures and responsibilities that are prerequisites to further evolution of management programmes.

Experimental pilot management projects have been set up by FAO in Cameroon and Gabon, but, unfortunately, project recommendations have never been put into general practice. Although silviculturally sound, management restrictions are sometimes resisted because they may:

(1) specify the utilization of too many second category species which are judged unprofitable in areas of poor access; and

(2) specify progressive harvesting per block, requiring utilization of all species in the annual allowable cutting area simultaneously, and disallowing repeated harvesting so as to let regeneration occur. This may create difficult marketing problems because of fluctuations in the tropical timber market.

Costs and benefits of moist forest silviculture

It has been difficult to measure accurately the productivity gains obtained by natural regeneration and stand improvement techniques. Uncertainty concerning the efficiency of the methods used has undoubtedly contributed to their progressive abandonment by most African countries. By contrast, the productivity and cost of forest plantations are relatively well-known.

In 1976, an important plan of action for the study of moist forest development relative to the different silvicultural interventions was set up in Côte d'Ivoire by the Societé Ivoirienne de Developpement des Plantations Forestieres (SODEFOR) with the technical support of the Centre Technique Forestier Tropical (CTFT). The advantage of this project, compared to previous experiments, was that it allowed accurate measurement of the impact of the silvicultural operations. The mensuration methods, size of trials and spatial replication of the treatments were designed so that the results obtained are reliable.

The project covers 1200 ha and three field stations characteristic of three ecological areas of the moist forest of Côte d'Ivoire: semi-deciduous forest, evergreen forest and transition forest (Maître 1986, this volume). The silvicultural practices used are the traditional harvesting of economic species and thinning by poison girdling. Two thinning regimes were tested (30% and 45% of the total basal area) by beginning systematically with the tallest trees in residual forest until the desired percentage of basal area was reached. The objective of the thinnings was to favour the valuable trees ≥ 10 cm dbh. No particular operation has been planned to aid regeneration through natural seeding.

After 4 years of observation, volume increment was 3 to 3.5 m^3 ha^{-1} yr^{-1} against 2 m^3 ha^{-1} yr^{-1} in the control stands, i.e. a gain in growth of 50 to 75% in stems of 73 main species ≥ 10 cm dbh. Measurements taken every year showed that volume increment increases with time, and that the impact of the thinning operations will probably be felt for at least 10 years. The largest yearly diameter increments, averaging 1 cm yr^{-1}, were found in species such as *Triplochiton scleroxylon*, (Sterculiaceae), *Terminalia superba* (Combretaceae), *Tarrietia utilis* (Sterculiaceae) and *Swietenia macrophylla* (Meliaceae).

A management system based on this practice would seem justified since the ratio of commercial volume to cost is greater than that of forest plantations set up in the same area. Over 30 years, the ratios for both systems may be estimated as follows:

(1) – Management of the natural forest: 25 m^3 ha^{-1}
140 US$

thus about 1 m^3 produced for every US$ 5.6 invested. (This assumes untreated natural forest would produce 60 m^3 and treated forest 85 m^3 of commercial volume).

(2) – Plantations: 250 m^3 ha^{-1}
1860 US$

about 1 m^3 produced for every US$ 7.4 invested.

It must be remembered that natural forest investments are usually compounded for a longer period, but the figures indicate the costs involved. If a wider range of species from the natural forest can be introduced in local markets, the economics of the operation improves.

The 10 000 ha Yapo forest is managed experimentally on the basis of results of the silvicultural research project set up by SODEFOR-CTFT. Other management programmes are planned in progressively larger areas to eventually cover all the forest lands of Côte d'Ivoire.

A comparison of Asia, America and Africa

It is often dangerous to extrapolate findings beyond the context from which they were derived, and this could be the case with too rigidly applying the experience from Asia to other tropical forest regions.

The ecological and socio-economic differences between the regions are quite marked. The dipterocarp forests of Malaysia, western Indonesia and the Philippines have significantly higher proportions of presently commercial species than the other regions, and these species have often regenerated abundantly after disturbing the forest. Institutionally, there is often significant change in the quality of management in very similar forests when one crosses a national boundary. Thus, it is unlikely that specific practices will have much direct application from one region to the other. However, to assume that there is nothing to be transferred from one region to another would be equally unconstructive. At the very least, the Asia experience shows that sustainable, integrated management of tropical mixed forest is technically feasible. In the other regions, faltering commitment and inadequate programmes, rather than silvicultural problems, have precluded success.

The complexity and variability of ecological dynamics and species composition in tropical rain forests means one thing for silviculture: flexibility, guided by common sense acquired through field experience. Although there may be debate about what constitutes an adequately stocked residual stand (Salleh and Baharudin 1985) or which specific silvicultural techniques should be employed, this should never impede the implementation of management programmes.

The common feature of discontinued natural forest management programmes in tropical America is that technical feasibility is never cited as the reason for abandoning them. In Colombia, the FAO project ceased operations because security broke down in the area. In Peru and Ecuador, the familiar pattern of landless farmers spontaneously occupying forest lands where roads had penetrated brought the management scheme to an end. In the Brazilian Amazon, the abundance of old growth timber resources, combined with publicly subsidized land clearing, saturated the existing conversion capacity, resulting in negative stumpage prices which eliminated the possibility of any economic management.

In Africa, political developments often meant that institutions dealing with forest management were given new directions, and new techniques, promising more rapid results than those achieved through the relatively slow growth of natural forests, were preferred. As these promises have often failed to materialize, there is some return of interest in managing natural forests; however, many institutions and national budgets do not have the strength to initiate new programmes.

These are not problems of species diversity, lack of understanding of ecosystem dynamics, inability to retain adequate regeneration, or lack of response to silvicultural treatment. The problems concern land-use policy, socio-economic conditions, and political realities (Palmer 1975). A large part of the problem is that the productive potential of resources found in abundance is undervalued. Certainly this chapter is not the first to point out that viable forest industries based on sustained yield could provide long term economic well-being for more people than are currently supported by opening up new areas of forest for unsustainable, short-term agricultural exploitation. The problem is also one of expanding the time horizons of policymakers. The tropical world already offers many examples of the consequences of continuing to neglect sustainable resource management policies. Once the original productive capacity of an area has been lost, efforts to restore it become very expensive.

PARADOXES AND CONTRADICTIONS

Although technical complexities have made it difficult to sustain successful management programmes, these factors have not been the principal stumbling blocks in areas where management based on natural regeneration had been attempted and subsequently discontinued.

High population density and established forest industries can threaten natural forest management by demanding production levels that the resources cannot sustain (FAO 1984). Conversely, sparsely settled areas with large expanses of unutilized forest resources, such as Amazonia or Central Africa, also present a difficult setting in which to initiate productive sustainable forest management. Since local timber demand is low, poorly organized industries can afford selective utilization, resulting in steady depletion of certain species in ever larger areas. Although selective utilization and no silviculture can result in forest depletion, harvesting of a large proportion of the species has also been very destructive.

Forest management is an economic activity. Investments must be justified in terms of their effectiveness to enhance future benefits to be derived from the resources they help develop (i.e. they must pay off). But the economic benefits of natural tropical forest management are largely unquantifiable. The future value of wood products is unpredictable and historically has usually been substantially different at the beginning than at the end of a rotation. For many years, work in natural tropical forest silviculture (Dawkins 1958, Wadsworth 1952) has attempted to establish a statistical foundation for evaluating quantitatively the effects of different treatments. The main result of these years of effort is the conclusion that the problem of adequate control and replication makes it very difficult to obtain reliable results in studies of the dynamics of tropical rain forests (Synnott 1980). Confident measures of response to treatment are essential for any economically quantitative evaluation of operations.

Economic analysis has not been able to evaluate adequately the larger social context, and the benefits of water quality, genetic conservation, and natural beauty that successful programmes of tropical forest management could generate. Consequences of alternative land-uses further complicate the validity of a strictly economic analysis. The clearing of forest land for unsustainable agriculture may not be profitable, even in the short-term, and discounting the inherent values of the destroyed forest. A long-term analysis certainly indicates that such a decision is unwise.

In areas where soil and vegetative cover have not been conserved, the costs of rehabilitation (reforestation on very poor soils, terracing, protection from fire, overgrazing, overcutting, etc.) are usually enormous. These are not investments which have any prospect of being repaid through direct financial benefits. Yet they are often justified as the only way of securing the welfare of the local inhabitants and preventing further deterioration of the life support system. A policy of focusing on already degraded areas while ignoring productive areas in a similar and continuous process of degradation, is a condemnation to relive the same disastrous history at ever increasing costs. Therefore, is natural forest management uneconomic, or is it the only economic alternative?

CONDITIONS AND CONSTRAINTS

The challenge and difficulty of implementing management is that any shortcoming, whether silvicultural, socio-economic, political or institutional (for example the lack of stable budgets or well-trained personnel), can effectively prevent success. Forest management systems must integrate the forest, the society and the means by which the needs of the latter are linked to the potentials of the former. Forest management programmes have traditionally focused on forestry and technology and given societal elements very simplistic treatment. Five conditions emerge for successful management programmes based on natural silviculture:

(1) The land must remain in forest use after timber harvest;

(2) Harvesting must not degrade the soil to the point where forest tree species cannot regenerate;

(3) Harvesting must not impair the environmental and social functions of the forest;

(4) Harvesting methods and silvicultural treatment must ensure that regeneration is adequate;

(5) The rate of timber harvest must not exceed the sustained yield capacity of the forest.

The first condition requires that the local population fully endorses the decision that the land remain under forest. Where poverty and population increase create pressures for crop land, official government land-use planning will be largely irrelevant. The local population will only endorse forest management if it helps to satisfy present day needs. Forestry programmes stressing community participation and useful social evaluation are becoming increasingly important in the tropics. Until now, these programmes have evolved primarily as reforestation efforts in degraded or deforested areas. The concept of applying participatory planning before deforestation and degradation occur is one that deserves close attention. Of course, where land-use policy actively encourages forest clearing for agricultural production which turns out to be unsustainable, a design for disaster is firmly in place.

The second, third and fourth conditions relate to harvesting operations. Although adequate logging controls have been an elusive goal in all tropical countries, the technical recommendation is remarkably simple. In a variety of situations, well-organized harvesting operations are more efficient and profitable than poorly organized and executed operations. And they leave tropical humid forests in a condition where the silviculturist has an adequate resource to work with (Mattson Marn and Jonkers 1981, De Bonis 1986). Control of logging has been difficult because it requires well qualified and motivated personnel to be on the logging site, and most forestry organizations in tropical areas do not adequately reward field foresters.

Good harvesting practices are more advantageous to all concerned, but the training, control and finance to implement such practices are not easily achieved. Institutional arrangements often divorce industrial and management activities. Although it is obvious that the two must be co-ordinated for successful operation, this is rarely the case. Timber management is academic without industrial operations, and industrial operations must continually shift to new forests unless adequate management is practised.

The fourth and fifth conditions are silvicultural. Analysis of silvicultural systems is abundant in the literature of tropical moist forest management. If regeneration cannot be achieved or assured in conjunction with timber harvesting, then one or the other has to give way. Historically, that has almost never been the harvesting. The last condition is that the removal of the growing stock must not exceed the regenerative and growth capacity of the forest. This presupposes a fairly accurate knowledge of the growing stock, its distribution by species, size classes and location, and of how these change with harvesting and treatment. Therefore, inventory and increment studies also figure prominently in management history.

Although more information would be useful in such areas as growth and yield statistics, annual increment and sustainable removals (FAO 1984), enough is now known to implement a sustainable natural system of management. Any claim that the silvicultural and yield regulatory elements are the limiting factors to the advancement of forest management would be hard to sustain. However, the void that exists for information on social components of management systems is much greater.

The five conditions imply a number of prerequisites. Effective legislation and national land-use planning which specifically identifies the forested areas to be managed are indispensable. National leadership at the highest level must be committed to current investment for direct economic benefits that will be realized 20 to 30 years into the future. National policy must insure that population development is in harmony with the optimum productive capacity of land available. Institutions responsible for management must have long-term programme stability, with stable leadership in key positions to provide continuity. Effective training institutions must prepare workers from technician to doctoral research level, and institutional arrangements must place knowledgeable and competent personnel in the field where management activities occur. Profitable management must be integrated with the national economy and the world timber market, and today's plans must assess the demand for products 20 years or more into the future. This is a formidable set of prerequisites interrelating with the entire fabric of national affairs.

Often it is difficult to commit scarce financial and human resources to unproven technical packages. There is therefore an urgent need to demonstrate feasibility through successful pilot management programmes. The difficulty is that these cannot operate in isolation, and they depend on the institutional strength and policy commitment referred to above. It has proven more effective to promote extensive management activities, such as the UFA system in Africa, than to initiate sophisticated and isolated intensive management projects.

CONCLUSIONS

It would be foolish to maintain that the current management picture in moist tropical forests is an encouraging one. In the three great centres of such forest, the Amazon basin, central Africa, and the islands of South-east Asia, substantial areas of forest are being cut over or converted to other uses, while significant programmes of silvicultural treatment are occurring only in Malaysia. There are several countries where reduction of productive moist forest areas has severely limited the potential for extensive natural forest management that once existed (Costa Rica, Nigeria and Laos are representative and diverse examples). However, Table 8.3 lists the 36 countries with at least one million ha of tropical forest suitable for productive management. They represent more than 90% of the world total. Countries with large areas of unlogged forest hold the highest potential.

Table 8.3 Largest national areas of productive closed broadleaved forest estimated for 1985 (x 10^6 ha)

Country	*Total**	*Logged over*	*Country*	*Total*	*Logged over*
Brazil	295.5	13.5	French Guyana	7.6	0.2
Zaire	79.2	0.4	Philippines	6.3	3.7
Indonesia	67.5	34.5	Madagascar	6.0	4.6
Peru	42.8	6.4	Kampuchea	5.0	0.5
India	37.8	3.9	Vietnam	3.5	2.3
Colombia	36.0	0.8	Central African Republic	3.4	0.4
Burma	21.8	5.5	Nicaragua	3.2	0.1
Gabon	19.8	9.9	Thailand	2.9	2.9
Venezuela	18.8	11.4	Panama	2.8	0.8
Bolivia	17.0	2.0	Laos	2.4	2.4
Cameroon	16.6	10.6	Angola	2.2	2.2
Malaysia	14.4	5.7	Paraguay	2.2	1.9
Papua New Guinea	13.9	0.4	Guatemala	2.8	1.2
Congo	13.6	3.4	Côte d'Ivoire	1.8	1.8
Guyana	13.5	1.4	Nigeria	1.6	1.5
Suriname	12.5	0.5	Honduras**	1.1	0.1
Mexico	11.4	0.3	Ghana	1.2	0.1
Ecuador	9.7	0.1	Sri Lanka	1.0	1.0

* The forest is not all lowland tropical in all cases. For example, in the Brazilian Amazon the estimate for tropical forest is (x 10^6 ha): 185.9 dense forest; 10.0 dryland forest; 76.6 open forest; 42.3 transitional and seasonal formations (total 314.8).

** Revised from FAO/UNEP 1981*a* based on personal communication (1987) from T.W.W. Wood.

Definitions: "Productive closed broadleaved forests ... cover ... a high portion of the ground and do not have continuous dense grass layer ... their characteristics ... allow (or might allow) for the production of wood for industry". "Logged over once or more times during the last 60 to 80 years".

From FAO/UNEP (1981*a*)

It will not be easy to achieve sound management, but failure would result in the disappearance of most tropical rain forests. There is every indication that economically unproductive areas in tropical countries will continue to be highly vulnerable to agricultural development or conversion to other uses, even if such development is unsustainable. There is no economically profitable alternative use known for large areas of biologically highly productive tropical forests.

It is hard to see how effective management programmes will be instituted without international expertise, co-operation and financial assistance, but such programmes will be carried out by national governments and institutions. National action and international assistance must be co-ordinated and comprehensive. Specific projects now being implemented in co-operation with FAO stress promotion of extensive management through compartment definition, inventory and control of harvesting. Projects also focus on demonstration areas and seminars assembling foresters, industrialists and policymakers. There is increasing support for natural forest management in the scientific and conservation communities, and this should be drawn upon to generate funding for management projects. One possibility is that naturally managed areas be established near protected areas in a combination that might enhance the effectiveness of both programmes.

The Tropical Forestry Action Plan (FAO 1985*a*, FAO Forestry Department 1986*b*), has a focus well suited to natural forest management in that it seeks to coordinate international assistance in priority areas based on comprehensive national forestry sector analysis. These are exactly the measures that are needed. The productive management of natural tropical forests, with the social and ecological benefits these forests can produce, is not the only necessary activity, but it is the key to combating irrational deforestation.

ACKNOWLEDGEMENTS

This contribution has been based on a series of studies on management of humid tropical forests undertaken in recent years by FAO's Forest Resources Division in collaboration with national institutes in tropical countries. Forestry Papers 53 and 55 presented case studies of management systems in India, Ghana, Honduras, and Trinidad. Two Forestry Papers in preparation will review management systems for tropical humid forests in Asia and Africa, and subsequent review of systems in Latin America is planned. In concentrating on management systems in selected areas of closed broadleaved tropical forests, the contribution draws heavily on studies of Malaysia by Thang Hooi Chiew and Abdul Rashmid bim Mat Amin of the Forestry Department of Malaysia and of the Philippines by the Bureau of Forest Development of the Philippines (subsequently compiled for FAO by A. Leslie), and on two recent papers on Sarawak by I. Hutchinson. Comprehensive studies of humid tropical forest management for French and English speaking countries of Africa, based *inter alia* on national studies by P. Karani and P.R.O. Kio *et al.*, were compiled by R. Catinot and M. Philip respectively. A review of these studies and other literature prepared by P. Mengin-Lecreulx and H. Maître, CTFT, constitutes the section on Africa in this paper. Finally, FAO field project reports concerning management of humid tropical forests were reviewed.

REFERENCES

Ashton, P.S. (1978).* The biological and ecological basis for the utilization of dipterocarps. *Proceedings of the Eighth World Forestry Congress,* Volume VIIA, pp. 819–30

Baur, G. (1968). *The Ecological Basis for Rain Forest Management.* (Forestry Commission of New South Wales: Sydney)

Boxman, O., Graaf, N.R. de, Hendrison, J., Jonkers, W.B.J., Poels, R.E.H., Schmidt, P. and Tjon Lim Sang, R. (1985). Towards sustained timber production from tropical rain forests in Suriname. *Netherlands Journal of Agricultural Science,* **33**, 125–32

Catinot, R. (1965).* Sylviculture tropicale en foret dense africaine. *Revue Bois et Forets des Tropiques,* **100–104**, (100, 5–18. 101, 3–16. 102, 3–16. 103, 3–16. 104, 17–29)

Chai, D.N.P. and Udarbe, M.P. (1977). The effectiveness of current silvicultural practice in Sabah. *Malaysian Forester,* **40(1)**, 27–35

Chapman, G.W. and Allan, T.G. (1978). *Establishment Techniques for Forest Plantations.* FAO Forestry Paper 8. (FAO: Rome)

Christakis, A. and McDonald, R. (1982). *Creating the Forestry Future – an Interactive Approach to Planning.* Manuscript. (University of Virginia: Charlottesville)

Correa de Lima, J.P. and Mercado, R.S. (1985).* The Brazilian Amazon Region; Forestry industry opportunities and aspirations. *Commonwealth Forestry Review,* **64(2)**, 151–6

Dawkins, H.C. (1958). *The Management of Natural Tropical High Forest With Special Reference to Uganda.* (Imperial Forestry Institute. University of Oxford: Oxford)

De Bonis, J.N. (1986). Harvesting tropical forests in Ecuador. *Journal of Forestry,* **84**, 4

Dixon, R.G. (1971). *Estudio de Preinversion para el Desarrollo Forestal del Noroccidente.* Ecuador. Informe Final Tomo III. Manejo Forestal. Document FAO/SF: 76/ECU 13. (FAO: Rome)

Eden, M.J. (1982).* Silviculture and agroforestry developments in the Amazon Basin of Brazil. *Commonwealth Forestry Review,* **61(3)**, 195–202

FAO. (1984). *Intensive Multiple-Use Forest Management in Kerala.* FAO Forestry Paper 53. (FAO: Rome)

FAO. (1985*a*). *Tropical Forestry Action Plan,* (FAO: Rome)

FAO. (1985*b*).* *Intensive Multiple-Use Forest Management in the Tropics. Analysis of Case Studies from India, Africa, Latin America and the Caribbean.* FAO Forestry Paper 55. (FAO: Rome)

FAO. (1986*a*).* *Tropical Silviculture.* Reissue of Forestry and Forest Products Study No. 13 (1954). (Published by arrangement with FAO. Periodical Experts Book Agency: Delhi)

FAO. (1986*b*). FAO's Tropical Forestry Action Plan. *Unasylva,* **152**, 37–64

FAO/UNEP. (1981*a*). *Proyecto de Evaluacion de los Recursos Forestales Tropicales. Los Recursos Forestales de la America Tropical. Part I: Regional Synthesis. Part II: Country Briefs.* (FAO: Rome)

FAO/UNEP. (1981*b*). *Tropical Forest Resources Assessment Project. Forest Resources of Tropical Asia.* UN 32/6. 1301–78–04. Technical Report 3. (FAO: Rome)

FAO/UNE.P (1981*c*). *Les ressources forestieres de l'Africque tropicale. Premiere partie: synthese regionale. Deuxieme partie: resumes par pays* (Country Briefs written in French, English or Spanish according to the official language of each country). (FAO: Rome)

Fox, J.E.D. (1968).* Logging damage and the influence of climber cutting prior of logging in the lowland dipterocarp in Sabah. *Malayan Forester,* **31**, 326–47

Fox, J.E.D. and Hepburn, A.J. (1972). *Code of Silvicultural Practice.* Forest Department, Sabah.

Gillis, M. (1984).* *Multinational Enterprises, Environmental and Resource Management Issues in the Tropical Forest Sector in Indonesia.* Development Discussion Paper No. 171. (Center for International Affairs, Harvard University: Cambridge)

Gómez-Pompa, A. (1985). *La funcion protectora y los servicios ambientales de los bosques: el problema de la reforestacion en el tropico.* Basic Document B-I:5, presented to Technical Commission 1, Theme 5, of the Ninth World Forestry Congress, Mexico.

Graaf, N.R. de. (1986). *A Silvicultural System for Natural Regeneration of Tropical Rain Forest in Suriname.* (Pudoc, Agricultural University: Wageningen)

Hadley, M. and Lanly, J.P. (1983).* Tropical forest ecosystems: identifying differences, seeking similarities. *Nature and Resources,* **19** (**1**), 1–19

Hutchinson, I.D. (1981). *Sarawak Liberation Thinning: Background and an Initial Analysis of Performance. A Practical Guide.* UNDP/FAO/ MAL/76/008. Field Document No.15. (Sarawak Forest Department: Kuching)

Hutchinson, I. (1987*a*). Improvement thinning in natural tropical forests: aspects and institutionalization. In Mergen, F. and Vincent, J.R. (eds.) *Natural Management of Tropical*

Moist Forests. Silvicultural and Management Prospects of Sustained Utilization, pp. 113–33. (School of Forestry and Environmental Studies, Yale University: New Haven)

Hutchinson, I. (1987*b*). The management of humid tropical forests to produce wood. In Figueroa, J.C., Wadsworth, F.H. and Branham, S. (eds.) *Management of the Forests of Tropical America: Prospects and Technologies*, pp. 121–55. (US Department of Agriculture Forest Service, Institute of Tropical Forestry: Rio Pedras: Puerto Rico)

Jonkers, W.B.J. and Schmidt P. (1984). Ecology and timber production in tropical rain forest in Suriname. *Interciencia*, **9(5)**, 290–7

Kio, P.R.O. (1981).* Regeneration methods for tropical high forests. *Environmental Conservation*, **8(2)**, 139–47

Kio, P.R.O., Ekwebelam, S.A. Oguntala, A.B. Ladipo, D.O. and Nwonwu, F.O.C. (1985).* *Management Systems in Tropical Mixed Forests of Anglophone Africa (With Special Examples from Nigeria)*. Report prepared for FAO.

Lanly, J.P. (1982). *Tropical Forest Resources*. FAO Forestry Paper No.30. (FAO: Rome)

Lee, H.S. (1982). The development of silvicultural systems in the hill forests of Malaysia. *Malaysian Forester*, **45(1)**, 1–9

Lee, P.C. (1977).* Objectives for forest management. In Sastry, C.B., Srivastava, P.B.L., Abdul Manap Ahmad (eds.) *A New Era in Malaysian Forestry*, pp. 139–42. (University Pertanian Malaysia: Serdang)

Leslie, A.J. (1985). *Study of Management Systems in the Tropical Mixed Forests of Asia*. Report prepared for FAO.

Leyton, J.I. (1980).* *Management and Utilization of the Latin American Moist Tropical Forests*. Paper presented to FAO Technical Consultation on Latin American Forests, 1980.

Maître, H.F. (1982). *Projet de recherches sylvicoles sur les peuplements naturels en fôret dense guyanaise*. Ministère de la recherche et de la technologie Secretariat d'Etat aux DOM-TOM.

Maître, H.F. (1986). Recherches sur la dynamique et la production des peuplements naturels on fôret dense tropical d'Afrique de l'Ouest. *Proceedings of 18th World Congress of IUFRO (Ljubljana. September 1986), Division I*, Volume 2, pp. 438–50. (Yugoslav IUFRO World Congress Organizing Committee: Ljubljana)

Mattson Marn, H. and Jonkers, W.B.J. 1981. *Logging Damage in Tropical High Forest*. UNDP/FAO/76/008 Working Paper No.5. (Forest Department: Kuching)

Mengin-Lecreulx, P. and Maître, H.F. (1986). *Les systemes d'aménagement dans les fôrets denses humides d'Afrique*. Prepared for FAO contribution to International Workshop on Rain Forest Regeneration and Management. Guri, Venezuela. (Centre Technique Forestier Tropical: Nogent-sur-Marne)

Mok, S.T. (1979). *Management of Timber Resources in Integrated Timber Complexes*. Persindangan perhutanan ketiga wilayah pantai timur, Kuantan Dec. 1979. Jabatan Hutan Negeri Pahang, Kuantan. (Chapter 8) cited in Whitmore 1984.

Moore, D. (1978).* *Analisis de alternativas de sistemas de manejo aplicables en las reservas forestales de Venezuela, en relacion a la produccion de materia prima para la industria mecanica de la madera*. UNDP/FAO/VEN/78/008. (FAO: Caracas)

Neyra-Roman M.G. (1979). *El bosque de Guandal y sus posibilidades de manejo silvicultural*. Proyecto de Investigaciones y Desarrollo Industrial Forestales, COL/74/005, PNUD/FAO. (FAO: Bogota)

Nillson, H.E., Marsch, H.E. and Singh, H.W. (1978). *Identification and Planning of a National Forest Inventory for the Philippines*. FAO, Working Paper No.15, PHI/72/006. (FAO: Rome)

Palmer, J.R. (1975). Toward more reasonable objectives in tropical high forest management for timber production. *Commonwealth Forestry Review*, **54(3/4)**, 273–89

Philip, M.S. (1986).* *Management Systems in the Tropical Moist Forests of the Anglophone Countries of Africa*. Report prepared for FAO.

Rankin, J.M. (1985).* Forestry in the Brazilian Amazon. In Prance, G.T. and Lovejoy, T.E. (eds.) *Amazonia*, pp. 369–92. (Pergamon Press: Oxford)

Roche, L. and Dourojeanni, M.J. (1984). *A Guide to* in situ *Conservation of Genetic Resources of Tropical Woody Species*. FAO/UNEP 1108–75–05 with collaboration of Unesco and IUCN. (FAO: Rome)

Romero Mejia, R. (1983). *La selva central: situacion actual y perspectivas para su desarrollo forestal*. PNUD/FAO/PER/81/002, Doc. de Trabajo No.11. (FAO: Lima)

Salleh, M.N. and Baharudin, J. (1985). Silvicultural practices in Peninsular Malaysia. In Davidson,

J., Tho Yow Pong and Bijleveld, M. (eds.) *Future of Tropical Rain Forests in South East Asia*. pp. 85–90. Commission of Ecology Papers No.10. (IUCN: Gland)

Smith, D.M. (1962). *The Practice of Silviculture*. Seventh Edition, (John Wiley and Sons: New York)

SODEFOR/CTFT (1985).* *Dispositifs d'étude de l'évolution de la fôret dense ivoirienne suivant différentes modalités d'intervention sylvicole. Présentation des principaux résultats après quatre années d'expérimentation.* (SODEFOR/CTFT: Abidjan)

Soepadmo, E. and Kira, T. (1977).* Contribution of the IBP-PT research project to the understanding of Malaysian forest ecology. In Sastry, C.B., Srivastava, P.B.L., Abdul Manap Ahmad (eds.) *A New Era in Malaysian Forestry*, pp. 63–90. (University Pertanian Malaysia: Serdang)

Soerianegara, I. and Kartawinata, K. (1985).* Silvicultural management of the logged natural dipterocarp forest in South-east Asia. In Davidson, J., Tho Yow Pong and Bijleveld, M. (eds.) *Future of Tropical Rain Forests in South-east Asia*, pp. 81–4. Commission of Ecology Papers No.10. (IUCN: Gland)

Synnott, T.J. (1980). *Tropical Rain Forest Silviculture, a Research Project.* Commonwealth Forestry Institute Occasional Paper No.10. (Department of Forestry, Oxford University: Oxford)

UNDP/FAO. (1970*a*). *Demonstration and Training in Forest, Forest Range and Watershed Management, The Philippines*. Forest Management, based on the work of D.I. Nicholson. FO:SF/PHI 16. Technical Report 3. (FAO: Rome)

UNDP/FAO. (1970*b*). *Estudio de preinversion para el desarrollo forestal en los valles del Magdalena y del Sinu*. Informe Terminal FO:SF/COL 14. (FAO: Rome)

UNDP/FAO. (1970*c*). *Estudio de preinversion para el desarrollo forestal de la Guayana Venezolana*. FAO/SF:82/VEN 5, Informe Final Tomo I, General. (FAO: Rome)

UNDP/FAO. (1973).* *A National Forest Inventory of West Malaysia* (1970–1972). Technical Report No.5. (FAO: Rome)

UNDP/FAO. (1977). *Fortalecimiento del Servicio Forestal*. FO:DP/ECU/71/527. (FAO: Rome)

UNDP/FAO. (1979). *Demostracion de manejo y utilizacion integral de bosques tropicales. Peru*. Resultados y recomendaciones del proyecto. FO:DP/PER/71/551. Informe Terminal. (FAO: Rome)

UNDP/FAO. (1980). *Management Plan for the Tapajos National Forest. Brasilia.* FO:BRA/78/003 Technical Report 2, based on work of Wood, T.W.W., (FAO: Rome)

UNDP/FAO. (1982*a*). *Forestry Development Project Sarawak Terminal Report*. Project Result, Conclusions, Recommendations. FO:MAL/76/008. (FAO: Rome)

UNDP/FAO. (1982*b*). *Report of the Joint UNDP/FAO Review Mission for BRA/82/008.* (UNDP: Brasilia)

UNDP/FAO. (1982*c*). *Desarrollo Forestal en Ecuador*. Resultados y Recomendaciones del Proyecto. FO:DP/ECU/77/005. (FAO: Rome)

UNDP/FAO. (1983*a*). *Roading, Logging and Transport in the Tapajos National Forest*. Technical Report Draft BRA/82/008. Based on work of J. Overgaard.

UNDP/FAO. (1983*b*). *Forestry Development Brazil: Project Findings and Recommendations*. FO:DP/BRA/78/003 Terminal Report. (FAO: Rome)

UNDP/FAO. (1983*c*). *Forestry Research in the Tapajos National Forest, Santarem, Brazil*. FO:BRA/82/008. Report based on work of T.W.W. Weaver (FAO: Brasilia)

UNDP/FAO. (1985*a*). *Forestry Development in Northeast Brazil; Terminal Statement*. FO:TCP/BRA 2202. (FAO: Rome)

UNDP/FAO. (1985*b*). *Investigacion en silvicultura tropical basado en la labor de J.T.Sterringa*. Documento de Trabajo No.38. (FAO: Rome)

Unesco/UNEP/FAO. (1979).* *Tropical Forest Ecosystems*. A state-of-knowledge report prepared by Unesco, UNEP and FAO. Natural Resources Research Series, 14. (Unesco: Paris)

Vadya, A.P., Colfer, C.J.P. and Brotokusumo, M. (1980).* Interactions between people and forests in East Kalimantan. *Impact of Science on Society,* **30**, 179–90

Vanniere, B. (1974).* *Management Possibilities of Tropical High Forest in Africa*. (FAO: Rome)

Wadsworth, F.H. (1952). Forest management in the Luquillo Mountains. III. Selection of products and silvicultural policies. *Caribbean Forester,* **13**, 93–142

Wadsworth, F.H. (1969). *Posibilidades futuras de los bosques del Paraguay*. FAO:SF/PAR 15. Documento de Trabajo No.2. (FAO: Rome)

Wadsworth, F.H. (1978).* *Silvicultural Aspects of Integrated Development of the Tapajos*

National Forest, Pará, Brazil. Report on FAO Consultancy 2/27–3/14, 1978. (FAO: Rome)

Walton, A.B. (1955).* Forest and forestry in North Borneo. *Malayan Forester,* **18**, 20–3

Whitmore, T.C. (1984). *Tropical Rain Forests of the Far East*. Second Edition. (Clarendon Press: Oxford)

Wood, T.W.W. (1982). *Suriname Forest Management. Operational Aspects*. FO:DP/SUR/81/004, Project Working Document No.22. (FAO: Paramaribo)

Wyatt-Smith, J. (1963). *Manual of Malaysian Silviculture for Inland Forests*. Malaysia Forest Record No.23. Two volumes. (Forestry Department of Malaysia: Kuala Lumpur)

*Documents not cited specifically in text which were reviewed for this paper and which are excellent sources of information on the topic.

Section 3

Case studies

CHAPTER 9

SILVIGENESIS STAGES AND THE ROLE OF MYCORRHIZA IN NATURAL REGENERATION IN SIERRA DEL ROSARIO, CUBA

R.A. Herrera, R.P. Capote, L. Menéndez and M.E. Rodríguez

ABSTRACT

Floristic and structural characteristics of the successional stages of the tropical evergreen submontane forest in the Sierra del Rosario Biosphere Reserve in Cuba are linked to natural and anthropogenic factors – their inter-relationships being expressed in a model of silvigenesis. The groups of plants that predominate within each phase of the successional sequence can be classified according to sclerophylly, subsclerophyllous plants being characteristic of the initial successional phases. A functional classification of the forest systems at Sierra del Rosario is being elaborated, based on five major characters:

(1) distribution of litterfall over the entire year,

(2) average degree of sclerophylly,

(3) decomposition and nutrient cycling rates,

(4) distribution of rains and total precipitation, and

(5) edaphic characteristics.

The role of mycorrhizae in forest regeneration is briefly discussed, based on measures of mycorrhizal infection levels of five tree species at different growth stages.

INTRODUCTION

Since 1974, researchers at the Institute of Ecology and Systematics of the Cuban Academy of Sciences have carried out studies at the Sierra del Rosario Biosphere Reserve in the province of Pinar del Rio of western Cuba. The predominant vegetation type is tropical submontane evergreen forest, with areas of thorny xeromorphic and herbaceous communities. The topography is dissected, with an altitude range of 50–600 m. Based on 20 years' data, annual rainfall and temperature average 2014 mm and 24.4°C, respectively. A detailed synthesis, in Spanish, of 10 years' research at Sierra del Rosario has been completed (Herrera *et al.* 1986, 1988). Here, we present insights on three aspects related to rain forest regeneration:

(1) relations between sclerophylly and silvigenesis,

(2) the search for a functional categorization of different forest types, and

(3) the role of mycorrhizae in regeneration processes.

SCLEROPHYLLY AND SILVIGENESIS STAGES

Scleromorphism has long been considered a parameter useful in evaluating the characteristics of plant formations. From the viewpoint of geobotany and plant community classification, the textural characteristics of leaves have been variously considered as membranaceous, papiraceous, chartaceous, and coriaceous. However, the classification of leaves has always somewhat depended upon the perception and judgement of individual observers.

During the last few years, sclerophylly has started to be reconsidered in functional ecosystem analysis. Various measures of the degree of scleromorphism have been developed, including determination of leaf hardness using penetrometers, lignin content, and other methods. Sclerophylly is especially important in considerations of leaf decomposition rates. Given the evidence that all the leaf material will decompose in time, methods based only on the determination of a part of such leaf material (e.g. use of penetrometers for leaf hardness) have to be discarded. One solution lies in determining the dry weight:fresh weight (DW:FW) ratio. This enables account to be taken of all the dry matter content, and is, in addition, an easy method to follow.

Another need is to recast or readapt the concept of sclerophylly as a continuous parameter which it is possible to measure by physical methods. Sclerophylly is thus hereby defined as the total dry content or hardness of leaves measured by the DW:FW ratio, and the degree of sclerophylly as the variation of DW:FW ratio as affected by cuticle thickness, stratification of epidermal tissue, constitution of mesophyll tissue, quantities of sclereids and latex tubes, density and hardness of conductive tissue, and, chemically, by the content of nutrients, lignin and other components.

Results obtained at the Venezuelan Institute of Scientific Research (IVIC) and other institutes have revealed differences between fast-growing trees – whose leaves are generally broader and more hygrophyllous – and slow-growing trees, whose leaves are usually smaller and more sclerophyllous. Negative correlations have been demonstrated between leaf hardness and decomposition rate (given similar environmental conditions for a given place) and also between leaf hardness and specific leaf area. In both groups of tree species, the highest degrees of sclerophylly have been correlated with lowest nutrient concentrations in leaves (Cuevas and Medina 1986).

Successional phases in the evergreen submontane forest at Sierra del Rosario are related to the magnitude and evolution of natural and anthropogenic factors, which give rise respectively to the so-called chablis or man-made transformations of the landscape. The following results refer only to naturally occurring plant succession, considering that human influence is restricted to the elimination of emergent individuals in the forest.

In line with the descriptions of Hallé *et al.* (1978) on the natural or man-made elimination of a large tree in a forest, an immediate phase arises in which graminous and herbaceous species predominate. This short-term phase is quickly followed by a first phase of vigorous secondary growth species which competitively give rise to the second growth forest (Homeostasis I). Subsequent phases comprise a second phase of vigorous growth which is followed by the development of a mature or primary forest (Homeostasis II). These successional phases can be recognized in the Sierra del Rosario forests. Moreover, the groups of plants that predominate within each phase (considering 75% or more of individuals) can be classified according to their sclerophyllous characteristics.

The initial immediate phase includes species with a range of sclerophylly (Table 9.1). In contrast, subsclerophyllous plants are characteristic of the first successional (i.e. second growth) phases, while most of the plants in the Regrowth I and Homeostasis I phases have leaves which are broad, hygrophyllous, soft and hairy, and normally decompose very rapidly; these are short-term phases, which may last for 15–20 years at Sierra del Rosario.

From consideration of the successional phases and sclerophylly values given in Table 9.1, three main groups of forest tree species can be recognized at Sierra del Rosario:

Table 9.1 Sclerophylly and leaf area for different successional phases at Sierra del Rosario

Species	*DW:FW*	*Specific Leaf Area (SLA)* (cm^2. g^{-1})
IMMEDIATE PHASE		
Olyra latifolia L.	0.447	
Laciacis divaricata (L.) Hitch.	0.338	
Pharus glaber H.B.K.	0.303	
Cestrum laurifolium L.Her	0.202	
Solanum erianthum D.Don.	0.333	
Solanum schlechtendalianum Walp.	0.272	
Muntingia calabura L.	0.388	
Phytolacca rivinoides Kunth et Bouché	0.167	
Ipomoea tuba (Schlcht.) G.Don.	0.166	
Pavonia fruticosa (Mill.) Fawx. et Rendle	0.263	
Piper aduncum L.	0.243	
Pisonia aculeata L.	0.165	
Trema micrantha (L.) Blume	0.365	
Average degree of sclerophylly ($\bar{x}$)	0.280	
Standard deviation ($s\bar{x}$)	0.091	
Coefficient of variation	32.5 %	
HOMEOSTASIS I. SECOND GROWTH FOREST		
Cecropia spp.	0.256	
Ipomoea tuba (Schlcht.) G.Don.	0.166	
Phytolacca rivinoides Kunth et Bouché	0.167	
Sapium jamaicense Sw.	0.194	
Vitis tiliaefolia (Hum.) C.Bonpl.ex R. et S.	0.228	
Cissampelos pareira L.	0.271	
Smilax mollis Willd.	0.272	
Piper aduncum L.	0.243	
Average degree of sclerophylly ($\bar{x}$)	0.224	
Standard deviation ($s\bar{x}$)	0.043	
Coefficient of variation	19.2 %	
HOMEOSTASIS II. INTERMEDIATE OR RESTORING PRIMARY FOREST		
Hibiscus elatus Sw.	0.268	190.7
Guarea guidonia (L.) Sleumer	0.266	192.0
Dendropanax arboreus (L). Dec. et Planch.	0.212	218.9
Zanthoxylum martinicense (Lam.) D.C.	0.332	200.0
Trichospermum grewiifolius (A.Rich.) Kosterm	0.216	307.4
Erythrina poeppigiana (Walp.) O.F. Cook	0.216	196.8
Ficus subscabrida Warb.	0.276	196.8
Ficus aurea Nutt.	0.231	196.9
Cedrela mexicana J.M. Roem	0.273	196.7
Alchornea latifolia Sw.	0.349	-

(continued)

Table 9.1 *continued*

Species	*DW:FW*	*Specific Leaf Area (SLA)* ($cm^2 . g^{-1}$)
Average degree of sclerophylly or EFA ($\bar{x}$)	0.266	209.3
Standard deviation ($s\bar{x}$)	0.046	35.3
Coefficient of variation	17.3 %	16.9%
HOMEOSTASIS II. PURE PRIMARY FOREST		
Pseudolmedia spuria (Sw.) Griseb.	0.402	142.3
Oxandra lanceolata (Sw.) Baill.	0.405	159.6
Matayba apetala (Macf.) Radlk.	0.443	134.1
Trophis racemosa (L.) Urb.	0.359	152.7
Prunus occidentalis Sw.	0.406	116.5
Amaioua corymbosa H.B.K.	0.413	122.0
Coccoloba retusa Griseb. Cat.	0.309	74.4
Guettarda valenzuelana A.Rich.	0.248	83.5
Calophyllum antillanum Britt.	0.417	88.5
Casearia sylvestris Sw.var.myricoides Griseb.	0.487	75.3
Bursera simaruba (L.) Sargent	0.440	-
Mastichodendron foetidissimum (Jacq.) Cronquist	0.410	-
Average degree of sclerophylly or EFA ($\bar{x}$)	0.395	114.9
Standard deviation ($s\bar{x}$)	0.064	32.5
Coefficient of variation	16.1 %	28.3%

(1) second growth species;

(2) intermediate or restoring species; and

(3) primary species.

According to this classification, regrowth and homeostatic phases (numbered II) may follow different successional paths depending upon environmental conditions, the most important of which is exposure to sunshine.

After the short-term secondary phases (I), there appear restoring species that are fast growing, with low sclerophylly, so that a Regrowth Phase II arises where competitive mechanisms are finally detrimental for *Cecropia* sp. individuals, perhaps because of the architectural characteristics of this species whose small number of very large leaves reduces its competitive abilities.

Depending upon sunshine exposure, different plant formations might develop from the Regrowth Phase II. Thus, in small humid valleys between mountains, and on humid north or west-facing slopes, which are somewhat protected from exposure to direct sunshine, the Regrowth Phase II is followed by an Homeostatic Phase II or Primary (Mature) Forest, where restoring forest tree species are predominant, giving rise to an Intermediate or Restoring Primary Forest (ombrophyllous or pluvisilva-like forest). On hilltops, and less humid north to east, or south-facing slopes, characterized by high levels of direct sunshine and wind, the Regrowth Phase II is of much longer duration; second growth species are substituted by a mixture of restoring and primary species, which might finally develop into a Mixed Primary Forest (restoring plus primary species) or a Pure Primary Forest (only primary species).

Thus, depending upon environmental factors, the so-called tropical evergreen submontane forest at Sierra del Rosario appears as a mosaic of mixed or pure primary forests, interspersed with semi-deciduous forests which occur exclusively on south-facing slopes or others with a slope greater than 45° which have no topographic protection against the wind.

FUNCTIONING OF FOREST ECOSYSTEMS AT SIERRA DEL ROSARIO

Over a period of about 10 years, data have been collected at Sierra del Rosario from different stands which could be classified as Restoring Primary Forest or Mixed Primary Forest – the former subsclerophyllous at a site called locally Yagrumal-Majagual, the latter mesosclerophyllous at Vallecito.

Two different groupings of functional characteristics and performance can be recognized (Table 9.2), the mesosclerophyllous forest being less productive and more fragile than the subsclerophyllous one. Caution is expressed in presenting such a comparison. From the forest management standpoint, there is no intention to suggest that the subsclerophyllous type should be imitated everywhere, since it is a forest type that is environmentally restricted. Also, the results obtained from plant formations functionally associated with the tropical evergreen submontane forest system cannot be extrapolated to the semi-deciduous forest systems occurring in the Cuban plains.

These caveats notwithstanding, a functional classification of tropical forest ecosystems is being refined by our group. Such a classification considers most of the parameters referred to in Table 9.2, in describing the main groups of forests and their variations. The classification is based upon five major characters:

(1) distribution of litterfall over the entire year;

(2) average degree of sclerophylly;

Table 9.2 Functional characteristics of mesosclerophyllous forest (Vallecito) and subsclerophyllous forest (Yagrumal-Majagual) at the Sierra del Rosario Biosphere Reserve (a)

Functional character	*Vallecito*	*Yagrumal-Majagual*
Root mat	Seasonal to permanent	Absent
Layer of dead leaves	Constant	Seasonal (only during "dry" season)
Hardness of rootlets	Higher	Lower
Biomass of rootlets	Higher	Lower
Living rootlets (%)	Lower	Higher
Vertical distribution of rootlets	Concentration in the top layers	More uniformly distributed
Decomposition rate of leaves	Lower	Higher
Decomposition rate of rootlets	Lower	Higher
Vertical distribution of VA mycorrhizal extramatrical mycelia (MEVA)	Approximately the same	Approximately the same
Vertical variation of MEVA: rootlet ratio (μg mg^{-1})	Increasing with depth	Uniformly distributed
Production of MEVA	Lower	Higher
Production of rootlets	Lower	Higher
Growth rates of trees	Slow growth	Fast growth
Nutrient content of leaves	Lower	Higher
Organic detritus in soil	Higher	Lower
Decomposer (fauna) populations	Lower	Higher
Productivity (b)	Lower	Higher
Turnover rates	Lower	Higher (about double)

(a) Actual figures corresponding to each character can be obtained from the authors.
(b) Considering stands of the same age.

(3) decomposition and nutrient cycling rates;

(4) distribution of rains and total precipitation;

(5) edaphic characteristics.

In addition, in considering forest regeneration and management, account also has to be taken of the original plant formations that were able to grow under the environmental conditions of eight different localities studied at Sierra del Rosario. A tentative conclusion is that silvigenesis can be closely related to the succession of plant associations characterized by groups of species with similar scleromorphic characteristics.

Table 9.3 Some characteristics of the seedling bank occurring on the forest floor of Vallecito at Sierra del Rosario. The data correspond to 48 months of observations between 1981 and 1985

Plant species	*Up to date observed number of seedlings*	*Mortality* (%)	*Average initial height* (cm)	*Mortality during the dry season*	*Herbivory* (a) (%)	*Growth rates* (cm yr^{-1})
T.racemosa	14	64.3	18.9 (1.1)	100.0	0.25 (0.09)	0.81 (0.47)
P.spuria	9	44.4	17.6 (0.9)	50.0	0.20 (0.10)	1.45 (0.47)
P.occidentalis	16	68.8	21.1 (1.3)	72.7	0.56 (0.02)	1.67 (0.74)
M.apetala	5	0.0	11.0 (1.3)	0.0	0.41 (0.06)	0.83 (0.11)
O.lanceolata	5	0.0	13.0 (2.1)	0.0	0.58 (0.08)	1.06 (0.35)

(a) Number of attacks by herbivores (including gasteropods) per plant per year.
Note: figures between parenthesis represent the standard errors.

ROLE OF MYCORRHIZA IN NATURAL REGENERATION

Regeneration of the mixed primary forest was studied during a 48-month period at Vallecito, by observing seedlings belonging to five predominant tree species (Table 9.3). Resistance of seedlings to different types of environmental stress was greater for *Matayba apetala* (Sapindaceae) and *Oxandra lanceolata* (Annonaceae) than for *Pseudolmedia spuria* or *Trophis racemosa* (Moraceae) and *Prunus occidentalis* (Rosaceae). Mortality occurred mostly during the relatively dry season, and seemed to be correlated with the characteristic vulnerability of each tree species, as measured by the so-called vulnerability index (this considers the tangential diameter of xylem fibres and the frequency of the latter per mm of wood tissue, the less vulnerable tree species being more resistant to drought). The incidence of seedlings affected by herbivory was larger for those species with a higher content of essential oils. Growth rates were very low for all species.

In a second (destructive) experiment, seedlings of different tree species and growth stadia were collected in order to analyze morphology and mycorrhizal infection levels. Growth stadia were differentiated in three main categories: cotyledonal (I), seedling (II), and small trees (III). Collected individuals belonged to *O. lanceolata* (I, II, III), *P. spuria* (I, III), and *M. apetala* (I, II, III) with relatively small seeds (0.5–1.0 cm); and *P. occidentalis* (I) and *Calophyllum antillanum* (I) with relatively large seeds (1.0–2.0 cm). For each species and stage, six to 12 individuals were

Table 9.4 Growth characteristics and mycorrhizal infection levels for different tree species and growth stadia in Vallecito, Sierra del Rosario

Tree species and growth stadia	*Plant height* (cm)	*Probable age* (years)	*Above-ground weight* (g)	*VA Mycorrhizal infection* %
OL I	12.3(0.8)	< 1	0.12(0.01)	97.8
OL II	19.8(1.0)	4.76	0.57(0.05)	93.6
OL III	25.5(1.1)	7.81	1.55(0.09)	90.0
PS I	10.4(0.5)	< 1	0.05(0.01)	15.9 (a)
PS III	51.4(4.2)	4.60	1.69(0.12)	30.3 (a)
PO I	16.5(1.1)	< 1	0.40(0.02)	62.0
CP I	16.0(1.0)	< 1	0.67(0.05)	66.2
MA I	7.1(0.4)	< 1	0.05(0.01)	45.4
MA II	14.9(1.0)	2.45	0.57(0.05)	45.0
MA III	45.6(3.6)	5.90	6.46(0.61)	65.6
MA IV	119.2(8.2)	7.50	38.50(3.35)	
MA V	157.0(4.6)	11.25	76.80(5.60)	68.4 (b)
MA IV S	175.0(25.5)	7.25	103.60(27.10)	
MA V S	196.0(17.7)	10.75	151.30(17.40)	72.8 (b)

Data are given as medians and (standard errors).
(a) mycorrhizal infection determined as percentage of mycorrhizal rootlet nodules;
(b) mycorrhizal determination was realized mixing rootlets from IV and V stadia individuals;
S individuals collected from the road slope.
OL *O. lanceolata*;
PS *P. spuria*;
PO *P. occidentalis*;
CP *C. antillanum*;
MA *M. apetala*.

collected. For *M. apetala*, two additional stadia belonging to the small tree bank were collected from the forest floor or a neighbouring open place on road slope (S), and named IV and V (see Table 9.4 for descriptions); in this case, three to five individuals were collected.

Since thickness of growth rings formed during the "dry" season was highly correlated with water deficits during drought periods, it was possible to estimate the approximate age of *M. apetala* seedlings. Data on seedling growth during the 48-month observation period were also used in calculating the age of all collected individuals. Vesicular-arbuscular (VA) mycorrhizal infection levels were determined by standard procedures for forest tree species carried out in our laboratory (Herrera *et al.* 1988).

Seedlings at the cotyledonal stadium were highly mycorrhizal only in *O. lanceolata*, *P. occidentalis* and *C. antillanum* (Table 9.4), which could be attributed to the large seed reserves in the two latter species. The explanation for *O. lanceolata* might be related to the light habits of this species, whose ecological plasticity is very large and which is able to live, flower and fruit in shady conditions. This is in line with observations that an heliophyllous plant species tends to have low levels of infection when growing in shade; this fact could be responsible for the low infection levels for *P. spuria* and *M. apetala*, where photosynthetic rates were probably affected by the reduced reception of light at the forest floor.

Increase in infection levels from stadia II to III in the case of *P. spuria* and *M. apetala* suggest that some metabolic changes are probably occurring during this change in seedling morphology, most likely due to a better reception of light which helps the development of the VA endophyte in rootlets.

The role of mycorrhiza in forest regeneration seemed not to be restricted to the seedling and small tree banks. In a third and final experiment, the VA mycorrhizal potential (infectivity: number of propagules per dm^3 of soil) was measured in soil samples collected from various plant formations. Table 9.5 shows these infectivity values, and corresponding information on VA spore populations and VA mycorrhizal infection levels.

Table 9.5 Vesicular-arbuscular mycorrhizal potential (infectivity) and other mycorrhizal characteristics of different plant formations at Sierra del Rosario

Site	*Mycorrhizal No. of propagules* dm^3	*Potential fiducial limits*	*Spores populations (a)*	*VA Mycorrhizal infection*
SECOND GROWTH FORESTS				
Natural chablis	3452	1362-8750	+++	56.8
Road slope	5408	2133-13710	n.d.	72.8
RESTORING PRIMARY FOREST				
Yagrumal-Majagual	345	136-875	++	40.4
Pluvisilva El Mulo	345	136-875	++	40.2
MIXED PRIMARY FOREST				
Vallecito	541	213-1371	+++	27.4
PURE PRIMARY FOREST				
Las Peladas	541	213-1371	+++	30.4
Semi-deciduous El Mulo	228	90-577	+	55.8

a, spore population classified as few (+), several (++), or many (+++) spores.
n.d., not determined.

The number of mycorrhizal propagules is higher in plant associations and environmental conditions typical for small open spaces in the forest system characterized by heliophyllous fast-growing and sub-sclerophyllous plant species with probably higher photosynthetic rates and fine root proportions. In these open places, the role of organic matter has also to be considered, since rapid decomposition rates could perhaps influence the infection development both in chablis and road slopes.

On the other hand, we consider that the actual importance of a larger mycorrhizal potential in the chablis (10 to 15 times greater than in the primary forest) rests on: first, the considerable ecological benefits represented by the opportunity for rapid and effective VA infection starting on the germination of primary species seeds (distributed by birds or other animals); and, second, the additional boost provided by the highly infective propagules surrounding the previously infected seedlings and small trees which remained after chablis opening. Taken together, the newly infected seedlings and improved previously infected seedlings and small trees are important influences on the future succession phases.

REFERENCES

Cuevas, E. and Medina, E. (1986). Nutrient dynamics within Amazonian forest ecosystems. I. Nutrient flux in fine litterfall and efficiency of nutrient utilization. *Oecologia,* **68**, 466–72

Hallé, F., Oldeman, R.A.A. and Tomlinson, P.B. (1978). *Tropical Trees and Forests. An Architectural Analysis*. (Springer-Verlag: Berlin)

Herrera, R.A., Menéndez, L. and Rodriguez, M.E.. (1986). *Ecologia de los bosques tropicales de la Sierra del Rosario, Cuba. Proyecto MAB 1. 1974–1986.* Libro de Resúmenes. (Academia de Ciencias de Cuba: Havana)

Herrera, R.A., Menéndez, H., Rodríguez, M.E. and Garcia, E.E. (eds.) (1988). *Ecología de los bosques siempreverdes de la Sierra del Rosario, Cuba. Proyecto MAB No.1, 1974–1987.* (Unesco: Montevideo)

CHAPTER 10

COMPARATIVE PHYTOSOCIOLOGY OF NATURAL AND MODIFIED RAIN FOREST SITES IN SINHARAJA MAB RESERVE IN SRI LANKA

N.D. de Zoysa, C.V.S. Gunatilleke and I.A.U.N. Gunatilleke

ABSTRACT

*The vegetation structure, floristic richness and density-dominance of endemic and pioneer species in three differently modified lowland rain forest sites with varying gap sizes (i.e. a selectively logged site, skid trail and shifting cultivation site) have been compared with those of an adjacent relatively undisturbed forest site at Sinharaja Biosphere Reserve in Sri Lanka. At each site, all plants, excluding epiphytes and parasites, were examined by plot sampling. In the undisturbed site with continuous, closed canopy and in the selectively logged site, vegetation reached 30–35 m in height. In each of the three disturbed sites, the canopy was open and discontinuous. In all four sites, 86–94% of the total density was contributed by individuals less than 1 m tall. The proportion of pioneer species ranged between 9–11% in the undisturbed and selectively logged sites, and between 29–32% in the logging trail and shifting cultivation site. In the latter two sites, they dominated the vegetation. The proportion of endemic species in the vegetation ranged between 59–62% in the two former sites, and between 44–51% in the latter two. Some 90% of the species encountered in the survey are in either Endangered, Vulnerable or Rare categories of the IUCN Red Data Book and their regeneration in disturbed forest sites was poor. The impressive performance of mahogany (*Swietenia macrophylla*), planted as an exotic enrichment species on the logging trail, indicates its potential for silviculture in disturbed lowland rain forests. The canopy of the undisturbed forest reached 30–35 m in height and was closed and continuous. The selectively logged forest while reaching a similar height had an open discontinuous canopy. The vegetation in both the logging trail and abandoned shifting cultivation site was 10–15 m in height with patchy cover.*

In all sites, 86–94% of the toal density was contributed by individuals less than 1 m tall. The proportion of pioneer species in the undisturbed and selectively logged sites ranged between 9–11% while in the logging trail and shifting cultivation it was 29–32%. In general terms, this study has laid the foundation for periodic re-enumeration of sites with different extents of disturbance and gap sizes, so that information on the regeneration potential of their constituent, indigenous and introduced species, or rare endangered endemics with low population densities, may be obtained.

INTRODUCTION

The species-rich lowland rain forests of Sri Lanka, with a high proportion of endemics in them, are of considerable phytogeographic importance to South and South-east Asian floras (Gunatilleke and Gunatilleke 1985, Ashton and Gunatilleke 1987, Gunatilleke and Ashton 1987). Today, they are reduced to about 9% of the land area of the wet lowlands of the island (Nanayakkara 1983), and are much fragmented and interspersed with village settlements (Perera 1972). Even these small fragments of forests are being gradually released for selective logging to meet the increasing timber demand of the country, and, at the same time, are subjected to encroachment by the villagers living along the forest periphery (Anon. 1986).

The ecological effects, both long and short-term, of selective logging in the rain forests of Sri Lanka have been studied mostly from a silvicultural point of view and the scantily available results have not yet been successfully incorporated in the management of these forests for timber needs (Anon. 1986). The role of non-timber species has been largely ignored in these management plans (Andrews 1961). As a part of our long-term studies in understanding the structure and functioning of the lowland rain forests of Sri Lanka, we have commenced a study to monitor, at periodic intervals, the changes in the structure, floristic richness, dominance and performance of primary and secondary forest species of potential silvicultural importance as well as those of endangered species in both natural and modified forest habitats.

SITE DESCRIPTION AND FIELD SAMPLING METHODS

This study was carried out in the northwestern part of Sinharaja MAB Biosphere Reserve [The geology, climate and soil properties of the area have been described previously (Gunatilleke and Gunatilleke 1981, 1985, Maheswaran and Gunatilleke 1987)] where a patch of relatively undisturbed lowland rain forest (Site 2 of Gunatilleke and Gunatilleke 1985) adjacent to three differently modified forest sites was selected for comparison of vegetational changes in both time and space. The salient features of the modified forest sites with different canopy gap sizes are as follows:

> Selectively logged forest site (SL) in which over 60% of the canopy was opened between 1972 and 1977 by selective removal of choice timber trees using heavy machinery (Gunatilleke and Gunatilleke 1981, 1983).
>
> Logging trails or skid trails (ST) of 4–5 m width constructed during logging operations between 1972 and 1977 by complete removal of surface vegetation together with top soil. They were subjected to soil compaction due to frequent movement of heavy logging machinery during the logging period and soil impoverishment from direct exposure to strong insolation and heavy precipitation.
>
> A traditional shifting cultivation site (SC) which was clear-felled, burnt, cultivated and abandoned 25–30 years ago.

Entire vegetation, excluding epiphytes and parasites, was categorized into four appropriate sizes (height/girth) classes viz (i) < 1 m in height, (ii) > 1 m in height but < 10 cm gbh, (iii) 10–30 cm gbh, (iv) > 30 cm gbh. These were sampled using 4 plot sizes: 1 m x 1 m, 2 m x 10 m, 10 m x 100 m and 25 m x 10 m respectively. The vegetation above 10 cm gbh was enumerated and tagged for future monitoring in permanently marked plots.

PHYSIOGNOMY AND STRUCTURE OF THE VEGETATION

The undisturbed forest site (UF) had a continuous canopy cover of 30–35 m height with a well packed tree layer but with relatively poorly developed shrub and climber vegetation. Structurally, this forest site has reached a mature phase with some gaps due to tree-fall within it. The SL site, on the other hand, had a discontinuous and uneven canopy cover with abundant gaps of varying sizes and shapes. Herbaceous and woody climbers and dense stands of a few primary forest species, such as *Shorea trapezifolia* (Thw.) Ashton (Dipterocarpaceae), were also common in these openings. In ST sites, a discontinuous and often patchy canopy of 5–10 m height was found interspersed with a dense herbaceous ground cover, notably of *Schizostigma hirsutum* Arn. (Rubiaceae). The SC site had a maximum canopy height of about 10 m with a discontinuous canopy and a decreasing gradient in height from the periphery of the clearing towards the centre of the site.

A comparison of the total density of individuals per ha in the four height/girth classes of the vegetation (i) < 1 m in height, (ii) > 1 m in height but < 10 cm girth at breast height (gbh), (iii) 10–30 cm gbh, (iv) > 30 cm gbh showed that the SC site had almost twice as many individuals as the SL site, and 40% and 55% more than the UF and ST sites respectively (Table 10.1). However, in each site, as much as 86–94% of the total density is contributed by ground vegetation less than 1 m in height, which includes herbs plus seedlings and saplings of shrubs and trees. The relatively poor density of ground vegetation in the SL sites may have resulted from a combination of several factors, such as excessive damage to the tree stand resulting in the paucity of parent trees of many species and poor germination and establishment of seedlings in the modified environment.

The next larger size class (i.e. >1 m height – 10 cm gbh) was best represented in the SL site followed by SC, ST and UF sites. The greater density of individuals in this size class in the SL site may result from the 'release effect' by canopy opening, fuelled by a flush of nutrients made available from the decomposing debris left behind after logging operations carried out 5–8 years before the sampling. Cohorts of individuals of species already established and awaiting a canopy gap in order to shoot up, as well as those species which are prolific seed producers at frequent intervals and capable of growing relatively faster at least during their juvenile stages, have taken full advantage of this artificial creation of varying gap sizes over a short period of time. A number of timber species, as well as several understorey

Table 10.1 Comparison of the density of individuals in each size class of the relatively undisturbed forest with those of the disturbed sites

	No. of individuals per ha and percentage of individuals in each size class			
Sites sampled *Size classes*	*Undisturbed forest (UF)*	*Selectively logged (SL)*	*Skid trail (ST)*	*Shifting cultivation (SC)*
1) More than 150 cm gbh	43	14	0	0
2) Less than 150 cm gbh and more than 90 cm gbh	115 (0.3%)	45 (0.3%)	0	0
3) Less than 90 gbh and more than 30 cm gbh	580	446	500 (0.2%)	120 (0.03%)
4) Less than 30 cm gbh and more than 10 cm gbh	1487 (0.6%)	2823 (1.7%)	4100 (1.8%)	5060 (1.5%)
5) Less than 10 cm gbh and more than 1 m in height	12 735 (5%)	20 300 (12%)	13 225 (6%)	15 580 (4.5%)
6) Less than 1 m in height	226 950 (94%)	144 020 (86%)	200 750 (92%)	318 600 (94%)
Total in all size classes	241 190	167 644	218 575	339 360

tree species, exhibit greater recruitment in pole size classes (> 1 m height – < 30 cm gbh). The greater densities of individuals of 10–30 cm gbh class in SC and ST sites can be attributed to individuals of light-loving and fast-growing species with a steady and abundant supply of propagules. However, the next higher girth class (30–90 cm gbh) is poorly represented in the SC site compared to the other three sites, although it had a fallow period of 25–30 years. Intense competition from herbaceous and semi-woody climbers for space, light and nutrients may have suppressed the entry of more individuals into the larger size classes at this site. Individuals over 9 cm gbh were not represented in either the SC or ST sites, probably because of the same reasons. On the other hand, the same girth class was better represented in the undisturbed forest, and only 35% of it was found in the selectively logged site 5 to 8 years after logging was stopped (Table 10.1). The extent to which the forest will recover following this degree of disturbance is being studied by recurrent sampling of these plots at regular intervals.

FLORISTIC RICHNESS AND DOMINANCE

The total floristic richness in the UF site was 267 species belonging to 180 genera and 80 families in the 5 ha sampled, in the SL site it was 180 species belonging to 120 genera and 70 families in 1.25 ha, in the ST site 115 species in 100 genera and 54 families and in the SC site, it was 134 species in 109 genera and 66 families, in the last two sites the sample area being 0.04 ha.

Both the UF and SL sites had Dipterocarpaceae, Sapotaceae, Clusiaceae, Myrtaceae and Euphorbiaceae dominating the canopy and subcanopy, with *Shorea trapezifolia*, *Mesua nagassarium* (Burm. f.) Kosterm. *Anisophyllea cinnamomoides* (Gardn. and Champ.) Alston and *Myristica dactyloides* Gaerth. among the dominant species. The understorey tree stratum of these two sites was dominated by Euphorbiaceae, Melastomataceae and Clusiaceae, with species such as *Xylopia championii* (Wight) Hook. f. and Thoms, and *Aporusa lanceolata* (Tul.) Thw. Woody climbers in these sites belonged to Leguminosae and shrubs to Rubiaceae and Loganiaceae. In the ST and SC sites, the canopy was dominated by Euphorbiaceae, while one of the dominants in the canopy of the ST site was the exotic species *Swietenia macrophylla*, which had been planted as an enrichment species after logging.

ENDEMIC AND ENDANGERED SPECIES

In both UF and SL sites, with a few exceptions, endemic species were the dominants in each stratum and life form category. The proportion of endemic species ranged between 59–62% of the total number of species enumerated. In contrast, in the ST and SC sites, their proportions were 51% and 44% respectively. These latter sites were dominated by widespread pioneer species such as *Macaranga peltata* (Roxb.) Muell. Arg (Euphorbiaceae) and localized pioneer species such as *Schumacheria castaneifolia* Vahl. (Dilleniaceae), the latter being a genus endemic to Sri Lanka.

The density or abundance of the woody species (> 30 cm gbh), which were grouped into different categories of rare species (Gunatilleke *et al.* 1987) along the guidelines of the IUCN Red Data Book (Synge 1981), was examined in the modified as well as undisturbed forest sites (Table 10.2). The majority of the species belonging to the endangered category showed a distinctly lower density distribution in the modified forest sites compared to that of the UF site. The majority of the vulnerable and rare species also showed a positive stand table in the UF site with more individuals in small size classes and are, therefore, maintaining themselves (Whitmore 1984). They were also represented in the SL site but not in ST and SC sites, clearly indicating their vulnerability to disturbance.

The proportions of pioneer species in the UF, SL, ST and SC sites were 9%, 11%, 29% and 32%, respectively, of the total number of species recorded from each site. The increasing floristic richness and the abundance of pioneers seem to have a suppressing effect on the performance of the rare species, at least during the early phases of recovery. Pioneer gap species viz: *Schumacheria castaneifolia*, *Gomphia*

Table 10.2 Distribution of woody plant species (>30 cm gbh) encountered in natural and modified forest sites at Sinharaja according to IUCN Red Data Book categories (E = Endangered, V = Vulnerable, R = Rare, K = Unknown and O = Out of Danger). See Table 10.1 for habitat category abbreviations

	Number of species in each RDB category					
Habitat categories	*E*	*V*	*R*	*K*	*O*	*Total*
1) Better growth in UF than in modified sites	9	9	12	1	0	31
2) Equal or better growth in SL but not in ST or SC	4	15	27	5	0	51
3) Better in any one or more modified forest sites than in UF	0	1	2	0	1	4
4) Growth in all four sites	0	7	10	0	3	20
Total number of species	13	32	51	6	4	106

serrata, *Gaertnera vaginans* (DC.) Merr., *Wendlandia bicuspidata* Wight and Arn. and *Lophatherum zeylanicum* Hook. f. were better represented in smaller, narrow gaps in the ST site, and also at the edges of the larger gaps of the SC site. Based on their restricted distribution in forest gaps, these species are termed localized pioneers in gap regeneration. Among the more widespread pioneers which are more common in larger gaps or in more exposed parts of the smaller gaps, *Hedyotes fruiticosa* L., *Melastoma malabathrica* L. and *Macaranga peltata* (Roxb.) Muell. Arg. were the dominant species. This latter group of species, being indicators of greater degrees of disturbance, were more abundant in number and species in the skid trail and shifting cultivation sites. In the undisturbed and selectively logged sites, however, there were only a few widespread pioneer species, represented in relatively low densities.

REGENERATION

The population structure of a few selected canopy and subcanopy species of merchantable timber value was examined in each site (Table 10.2). A comparison of stand tables for each species, in each of the four sites, indicates that some of the density-dominant primary forest species – such as *Shorea trapezifolia*, *Syzygium makul* Gaertn., *Syzygium rubicundum* Wight and Arn. and *Shorea stipularis* Thw. – provide, on the whole, a better representation of size classes 2 and 3 (> 1 m in height, < 30 cm gbh, mostly pole-size individuals) in the SL site. Their representation in the ST and SC sites is more or less restricted to size classes 1 and 2. These results indicate that the above species perform better in canopy gaps than under the shade in the undisturbed sites. Their relatively poor recruitment into pole size classes in SC and ST sites may result from their poor competitive ability as compared with that of pioneer species abundant in these sites. Continued monitoring of the performance of these species in the disturbed sites would provide valuable information for silvicultural management of these forests.

The juveniles of some canopy and subcanopy species, on the other hand, were better represented in the undisturbed forest than in any of the modified forest sites, indicating either their extreme shade loving nature or restricted distribution related to site specificity and fruit dispersal pattern and therefore not represented in our sample plots. Although *Mesua nagassarium* var. *pulchella* is a dominant and widespread species in the undisturbed forest, it was not represented among the large size classes in the SL site and poorly represented in smaller size classes. *Mesua nagassarium* is often associated with ridge tops within the undisturbed forest, and is well represented in all size classes in such habitats. Its poor performance in our disturbed study sites may be because the selective

Table 10.3 Population densities of a few selected density dominant tree species in the four different forest sites in Sinharaja biosphere reserve, Sri Lanka. Size classes 1 = < 1 m in height, 2 = > 1 m in height – 10 cm gbh, 3 = 10–30 cm gbh, 4 = 30–90 cm gbh, 5 = 90–150 cm gbh and 6 = >150 cm gbh

	Individuals ha^{-1} in each size class					
	Undisturbed forest (UF)					
	1	*2*	*3*	*4*	*5*	*6*
A. Density dominant canopy species with best growth in selectively logged site and also established in larger gaps (ST and SC)						
1. *Shorea trapezifolia*	21600	70	6	1	2	3
2. *Shorea stipularis*	5700	65	7	3	2	3
3. *Cryptocarya wightiana*	4550	270	2	8	12	0
4. *Syzghium rubicundum*	7300	175	23	18	12	6
5. *Syzyghium makul*	10550	15	3	4	4	3
6. *Mastixia tetrandra*	750	65	3	3	6	1
B. Density dominant canopy species with better growth in selectively logged site but with limited establishment in larger gaps (ST and SC)						
1. *Myristica dactyloides*	750	270	46	24	9	1
2. *Anisophyllea cinnamomoides*	700	165	30	26	3	1
C. Density dominant canopy species with better growth in undisturbed forest than in disturbed sites						
1. *Mesua nagassarium*	22250	290	59	17	8	3
2. *Shorea worthingtonii*	15000	175	53	21	4	2
3. *Palaquium thwaitesii*	4050	595	64	8	3	1
D. Density dominant localized pioneer treelet species better established in larger gaps (ST and SC)						
1. *Schumacheria castaneifolia*	500	80	40	2	0	0
2. *Aporusa lanceolata*	2100	875	151	1	0	0
3. *Gaetnera vaginans*	2650	705	14	0	0	0
4. *Gomphia serrata*	900	65	43	29	1	0

logging was mostly concentrated in lower slopes and valleys, and rarely on ridge tops. However, its seed dispersal, germination and survival also need to be studied before coming to a definite conclusion on its poor performance in disturbed sites, in spite of being the single, most dominant species in the whole forest.

Table 10.3 *Continued from facing page*

Individuals ha^{-1} in each size class													
Selectively logged Forest (SL)						*Logging Trail (ST)*				*Shifting cultivation Site (SC)*			
1	2	3	4	5	6	1	2	3	4	1	2	3	4
14400	2580	442	10	8	6	0	250	0	0	16750	500	25	0
1000	420	98	10	2	0	17500	75	0	0	250	0	0	0
2000	820	22	0	0	0	1500	75	0	0	1500	100	0	0
2200	500	32	8	5	0	7500	1150	50	0	2500	175	0	0
600	340	36	2	0	0	750	150	0	0	4000	100	25	0
600	140	66	2	1	0	750	25	0	0	1250	0	0	0
0	300	42	35	5	1	1500	75	50	0	0	0	0	0
200	400	42	17	2	2	0	0	0	0	0	0	0	0
0	0	0	1	0	0	47500	75	0	0	0	0	0	0
0	0	0	1	2	0	0	0	0	0	0	0	0	0
0	40	44	6	0	0	0	100	0	0	250	0	0	0
1800	1080	514	21	0	0	750	50	200	0	24500	3325	1825	0
1800	1440	162	1	0	0	0	25	0	0	500	100	150	0
1200	960	40	0	0	0	0	50	0	0	50750	1800	50	0
80	220	10	14	0	0	0	25	0	0	1250	425	150	0

A third group of species, which are smaller in both height and girth at maturity (< 90 cm gbh), showed strongly 'positive' stand tables in larger gaps such as those within the ST and SC sites (Table 10.3). Amongst these are *Schumacheria castaneifolia*, *Gomphia serrata* and *Wendlandia bicuspidata* Wight and Arn., all of which are also good coppicing species.

This spatial distribution of tree species in different sized gaps and grown over different periods of time could be utilized in developing silvicultural systems suitable for degraded forests in the region. Mixed planting of species of different growth rates and age structures and having different site specificities, germination strategies and utility values have been considered in several reforestation models prepared for the lowland wet zone of Sri Lanka (Gunatilleke *et al.* manuscript).

Continued monitoring of these plots over a long period of time would provide valuable information on the overall regeneration pattern as well as the performance of individual species of both utility and conservation value. It can be clearly seen that the selective logging alters the floristic composition and density of individual species. In the SC site, even after 25 years of regrowth since cultivation was abandoned, very few primary forest species were able to establish, and no species has been able to reach a size exceeding 90 cm gbh. Performance of mahogany as an enrichment species was extremely good in logging trails and therefore useful in silvicultural management of these forests. However, its establishment in biosphere reserves such as Sinharaja could result in the reduction of species diversity because of its ability for prolific seed production and good seedling establishment.

ACKNOWLEDGEMENTS

This research project was funded by a grant from Natural Resources, Energy and Science Authority (NARESA) of Sri Lanka (RG/83/13) given to Drs. I.A.U.N. Gunatilleke and M.D. Dassanayake. We are also grateful to NARESA for providing funds for a field station at Sinharaja that facilitated this work tremendously, and to the Conservator of Forests and his staff for permission and assistance in carrying out the field work. Finally, we thank Dr.M.D. Dassanayake for his advice and assistance during various stages of this study.

REFERENCES

Andrews, J.R.T. (1961). *Forest Inventory of Ceylon*. (A Canadian-Ceylon Colombo Plan Project). (Ceylon Government Press: Colombo)

Anon. (1986). *Forestry Master Plan for Sri Lanka*. (Jakko Poyry International: Helsinki)

Ashton, P.S. and Gunatilleke, C.V.S. (1987). New light on the plant geography of Ceylon. I. Historical plant geography. *Journal of Biogeography,* **14**, 249–85

Gunatilleke, C.V.S. and Ashton, P.S. (1987). New light on the plant geography of Ceylon. II. The ecological biogeography of the lowland endemic tree flora. *Journal of Biogeography,* **14**, 295–327

Gunatilleke, C.V.S. and Gunatilleke, I.A.U.N. (1981). The floristic composition of Sinharaja – a rain forest in Sri Lanka with special reference to endemics and dipterocarps. *Malaysian*

Forester, **44**, 386–96

Gunatilleke, C.V.S. and Gunatilleke, I.A.U.N. (1983). A forestry case study of the Sinharaja rain forest in Sri Lanka. In Hamilton, L.S. (ed.) *Forest and Watershed Development and Conservation in Asia and the Pacific*, pp. 289–358. (Westview Press: Colorado)

Gunatilleke, C.V.S. and Gunatilleke, I.A.U.N. (1985). Phytosociology of Sinharaja – a contribution to rain forest conservation in Sri Lanka. *Biological Conservation,* **31**, 21–40

Gunatilleke, C.V.S., Gunatilleke, I.A.U.N. and Sumithraarachchi, B. (1987). Woody endemic species of the wet lowlands of Sri Lanka and their conservation in botanic gardens. In Bramwell, D., Hammann, O., Heywood, V. and Synge, H. (eds.) *Botanic Gardens and the World Conservation Strategy*, pp. 183–96. (Academic Press: London)

Maheswaran, J. and Gunatilleke, I.A.U.N. (1987). Soil physiocochemical properties of Sinharaja forest, Sri Lanka and the effect of deforestation. In Kostermans, A.J.G.H. (ed.) *Proceedings of Third Round Table Conference on Dipterocarps*. Samarinda, Indonesia, 16–20 April 1985, pp. 121–34. (Unesco: Jakarta)

Nanayakkara, S.D.F.C. (1983). Sri Lanka Forest Cover Map – Forestland and Management. (Centre for Remote Sensing, Survey Department: Colombo)

Perera, W.R.H. (1972). A study of the protective benefits of the wet zone forestry reserves of Sri Lanka. *Sri Lanka Forester,* **10**, 87–102

Synge, H. (1981). *The Biological Aspects of Rare Plant Conservation*. (Wiley: Chichester)

Whitmore, T.C. (1984). *Tropical Rain Forests of the Far East*. (Second Edition.) (Clarendon Press: Oxford)

CHAPTER 11

PATTERN AND STRUCTURE ALONG GRADIENTS IN NATURAL FORESTS IN BORNEO AND IN AMAZONIA: THEIR SIGNIFICANCE FOR THE INTERPRETATION OF STAND DYNAMICS AND FUNCTIONING

E.F. Bruenig

ABSTRACT

Two areas of tropical rain forest on strongly developed edaphic gradients have been analyzed for pattern; San Carlos de Rio Negro in Venezuela and Sabal Forest Reserve in Sarawak, Malaysia. The two areas have almost identical soil types and show distinct but differently patterned catenas of micro-relief and soil types. The climatic conditions and the correlated physiognomic features of the forest associations are almost identical. Tree species distribution patterns are, however, different between the two areas, indicating differences in dynamics and regeneration. The implications for research and management are discussed. The hypothesis that pattern is related to dynamics, and consequently should be a useful indicator for regeneration mechanisms and phasic developments, is being tested through continuing comparative research in Sabal (Sarawak), Danum (Sabah), Bawang Ling/Hainan (China) and San Carlos de Rio Negro (Venezuela).

STUDY AREAS

The permanent natural forest sample plot in Sabal Forest Reserve lies just north of 1°N in West Sarawak, Malaysia, at about 100 m altitude in a perhumid tropical lowland climate. Mean annual rainfall is about 4000 mm, with a seasonal peak in January. San Carlos de Rio Negro is similarly situated just south of the 2°N parallel at about 60 m altitude in the Territorio Amazonas, Venezuela. Mean annual rainfall is about 3500 mm, with a peak in May–June. The climate at both locations is characterized by irregular, sporadic periods of drought (Bruenig 1966, 1971, Heuveldop 1977) and supersaturation with intense soil flushing (Bruenig and Schmidt-Lorenz 1985).

The soil types in both areas are somewhat similar, except for the soil in the *bana* vegetation type in San Carlos, which has no equivalent in Sabal. In both areas, soil types are distributed in broad bands diagonally NW–SE, but the bands are more clearly distinguished in San Carlos while the soil type distribution pattern is more complex with scattered small enclaves in Sabal. The hydrology in both areas is characterized by alternating water saturation and dryness, but the more sloping (15–20%) Sabal soils on sandstone and bouldery conglomerate are better drained and more evenly supplied with water than the flat to almost flat (0–5%) San Carlos soils. The soil types are sufficiently similar to make comparison of the vegetation feasible.

The vegetation in both areas is a mixture of species-rich, complex forest and species-poor, simple forest – mixed dipterocarp forest and Kerangas in Sabal and complex mixed forest (*tierra firme*), *caatinga* and *bana* in San Carlos. In Sabal, the vegetation in the 20 ha plot is as heterogeneous as the soil. While species distribution in Sabal could be related to soil Munsell colour notation using a simple sorting programme on an IBM computer (Bruenig 1966), it was not possible to show such relations subsequently using special correlation programmes MUNCSELL, MODCRO and MUNTAB (Weischke 1982). Analysis of variance of the pattern of basal area distribution indicated the existence of distinct populations, but their distinction and mapping in the very complex mixture of micro-sites and soil types proved difficult (Bruenig 1973, Weischke 1982).

In San Carlos, correlation of tree species distribution with soil types is more distinct and the classification by FANTASMB produced distinct tree association groups (= forest vegetation types) which can be mapped and explained (Bruenig *et al.* 1979). This was certainly the case, if the chosen minimum tree diameter was 20–30 cm and the stopping rule for information reduction rate > 3%, but the pattern of association groups became complex and confused with no discernible association with soils if the diameter and the stopping rule were reduced. Bruenig *et al.* (1979) concluded tentatively that this was possibly due to the effect of stand dynamics at small areal scale (0.01–0.1 ha). This effect would blur the association of the successful upper canopy individuals of the various species, since the inclusion of smaller individuals introduces a stronger element of chance.

In Sabal, the results of classifying with FANTASMB were much less satisfactory, presumably as a result of the interaction of the greater complexity of soil-type distribution and the somewhat larger total number of tree species present in each group used in the similarity analysis (Weischke 1982).

SPECIES RICHNESS AND DIVERSITY

The total number of tree taxa recorded was 453 in Sabal (in 16063 individuals > 9.7 cm diameter on 20 ha, exact 20 x 25 chains) and 465 in San Carlos (13931 individuals > 1.5 cm, > 5 cm and > 13 cm d on 10 ha). The species richness (number of species per 100 individuals) is also very similar in Sabal and San Carlos under the same soil conditions. In both areas, the species richness is not consistently different between the top, intermediate and lowest canopy layers. Any of the canopy layers can have the highest, intermediate or lowest value. The same applies to diversity.

Species diversity, as expressed by the McIntosh index, tends to be lower on the less favourable sites, especially in the top-canopy. Diversity is higher on the better sites, which also carry a more species-rich, complex and taller vegetation. Species richness within a small area (interacting group of trees at the level of an eco-unit, small gap size up to about 0.1–0.3 ha) seems to be controlled by edaphic conditions, within limits which appear to be little affected by the size of the regional pool of available species. Species diversity, on the other hand, seems to be more strongly influenced by phase dynamics and the relative competitiveness and biochemical aggressiveness of tree species which are capable of gregariousness (for example Bruenig 1966, Bruenig and Sander 1983). It also seems to be influenced by the conditions in the lower canopy for top-canopy species to regenerate, and by the chances of gaps becoming available as a result of mortality, wind-throw and lightning.

Generally, in both Sarawak and in Sabal, species diversity and species richness are higher in transitional association groups where one vegetation type changes into another with concurrent change of soil conditions in time and space (i.e. there is a dynamically changing ecotone). They are also higher in association groups which have a high frequency of small-gap size disturbance from lightning and windfall as a result of soil and terrain (steep slopes, narrow valley bottoms, pronounced microrelief) or of forest stature (tree height, stem diameter, crown volume, aerodynamic roughness).

PHYSIOGNOMY

Assumptions and hypotheses on the possible ecological significance of plant and vegetation physiognomy date back to Humboldt (1847). Schimper (1903) took the subject up in more detail and at a global scale, while the topic has received new interest in respect to humid tropical evergreen trees and forests (e.g. Bruenig 1966, 1970, 1976, Medina *et al.* 1978). In terms of regeneration and management, the key exchange processes between the atmosphere and the plant or vegetation surface concern the input, storage, transformation and output of radiative, thermal and kinetic energy and various kinds of matter, especially carbon dioxide and other gases, water, aerosols, and organic and mineral particles and solutes. The kind and intensity of these exchanges are influenced by atmospheric conditions and vegetation canopy.

Controlling canopy elements are:

(a) leaf morphology, optical properties, mass and arrangement in the crown;

(b) angle (ideal angle for static strength 45°) and elasticity of branches and twigs to resist and modulate wind stress;

(c) the arrangement of the crowns in the canopy;

(d) the resulting aerodynamic roughness and the capacity to intercept, transmit and emit light, wind, particles, gases, water vapour and raindrops;

(e) subject to the external forces and gradients, the variable stomatal and internal diffusion resistances and conductances.

In this respect, crown and canopy architecture and branching models are ecologically relevant only with respect to the implications of crown and canopy absorptivity, reflectivity and transmissivity for energy and matter, their resistance against wind stress and their regulatory effects on regeneration, growth and mortality. This importance of canopy structure is dramatically demonstrated in the "forest decline" phenomenon in the Federal Republic of Germany.

Regression analysis with four different equations (Kurz 1982) showed that the "leaf surface area per tree crown" as dependent variable is only loosely correlated with the independent variable tree height ($r = 0.50$). This lack of correlation would explain the unexpected result from measurements of light extinction in four floristically and structurally different association groups by Heuveldop (1979). He found that the top canopies intercepted and extinguished light at the same rates, the main differences being attributed to the greater depth of canopy and the existence of undergrowth in the taller stands. The result is the existence of very different conditions for the regeneration of top-canopy species along site and vegetation structure gradients (Fig. 11.1).

The dependent variable "dry leaf mass" has been shown to be linearly correlated with the product of crown sectional area and the second power of breast-height diameter (equivalent to basal area). The same correlation

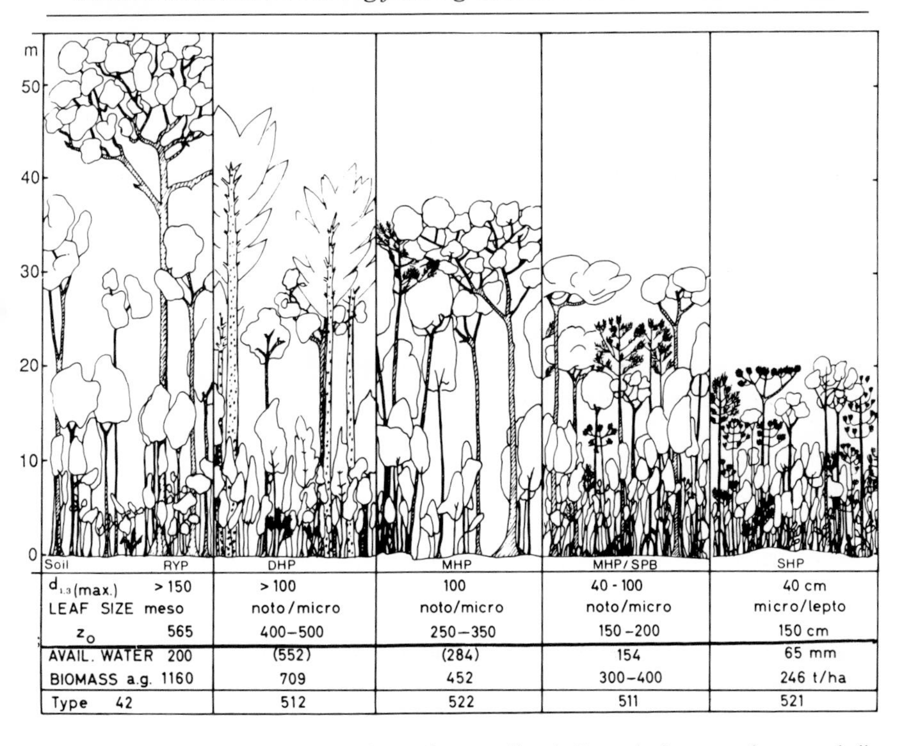

Soil	RYP	DHP	MHP	MHP/SPB	SHP
$d_{1.3}$(max.)	> 150	> 100	100	40 - 100	40 cm
LEAF SIZE	meso	noto/micro	noto/micro	noto/micro	micro/lepto
z_0	565	400–500	250–350	150 -200	150 cm
AVAIL. WATER	200	(552)	(284)	154	65 mm
BIOMASS a.g.	1160	709	452	300–400	246 t/ha
Type	42	512	522	511	521

Fig. 11.1 Change of physiognomy along a forest gradient in Sarawak, from complex mesophyll mixed dipterocarp forest on deep rooted, well structured acrisol through noto-microphyll *Agathis*–Kerangas on deep humic or orthic podzol and *Agathis–Casuarina* Kerangas on medium deep rooted, occasionally gleyic podzol (such as in San Carlos in Yevaro – Hiua ass. groups) to simple noto-microphyll mixed *Casuarina* Kerangas on moderately shallow rooted and simple, low *Casuarina–Dacrydium* Kerangas on very shallow rooted humic or orthic podzol. Z_0 is the dimensionless estimator of aerodynamic roughness of the canopy. The plant available water in biomass and soil is only partly accessible in pure sands because much of the B-horizon is devoid of roots. The types refer to Bruenig (1974). Type 512 corresponds to the Yevaro association groups in San Carlos, types 522 and 511 to the various Cunuri assoc. groups, which however lack leptophylls, phyllodials and coniferous needle-leaves; all types in San Carlos are about 30% less tall than the corresponding type in Sarawak

exists for leaf area. The ratio leaf mass to leaf area is smaller in small trees (0.115 kg m^{-2} for trees of diameter 1–5 cm) than in larger trees (0.162 kg m^{-2} for trees < 13 cm diameter), which are more sclerophyllous. The larger leaf area and leaf mass in the more complex, tall association groups with large crowns in the top-canopy is therefore not the result of a denser top-canopy, but of a denser subcanopy. The structurally very different association groups do not differ so much in absolute leaf surface area and leaf mass throughout the canopy, but rather in the arrangement of the leaves in the canopy. Correspondingly, the phytomass, which is mainly determined by the stem mass, varies much more and differs more strongly between association groups than the leaf area and leaf mass. The general patterns of distribution of above-ground phytomass, leaf area and leaf mass are broadly similar in the San Carlos area, but leaf area and leaf mass show a more moderate variation than phytomass (Kurz 1982).

SPATIAL PATTERN

Spatial pattern is the arrangement formed by the occurrence of certain defined, individual, distinct features in space and its mathematical description. In a forest, spatial pattern of all trees, and of a tree species or population, is accordingly the distributional arrangement by which individual trees of the whole collective, or of the species population, are distributed over an area. The patterns of the 64 most abundant species in Sabal showed that 30 species were clumped at an areal scale of 35–55 m (about 0.1–0.3 ha). These species were less abundant than the 34 species with a non-clumped pattern and they had relatively more individuals in the smaller (10–20 cm diameter) sizes and relatively fewer trees in the top-canopy (see Newbery *et al.* 1986 for more details on pattern trend, repetition and direction). Trends in the distribution pattern across the 20 ha area corresponded to the change of soils from acrisols (oxisols-ultisols) to podzols and peats. Trends in species distribution would seem to correspond to the soil type distribution pattern, even though conventional correlation and association analyses have not yet proved this correspondence conclusively.

A similar but tentative and only exploratory 2D-spectral analysis of the 14 most common species at San Carlos produced an unexpectedly deviating result. I had earlier surmised that the chaotic, individualistic pattern produced by FANTASMB at low minimum diameter and low information-reduction level stopping rule reflected a small-gap patch pattern of regeneration. Small gaps of any age are common at San Carlos, and the gaps show a trend related to the association group distribution (see Fig. 17 in Bruenig *et al.* 1979). The results of Newbery's spectral analysis (unpublished report, dated 13 February 1985) were in marked contrast to the expectations that gap-size clumps would be revealed and the results indicate a markedly different pattern to that at Sabal. For the 14 species examined, there was no evidence of repetition of basal area peaks and basal area lows at any regular interval across the "window" selected from the 10 ha area, to produce a visually even spread of tree locations similar to that over the whole area at Sabal (discounting any effect of particular soil-type patterns). If there exist any clumps, they are random aggregations and distributed at random. This is also unexpected insofar as the pattern of gap distribution over the area shows definite relationships to soil and vegetation types, with peaks of gap size and frequency in the transition *bana–caatinga* and a high frequency of small gaps in the Yevaro association groups (Bruenig *et al.* 1979).

CONCLUDING ASSUMPTIONS AND HYPOTHESES

Some tentative conclusions may be put forward with respect to regeneration and management, as follows:

Site conditions determine the limits for species richness;

Diversity is strongly influenced by phasic and successional development, including the effects of species competition (plus interactions with change of soil conditions by species producing acid, tannin-rich litter);

The physiognomy of the upper canopy is adapted to episodic climatic stresses which may occur and become influential on all sites, but especially so on favourable soils;

The subcanopy uses light and water resources left by the top-canopy trees, and consequently can be more luxurious on better soils with higher field capacity and more secure, reliable water supply;

Seedling establishment and growth of top-canopy species are continuous in the better illuminated vegetation stands on poor (oligotrophic-xeric) soils, but depend on mortality and catastrophic collapse of the top- and mid-canopy for success on better soils;

Patterns of species distribution are indicative of the dynamic state of the vegetation with respect to succession and to gap dynamics, and silviculture management systems must be designed accordingly in order to be successful;

Repetition in the clumped pattern in Sabal may be the result of a long period of small-gap regeneration during which patterns develop by self-organization under impact of external forces (wind, lightning).

A better and comprehensive understanding of these relationships and of the importance of canopy and structure for the survivability and functioning of the ecosystem as a whole would contribute to management. It would be particularly important for assessing the most probable reaction and performance of residual individual trees and forest stands to impacts such as selective or uniform logging, for comparing artificial complex-mixed forests and single species plantations, and for designing more efficient and site-adapted silvicultural management systems (Bruenig and Sander 1983, Bruenig 1984).

ACKNOWLEDGEMENTS

The field research in Sabal has been part of the research programme of the Forest Department, Sarawak, and has been financially supported by the German Research Foundation (DFG). The International Amazon Ecosystem Project is a MAB-pilot project co-ordinated and supported by the Instituto Venezolano de Investigaciones Cientificas (IVIC), Caracas; the German contribution has also been financially supported by DFG. Several research scientists contributed to the data analyses and evaluations, especially Drs. T.W. Schneider, D. Newbery, E. Renshaw, J. Heuveldop and H. Klinge, and research students W. Kurz and A. Weischke. The author is grateful to Mrs. M. Tarasenko for processing of the text.

REFERENCES

Bruenig, E.F. (1966). *Der Heidewald von Sarawak und Brunei – eine Studie seiner Vegetation und Oekologie*. (The heath-forest of Sarawak and Brunei – a study of its vegetation and ecology.) Thesis, University of Hamburg, Hamburg.

Bruenig, E.F. (1970). Stand structure, physiognomy and environment factors in some lowland forests in Sarawak. *Tropical Ecology*, **2**, 26–43

Bruenig, E.F. (1971). On the ecological significance of drought in the equatorial wet evergreen (rain) forest of Sarawak (Borneo). In Flenley, J.R. (ed.) *Transactions of the First Aberdeen-Hull Symposium on Malesian Ecology*, pp. 66–97. University of Hull Department of Geography Miscellaneous Series No. 11. (University of Hull: Hull)

Bruenig, E.F. (1973). Biomass diversity and biomass sampling in tropical rainforest. In *Proceedings of IUFRO Working Group on Biomass Studies*, pp. 269–93. (College of Life Science and Agriculture, University of Maine: Orono)

Bruenig, E.F. (1974). *Ecological Studies in the Kerangas Forests of Sarawak and Brunei*. (Borneo Literature Bureau, for Sarawak Forest Department: Kuching)

Bruenig, E.F. (1976). Tree forms in relation to environmental conditions: an ecological viewpoint. In Cannell, M.G.R. and Last, F.T. (eds.) *Tree Physiology and Yield Improvement*, pp. 139–56. (Academic Press: London-New York-San Francisco)

Bruenig, E. (1984). Designing ecologically stable plantations. In Wiersum, R.F. (ed.) *Strategies and Designs for Afforestation, Reforestation and Tree Planting*, pp. 348–59. (Pudoc: Wageningen)

Bruenig, E.F., Alder, D. and Smith, J.P. (1979). The International MAB Amazon Rain Forest Ecosystem Pilot Project at San Carolos de Rio Negro: Vegetation classification and structure. In Adisoemarto, S. and Bruenig, E.F. (eds.) *Transactions of the Second International MAB-IUFRO Workshop on Tropical Rain Forest Ecosystem Research*, pp. 67–100. Special Report N°. 2. (Chair of World Forestry: Hamburg-Reinbek)

Bruenig, E.F. and Sander, N. (1983) Ecosystem structure and functioning. In Huxley, P.A. (ed.) *Plant Research and Agroforestry*, pp. 221–48 (Pillan and Wilson: Edinburgh)

Bruenig, E.F. and Schmidt-Lorenz, R. (1985). Observations on the humic matter in Kerangas and caatinga soils with respect to their role as sink and source of carbon in the face of sporadic episodic events. *SCOPE/UNEP Special Issue,* 58, 107–22. (Hamburg University, Institute of Palaeogeology: Hamburg)

Heuveldop, J. (1977). Erste Ergebnisse bestandesmeteorologischer Untersuchungen im Regenwald von San Carlos de Rio Negro. In *Tropical Moist Forest* (**II**), Mitt. Bundesforschungsanstalt für Forst- und Holzwirtschaft, Hamburg-Reinbek, No. 115, 101–16

Heuveldop, J. (1979). The International MAB Amazon Rain Forest Ecosystem Pilot Project at San Carolos de Rio Negro: Micrometeorological studies. In Adisoemarto, S. and Breunig, E.F. (eds.) *Inernational Transactions of the Second International MAB- IUFRO Workshop on Tropical Rain Forest Ecosystem Research*, pp. 106–23. Special Report N°. 2. (Chair of World Forestry: Hamburg-Reinbek)

Humbolt, A.V. (1847). *Kosmos. Entwurf einer physischen Weltbeschreibung.* Vol **1–4** (Cotta'scher: Stuttgart-Tuebingen)

Kurz, W. (1982). *Biomasse eines amazonischen Feuchtwaldes: Entwicklung von Biomasseregressionen.* Diplomarbeit, Universität Hamburg, (Fachbereich Biologie: Hamburg)

Medina, E, Sobrado, M.A. and Herrera, R. (1978). Significance of leaf orientation for temperature in Amazonian sclerophyll vegetation. *Journal of Radiation and Environmental Biophysics,* **15**, 131–40

Newbery, D., Renshaw, E. and Bruenig, E.F. (1986). Spatial pattern of trees in kerangas forest, Sarawak. *Vegetatio,* **65**, 77–89

Schimper, A.F.V. (1903). *Plant-Geography upon a Physiological Basis* (Oxford: Clarendon Press)

Weischke, A. (1982). *Struktur und Funktionen in Waldökosystemen: Strukturvergleich zwischen Kerangas und Caatinga.* Diploma Thesis. Mimeographed. Chair of World Forestry, Department of Biology, University of Hamburg, Hamburg

CHAPTER 12

RIVER DYNAMICS AND NATURAL FOREST REGENERATION IN THE PERUVIAN AMAZON

J.S. Salo and R.J. Kalliola

ABSTRACT

Flat topography, high loads of suspended solids and easily erodable alluvial substrates are three factors underlying the important role of river dynamics in land transformation, forest regeneration and species diversity in Western Amazonia. Satellite analyses show that 26.6% of the modern lowland forest has characteristics of recent erosional and depositional activity, and that 12.0% of the Peruvian lowland forest is in successional stages along rivers. These findings contrast with some traditional views of Amazonian rain forest, which have tended to emphasize stability, with the dominant mode of forest regeneration occurring in light gaps created by fallen trees. The factors causing fluvial perturbance are discussed, the forests of various floodplain generations (contemporary and past) described, and the role of river dynamics in rain forest regeneration examined.

INTRODUCTION

Interest in the mechanisms of forest regeneration and succession in the lowland rain forests of Amazonia has grown due to the large-scale development programmes being carried out in this region (Lovejoy 1985). These activities have revealed a lack of information concerning the mechanisms of forest regeneration, i.e. primary and secondary succession in the area. We have earlier proposed (Salo *et al.* 1986) that river dynamics are a major mode of forest regeneration in western Amazonia and that they are an important factor in the high ß-type species diversity (between habitats) characterizing the area. The river dynamics modify large areas because of the flat general topography, their high load of suspended solids and the easily erodable alluvial substrate. Landsat imagery analyses demonstrate that 26.6% of the modern lowland forest shows characteristics of recent erosional and depositional activity (Salo *et al.* 1986); 12.0% of the forest is under the influence of the modern erosion–deposition cycle.

In this contribution, we outline a general theory of rain forest regeneration initiated by fluvial perturbance. This approach is based on the fact that the geological history of the Peruvian Amazon basin shows a dominance of continental sediments deposited by migrating floodplain generations (RADAMBRASIL 1977, Räsänen *et al.* 1987). These fluvial deposition dynamics have been present from the late Cretaceous to modern times. At present, the floodplain dynamics are especially active in the area generally referred to as the western periphery of the Amazon basin (Fittkau 1971). The major Amazonian whitewater rivers (Sioli 1950) with a broad and complex floodplain system (*várzea*) have their origin in this region (Fig. 12.1).

The few works which deal with the factors behind lowland forest succession in Amazonia only describe gap-phase regeneration (Hartshorn 1980) or secondary succession in man-induced clearings (Saldarriaga 1986,

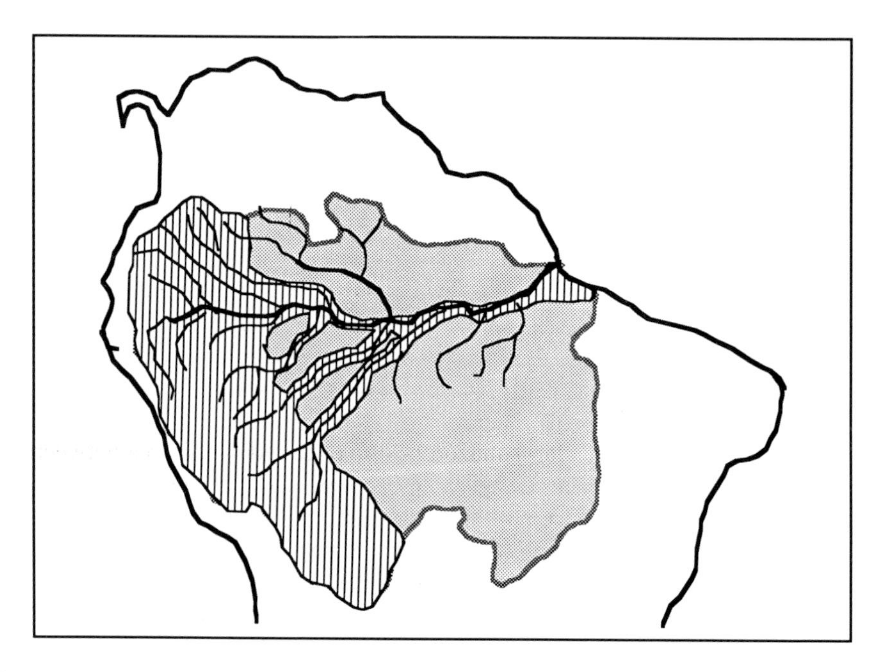

Fig. 12.1 The area of western peripheric Amazonia (Fittkau 1971), Peruvian Amazonia and major river systems

Uhl *et al.* 1982, Uhl and Jordan 1984). The primary succession following the formation of meander scrolls has been described only locally (Foster *et al.* 1986, Kalliola *et al.* 1987, Ule 1908). Most of the works which describe the habitat heterogeneity of the Amazonian lowland forest have been carried out in the central or eastern Amazonia. Here, the floodplain areas have remained relatively stable during the Holocene (Sternberg 1960). The works which analyse the structure of the western Amazonian whitewater floodplain forests (Balslev *et al.* 1987, Malleux 1982) have revealed highly mosaic forest structures. Outside Amazonia, White (1975) has described fluvial disturbance causing mosaic forest structure in the lowlands of Papua New Guinea.

FACTORS CAUSING FLUVIAL PERTURBANCE

The geological history of the Amazon basin is poorly known and its interpretation is subject to severe discrepancies. Even the interpretation of basic sedimentological processes, i.e. whether the basin has been dominated by semi-marine, lacustrine or fluvial sedimentation, has remained without comprehensive and large-scale generalizations. Interpretations of present-day floristic and biogeographical patterns have largely relied on the Pleistocene refuge hypothesis, although the refuge theory is predominantly based on present biogeographical data without adequate backing by biostratigraphical and geological evidence (Salo 1987).

Recent findings in the Amazon rain forest history research strongly suggest that, in geologically more active western Amazonia, perturbance processes promoted by fluvial activity have been the major factor causing large scale forest destruction and the subsequent primary succession (Räsänen *et al.* 1987).

In Peru, the forest areas of most active present day fluvial regeneration are the Pastaza (Huallaga-Marañon), Ucayali and Madre de Díos basins (Fig. 12.2). In these areas, the forests have gone through several complete regeneration cycles, each of which has followed the successional development pattern described here.

In this paper, the Peruvian Amazon basin is defined as the area between the Cordillera of the Andes in the west, and the Jutaí (Iquitos) arch in the east, including the major upper catchment areas of the rivers Madre de Díos, Purus, Juruá, Ucayali, Huallaga, Marañon, Napo, Putumayo, and Japurá, The geomorphology of these areas shows the piedmont-type alluvial fans and terraced deposits which form flat lowlands (80–350 m asl) in the eastern Andean forelands. These Tertiary and Quaternary fluvial deposits, entirely covered by diverse lowland rain forest, have been insufficiently studied, even though the sediments belong to one of the largest fluvial systems which has ever existed on earth. The Quaternary fluvial history of the area is primarily dominated by the basinal structures and the postorogenic tectonics of the Andes. A more detailed description of the Tertiary-Quaternary continental sedimentation of western Amazonia is given in Räsänen *et al.* (1987).

RIVERS AFFECTING SURROUNDING VEGETATION

River dynamics (channel processes) include the effects of channel migration (erosion, deposition, channel abandonment, etc.) and flooding, which create, destroy, and recreate fluvial landforms. The intensity of these influences varies between rivers, partly following the classification of channels into straight, sinuous, meandering, braided and anastomosing types. In western Amazonia, most of the rivers are meandering but they also have braided and anastomosing stretches.

The most active forest perturbance occurs within the present meander plain delimited by the outer meander curve points and oxbow lakes. Within this plain, the lateral channel migration causes under-cutting of the forest and subsequent point-bar sedimentation at the outer curve of the meanders, leading to the sequential meander scroll formation characterizing the meander loop. The present floodplain is further diversified by the formation of oxbow lakes, the successional development after the filling of these lakes by silting, and by the primary succession of floating macrophytes. Channel migration causes a mosaic structure of the floodplain forest ground. The heterogeneity is twofold:

(1) the patches within the present meander plain are of different age, thus creating patches of forest of different age,

(2) the sedimentation process causes differences in the developing forest bed due to the uneven deposition of the clay, silt, and sand fraction of the suspension according to the site of deposition.

The filling oxbow lakes form a distinct pseudo-layered stratigraphy characterized by the sequential sedimentation of a clay-gyttja during the low water period and a riverine silt-sand fraction during the period when the annual floods reach the lakes.

The major rivers present a series of former floodplain generations with more or less clear series of terraces. In eastern Amazonia, the formation of these terraces has been explained by the Pleistocene–Holocene oscillation of the sea-level (Lathrap 1968). However, these oscillations do not explain the terrace formations of Peruvian Amazonia, which predominantly fall between 80–350 m asl. Catastrophic floods after the melting of the Andean glaciers at the end of the Lujanian have also been proposed as a major event modifying the fluvial system of the area. However, Räsänen *et al.* (1987) have presented evidence that the floodplain dynamics are regulated by the long-term geological activity of the Sub-Andean fault system and the epeirogenic uplift of the eastern shield areas. This activity, together with the aggrading rivers, causes abandonments of floodplains and a series of lateral migrations of whole floodplain systems causing the distinctive high erosion banks (*cerros*). These former floodplain areas are detectable by radar imagery (SLAR) from the flat general topography promoted by the annual sedimentation of suspended sediments and by the presence of terraces and other obvious river marks (RADAMBRASIL 1977).

Most of the Peruvian Amazon basin is characterized by the dissected convexo-concave morphology which indicates denudation of the forest ground. We have proposed elsewhere that this formation is also of fluvial origin, evolved by the same sedimentation processes as the present-day floodplains (Salo *et al.* 1986, Räsänen *et al.* 1987). This led us to the conclusion that the whole Peruvian Amazon basin has been formed by a series of floodplain generations, from the present floodplain formation to the Pleistocene dissected *tierra firme* formations. Figure 12.2 gives a general model describing the fluvial regeneration cycle from present to old forests on former floodplains.

FORESTS OF THE FLOODPLAIN GENERATIONS

Forests on present floodplains

The present floodplain formation covers 62 000 km^2 (12.0%) of the Peruvian lowland forest (Salo *et al.* 1986). This is the area with the youngest forests which have all gone through primary succession. These forests are flooded annually and are subject to further fluvial perturbances. The formation can be further subclassified into sequential successional forests and mosaic forests.

Sequential successional forests

The sequential successional forests are located in the loops of meandering rivers and on the channel islands of braided rivers. The forest bed is characterized by parallel meander scrolls formed by primary sedimentation of the bedload and by suspended sediment of the channel at the inner bank of the meander. At the Manu River in the department of Madre de Dios (Peru), we measured an average lateral channel migration of 12 m yr^{-1} at the meander points during the years 1962–1976 (Table 12.1).

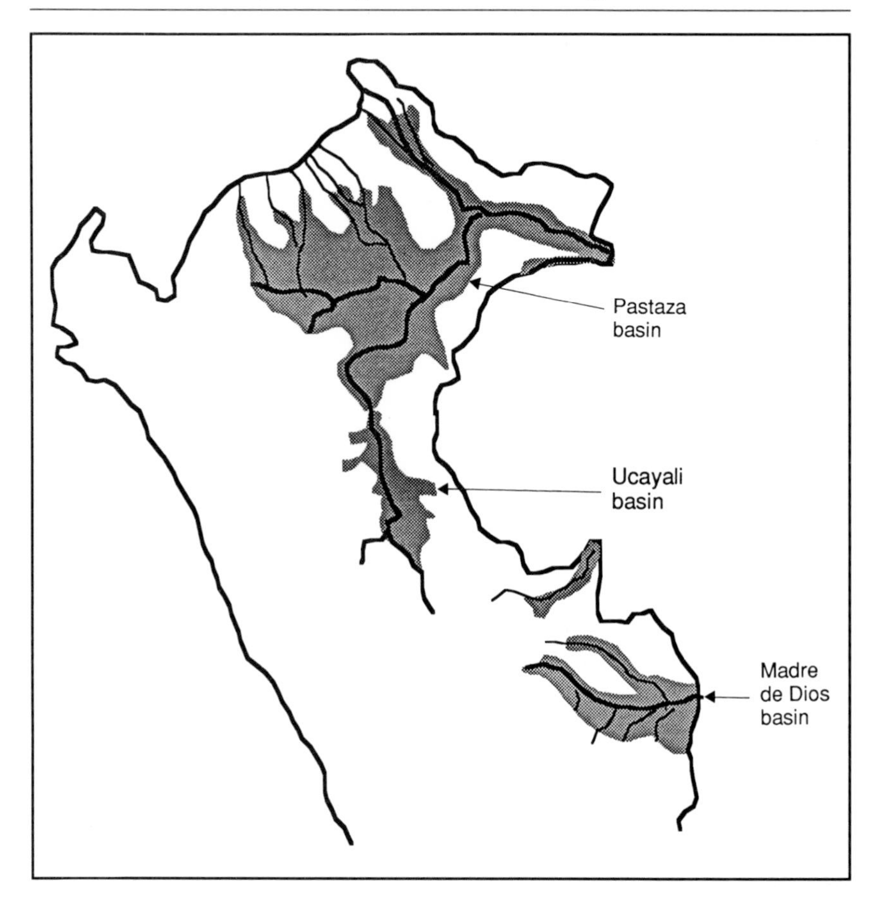

Fig. 12.2 The major depositional basins in Peru

The development of the parallel scroll-swale system gives rise to a sequential successional pattern in the forest, with increasing age towards the meander neck. The sequential pattern of meander loop forests has been documented in some fluvial systems of northern latitudes but a general treatment of the subject is still lacking.

The primary succession begins at the first point-bar "beaches" created by the deposition of river sediment on the convex bank of the bend. This sedimentation leads to the stabilization of the point-bar and to the addition of a new ridge to the developing meander bend. So far, there are no reliable data on the rate of ridge formation in the upper Amazonian floodplains.

The primary successional species of the first point-bar are subjected to climatic conditions dramatically different from those prevailing in the surrounding older forest. Growth is initiated during low water which is

Table 12.1 Rate of lateral channel erosion and areas subjected to primary succession in the Manú River, southwestern Peru, during the years 1962/3–1976

River section		*1*	*2*	*3*	*4*	*5*	*6*	*7*	*x*	Σ
River sinuosity	1963	2.1	2.5	1.4	1.9	1.6	2.0	3.3	2.1	
	1976	2.3	2.2	1.5	1.6	1.8	1.6	3.7	2.1	
Mean rate of lateral erosion in the meander bends (m yr^{-1})		11.3	11.6	8.4	17.2	15.7	9.8	10.0	12.0	
Area of new land created 1963-1976 subject to primary succession (km^2)		1.8	1.8	1.2	2.4	1.9	1.3	1.6		12.0
Number of point-bar "beaches"	1963	13	10	9	11	11	16	23		93
	1976	13	10	9	10	11	16	18		87
Number of oxbow lakes	1963	3	5	5	5	7	7	3		32
	1976	3	6	5	6	7	7	4		38

In the table, a 70 km section (air distance) of the river from the Cocha Cashu biological station (11°52'S, 71°23'W. 350 m a.s.l.) to Boca Manú (12°15'S, 70°55'W, 320 m a.s.l.) is divided into 7 sections. The analysis is based on manual comparison of aerial photographs from 1962/3 with Landsat MSS image 276296-135752-7.

usually also the driest period of the year. The primary successional community is dominated by fast growing, shade intolerant trees, such as *Alchornea castaneifolia* Juss. (Euphorbiaceae), *Cecropia membranacea* Trécul. (Cecropiaceae), *Ochroma lagopus* Swartz (Bombacaceae), *Salix martiana* Leybold (Salicaceae), and *Tessaria integrifolia* Ruiz and Pavón (Compositae), with a wide geographical distribution in western Amazonia. The seedlings of *T. integrifolia* and *A. castaneifolia* may repeatedly reach dominance with a frequency of > 90% in the point-bar community. In this environment, the dominance of trees over annual herbs is driven by their ability to bind the living sedimentary bed and to their persistence despite the up to 2 m annual deposition of river sediment during high water periods.

The ridge-swale system promotes zonated forest patterns during the later successional development. There are no data on the development of sequential successional forests beyond the *Ficus-Cedrela* association, which is a forest occurring on approximately 30–50 year-old meander loops, with *C. odorata* L. (Meliaceae) and *F. insipida* Willd. (Moraceae) as the dominant species. At present, we are conducting a line transect programme in these forests in the central Ucayali and Madre de Dios basins.

Mosaic forest

The result of meander cutting and the abandonment of river channels is the development of a highly complex forest. In general, the genesis of the Ucayali-Marañon-Solimões and the Madre de Dios floodplains follows the course of *várzea* formation along the main Amazonia trough. In the central part of the Amazon basin, a major factor affecting the increased sedimentation intensity during the Holocene is the prevailing maximum in the eustatic sea-level fluctuation (Klammer 1984). The Pleistocene sea-level minimum was reached during the pre-Flandrian regression causing the incision of the main Amazon channel up to the mouth of the Japurá. It is obvious that the upper Amazonian floodplain dynamics beyond the Iquitos (Jutaí) arch cannot be explained by the sea-level fluctuation cycle. The dynamics are mainly driven by tectonic activity in the area (Räsänen *et al.* 1987) causing similar geomorphic structures within the aggrading floodplains as the filling of the Pleistocene river valley in central Amazonia. These processes include the cutting of meander bends and the formation of oxbow lakes, lakes in abandoned channels, lateral levee lakes, lakes in depressions formed by uneven aggradation and crescentic levee lakes. Although the lakes formed in sediment-blocked river valleys (*ria*-lakes) are more common in central and eastern Amazonia, they are also present in the Ucayali and Pastaza-Marañon basins as a result of blocking by the aggrading floodplains of the main rivers.

The primary factor diversifying the geomorphology of the floodplains following the formation of meander bends is the cutting off of the meanders and the formation of oxbow lakes (Fig. 12.3). This process has several consequences for forest dynamics. First, it initiates a new type of primary succession, which takes place along the sites of the oxbow lakes. Most of the oxbow lakes are first filled through their upper mouth with the sand-gravel fraction of the channel bedload. This is followed by the silting of the rest of the lake during the annual floods. For this reason, the sedimentary bed for the primary successional development is already mosaic in structure during the early phases. Second, where channels have been abandoned, successional forests of different ages occur side by side.

Malleux (1982) divides the Peruvian whitewater floodplain forests into two categories: permanently and temporally inundated forests. Apparently, the largest areas of permanent inundation are located in the Pastaza basin, but there are also large areas in the Holocene floodplain of Ucayali. These areas are dominated by extensive and almost pure stands of *Mauritia flexuosa* L. f. (Arecaceae).

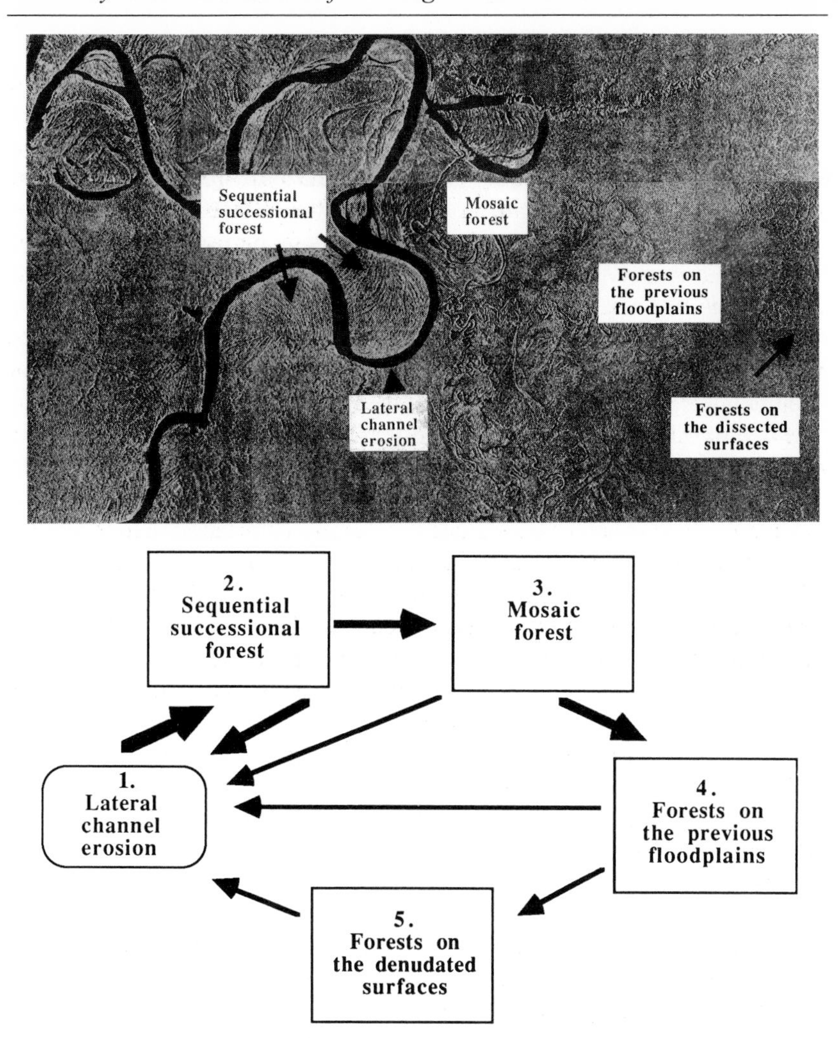

Fig. 12.3 The fluvial regeneration cycle of the western Amazonian forests on present and former floodplain generations in central Ucayali floodplain at Pucallpa, Peru. The forest regeneration cycle is initiated by lateral channel erosion and subsequent sedimentation of point-bars and migrating channel islands (above). The primary succession on these deposits follows a sequential pattern due to (i) further migration of the channel, (ii) formation of age structured sets of cohorts of the early successional trees according to the seasonal sediment accumulation and (iii) competition between species. The mosaic forest is formed on the present floodplain generation composed of cut-off meanders (sequential successional forests) and oxbow lakes. Transitional forests with increasing autogenic light-gap regeneration dominate the older floodplain generations outside the present active meander plain. Finally the colluvial processes cause dissection of the floodplain reliefs, resulting in the dominance of denudated forest beds. Sideways looking airborne radar (SLAR) image, courtesy of ONERN, Lima. Below, the relative width of the arrows indicates the areas most recently disturbed by the channel erosion. These are the areas most likely to be re-disturbed

Forests on previous floodplains

So far there are no extensive studies which cover the dynamics of the upper Amazonian floodplain generation system. It is possible to detect the most recent floodplain generations beyond the Holocene floodplain formation by means of radar (SLAR) or Landsat imagery (Fig. 12.3). In the Landsat MSS (multispectral scanner) images (channels 6 and 7), the former floodplain formations appear without the dissected surface erosion which characterizes the oldest type of alluvium. Their topography is flat due to the filling up of the primary floodplain morphology by annual flood sediments and colluvial processes. In the Peruvian lowland forest, previous floodplains cover 75 100 km^2 or 14.6% of the area (Landsat resolution, Salo *et al.* 1986). The area is undoubtedly much larger if the resolution of SLAR imagery is used to give additional information on the terrace system of the floodplains.

Forest regeneration in this formation takes place increasingly through gap-phase dynamics. However, as the erosional reliefs of the area are subjected to continuous tectonic change, the perturbance caused by channel processes may repeatedly modify the forest bed.

Forests on the denuded surfaces

Floristic studies and works describing the forest structure in unflooded Amazonian forests are numerous but deal mainly with the areas close to Manaus and Belem; thus there is inadequate geographical cover. In Peruvian Amazonia, the forests outside present or previous floodplain generations and with apparent surface dissection comprise 73% of the lowland rain forest (Salo *et al.* 1986). In the literature, this area is generally referred to as *tierra firme* as opposed to the flooded Holocene floodplain or *várzea*. The denudation of the surface is caused by colluvial processes driven by the high precipitation. The uppermost layered sediments are poorly consolidated and lie on slightly folded Tertiary sediments. At present, it seems clear that the uppermost beds which form the base of most of the Amazonian rain forest are of fluvial origin.

The western Amazonian *tierra firme* forests seem to have gone through perturbance and regeneration cycles, caused by changing floodplains. In addition, in the Peruvian lowlands we have observed several instances in *tierra firme* forests where the gap-phase dynamics are accelerated by small-scale landslides in deepening gullies. This leads us to believe that "classic" autogenic gap-phase dynamics, mainly driven by the senescence of mature trees, is only one mode of forest regeneration. Studies of the edaphic regeneration mechanisms should be given more emphasis.

CONCLUSIONS

The complex fluvial processes suggest that the western Amazonian rain forest is a mosaic of floristic and faunistic units which are larger than the eco-units created by tree falls and landslides, but not as large as the phytogeographical units of Amazonia. There also seems to be a difference in the present rate of fluvial perturbance between central and western Amazonia. The implications of this difference are still largely undocumented.

Also, there may be other differences between the areas of the Amazon basin related to forest dynamics. In the Rio Negro basin area of northern Amazonia, there is a long tradition of slash-and-burn agriculture, promoting regeneration processes (Saldarriaga 1986, Sanford *et al.* 1985).

The long-term rain forest regeneration cycle described here still lacks an exact timetable. At present, the lack of absolute Pleistocene datings and the unsatisfactory Holocene datings do not allow us to reconstruct the rate of forest regeneration promoted by fluvial disturbance, or to evaluate the relative importance of different regeneration mechanisms. The fact that the forests of the upper Amazon basin are growing on up to 4000 metres of predominantly continental sediments (Kummel 1948) derived from fluvial processes, similar to the present, immediately suggests that the dynamic processes described here offer a tentative research plan for modelling the mechanisms of biological diversification in the lowland Amazonia. The wide geographical range and long history of forest regeneration driven by fluvial perturbance also suggests that there is a need for studies of the life-histories of floodplain trees in order to manage their natural populations.

REFERENCES

Balslev, H., Luteyn, J., Olgaard, B. and Holm-Nielsen, L.B. (1987). Composition and structure of adjacent unflooded and floodplain forest in Amazonian Ecuador. *Opera Bot.*, **97**, 18–27

Fittkau, E.J. (1971). Ökologische Gliederung des Amazonas-Gebietes auf geogemischer Grundlage. Münster. *Forsch. Geol. Paläont.*, **20/21**, 35–50

Foster, R.B., Arce, B.J., and Wachter, T.S. (1986). Dispersal and the sequential plant communities in Amazonian Peru floodplain. In Estrada, A. and Fleming, T.H. (eds.) *Frugivores and Seed Dispersal*, pp. 357–70. (Dr. W. Junk Publishers: Dordrecht, Boston, Lancaster)

Hartshorn, G. (1980.) Neotropical forest dynamics. *Biotropica,* **12**, 23–30

Kalliola, R., Salo, J. and Mäkinen, Y. (1987). Regeneration natural de selvas en la Amazonia Peruana. 1: Dinamica fluvial y sucesion ribereña. *Memorias del Museo de Historia Natural "Javier Prado"* No.**19A**. (Universidad Nacional Mayor de San Marcos: Lima)

Klammer, G. (1984). The relief of extra-Andean Amazon Basin. In Sioli, H. (ed.) *Amazon. Limnology and Landscape Ecology of a Mighty Tropical River and its Basin*, pp. 47–83. (Dr. W. Junk Publishers: Dordrecht, Boston, Lancaster)

Kummel, B. (1948). Geological reconnaissance of the Contamana region, Peru. *Bulletin of Geological Society of America*, **59**, 1217–66

Lathrap, D.W. (1968). Aboriginal occupation and changes in river channel on the central Ucayali, Peru. *American Antiquity,* **33(1)**, 63–79

Lovejoy, T.H. (1985). Amazonia, people and today. In Prance, G.T. and. Lovejoy, T.H (eds.) *Amazonia*, pp. 328–38. (Pergamon Press: Oxford)

Malleux, J. (1982). *Inventarios forestales en bosques tropicales*. (Universidad Nacional Agraria la Molina: Lima)

RADAMBRASIL. (1977). *Levantamento de recursos naturais Vol. 15*. Folha S.B.19 Juruá. (Ministerio das Minas e Energia, Departamento Nacional da Producao Projeto Radambrasil: Rio de Janeiro)

Räsänen, M.E., Salo, J.S. and Kalliola, R.J. (1987). Fluvial perturbance in the Western Amazon Basin regulation by long-term sub-Andean tectonics. *Science,* **238**, 1398–401

Saldarriaga, J. (1986). Recovery following shifting cultivation. A century of succession in the upper Rio Negro. In Jordan, C.F. (ed.) *Amazonian Rain Forests. Ecosystem Disturbance and Recovery*, pp. 24–33 (Springer-Verlag: New York)

Salo, J. (1987). Pleistocene forest refuges in the Amazon: evaluation of the biostratigraphical, lithostratigraphical and geomorphological data. *Ann. Zoologici Fennici,* **24(3)**, 203–11

Salo, J., Kalliola, R., Häkkinen, I., Mäkinen, Y., Niemelä, P., Puhakka, M. and Coley, P.D. (1986). River dynamics and the diversity of Amazon lowland forest. *Nature,* **322**, 254–8

Sanford, R.L. Jr., Saldarriaga, J., Clark, K.E., Uhl, C. and Herrera, R. (1985). Amazon rain forest fires. *Science,* **227**, 53–5

Sioli, H. (1950). Das Wasser im Amazonasgebiet. *Forsch. Fortsch*, **26**, 274–80

Sternberg, H.O'R. (1960). Radio-carbon dating as applied to a problem of Amazonian morphology. *Comptes Rendus 18eme Congres International de Geographie*, 2, 299–424. (International Geographical Union: Rio de Janeiro)

Uhl, C., Clark, H. and Clark, K. (1982). Successional patterns associated with slash-and-burn agriculture in the upper Rio Negro region of the Amazon basin. *Biotropica,* **14(4)**, 249–54

Uhl, C. and Jordan., C.F. (1984). Succession and nutrient dynamics following forest cutting and burning in Amazonia. *Ecology,* **65(5)**, 1476–90

Ule, E. (1908). Die Pflanzenformationen des Amazonas-Gebietes. *Botanischer Jahresbuch,* **40**, 114–72, 398–443

White, K.J. (1975). *The Effect of Natural Phenomena on the Forest Environment*. Presidential address to the Papua New Guinea Scientific Society.

CHAPTER 13

THE FIRST 26 YEARS OF FOREST REGENERATION IN NATURAL AND MAN-MADE GAPS IN THE COLOMBIAN AMAZON

T. Walschburger and P. von Hildebrand

ABSTRACT

Twelve tree species of different demographies were studied at different stages of the successional sequence in the Colombian Amazon, both in cultivated areas abandoned by Indians and in natural gaps. The 13 selected study areas cover the first 26 years of regeneration, including four primary forest sites. All individuals belonging to the selected species were described within 1000 m² study plots in 1982 and 1984. Demographic strategies have been analyzed on the basis of variations in frequency, basal area and stem volume. The importance of the 12 species in the regeneration process is discussed in the light of complete floristic and structural descriptions of the study sites.

INTRODUCTION

An analysis has been made of variations in demographic pattern and survival strategies of 12 secondary growth species, occurring in natural gaps, in *rastrojos* (the Spanish term for secondary forest growth in abandoned slash-and-burn fields) and in mature forest in the Colombian Amazon. Survival strategies of these 12 species were studied for the first 26 years of regeneration. The main aim of the overall study, which is continuing, is to establish whether the processes of rain forest regeneration are substantially altered by the disturbance represented by slash-and-burn agriculture. Associated work of the Amazon Rain Forest Research Centre (Puerto Rastrojo Foundation 1987) includes studies on the interdependent relations of fishes and flooded forest, resource use by Indian communities in the Colombian Amazon region (Walschburger and von Hildebrand 1988) and the encouragement of Indian participation in the definition of management plans and protected areas (Colombia. Asuntos Indigenas-Ministerio de Gobierno 1988).

STUDY AREAS, SPECIES AND METHODS

The study was carried out in the Miriti-Parana mid-river course, the most northerly affluent of the Caqueta River in Colombian Amazonia. The area is gently undulating and is of Tertiary geological origin. The soils are basically oxisols (PRORADAM 1979). Annual precipitation is *c*.3500 mm, and the area is therefore one of the most humid in the Amazon region, except for the Andean piedmont.

Floristic descriptions of all individuals > 2 m high in 500 m^2 plots in 32 natural gaps and *rastrojos* of different ages were chosen using the following criteria: numerical frequency, basal area, canopy or subcanopy species, and duration of persistent occurrence during the first 30 years of regeneration. The selected species were: *Cecropia discolor*, *C. tessmannii* and *C. sciadophila*, *Vismia guianensis*, *V.* sp. and *V. macrophila*, *Solanum* sp.,*Cespedesia spatulata*, *Clathrotropis macrocarpa*, *Macrolobium* sp., *Miconia minutiflora* and *M.* sp. (von Hildebrand and Walschburger 1985).

Study sites comprised four areas of mature forest, six areas of abandoned slash-and-burn cultivation (*rastrojos*) and three natural gaps of different ages. In each site, eight 125 m^2 plots were established, thus providing a total sample area of 1000 m^2. The number of plots in natural gaps varied according to the gap area. Each individual belonging to the 12 selected species was described by diameter, height and general phenology. A first description was made in 1982, and was repeated 2 years later.

RESULTS AND DISCUSSION

Data were organized chronologically according to the age of the different study sites and separately for natural gaps, *rastrojos* and mature forest stands. A frequency analysis was established for different height and diameter classes. Within these classes, basal areas and stem volumes were calculated. Total stem volume was estimated by multiplying total basal areas by height. All analyses were made for individual species as well as for groups of species. Table 13.1 summarizes data on litterfall, decomposition rates and basal areas.

Table 13.1 Age, number of species, basal areas, litterfall and annual decomposition rates in % in the different study sites (von Hildebrand and Walschburger 1985)

Study site	*Age in years*	*Total species > 2 m*	*Number of studied species present*	*Litter fall t ha⁻¹*	*% Decomp. year*	*Basal area 12 studied species*	*Basal area For total of sp. > 2 m*
CH-OV-82	0		0			0.0	
CH-OV-84	2		7			1.0	
R20-82	3		11			7.9	
R9-82	4	31	11	7.6	80	9.6	15.6
R20-84	5		11			10.2	
R9-84	6		10			10.4	13.1
R10-82	7	59	10	7.9	76	18.8	
R10-84	9		10			18.0	21.6
R8-82	12		11			17.9	
R8-84	14	42	10			14.5	28.0
R17-82	22		9	11.1	75	10.8	
R17-84	24	56	8			4.0	19.3
CN10-82	4	65	8	4.1	73	1.4	3.9
CN10-84	6		7			1.5	
CN3-82	7	41	4	9.2	74	0.8	13.3
CN3-84	9		4			0.9	
CN9-82	24	97	2	7.2	75	1.9	8.6
CN9-84	26		2			2.6	
M1	> 100		3	9.3	75	3.8	
M2	> 100		4	8.9	78	2.2	
M3	> 100		3	9.0	74	0.7	
M4	> 100		3	9.3	73	5.8	
M6	> 100	116					34.0

The prefix CH refers to *Chagra* (Indian cultivated areas)
The prefix R refers to *Rastrojo* (Slash-and-burn secondary growth areas)
The prefix CN refers to *Claro Natural* (Natural gap)

The suffix 82 and 84 refer to the years of description of the sites

Invasion patterns of open spaces

The main differences between *rastrojos* and natural gaps are the size and the treatment before the regeneration process begins. Slash-and-burn agriculture includes burning, planting, weeding and harvesting of an area of approximately 1 ha. In contrast, natural gaps are generally not greater than 500 m^2, and are mainly created by falling trees. Perturbation of the root mat that covers the soil in natural gaps is mainly confined to the area affected by the roots of the large fallen trees. This disturbance pattern offers the opportunity for new individuals to invade the otherwise closed root mat. These differences in initial soil surface conditions lead to different strategies of space invasion and also provide numerous possibilities for variation in subsequent floristic composition (see also contributions by Whitmore and Bazzaz in this volume).

Thus, in the 3-year-old *rastrojo*, 80% of individuals of height >2 m belonged to the 12 studied species. In contrast, these species accounted for only 10% of the total number of individuals in natural gaps. Considering all individuals in all size classes for the 12 studied species in the 3-year-old *rastrojo*, eight species belonging to the genera *Cecropia*, *Vismia* and *Miconia* accounted alone for 8280 of the 9320 individuals ha^{-1} (87%). In natural gaps, the species of these same genera represented only 320 of the total 2100 individuals ha^{-1} (15%). *Cecropia* and *Vismia* were practically absent in mature forest stands.

During the first stages of regeneration, *rastrojos* were characterized by very high spatial population densities and growth rates. Most were 0–7 m tall, some were already > 12 m in height. General mortality between 1982 and 1984 in the same *rastrojo* showed that the pioneer species *Cecropia* spp., *Vismia* spp., *Miconia minutiflora*, must reach ≥ 7 tall in order to have a reasonable chance of survival. In the 3-year-old *rastrojo*, > 80% of all individuals < 7 m in height died, while almost all the taller individuals survived. The survival strategy of these species comprises a rapid invasion of an open area and rapid growth in order to remain in the canopy strata, so as to be able to overcome competition and subsequently to reproduce. In comparison, the space conditions offered in natural gaps to colonization by new individuals are less favourable so that the invasion patterns of typical *rastrojo* species do not occur in the same way.

Dominance periods of the 12 species

Temporal variations in numerical frequencies, as well as variations in total stem volume of the different species studied, reflect the strategy of area invasion and, later, the survival strategy of using the continuously changing vertical and horizontal space in different ways.

Chronosequences of dominant species in *rastrojos*

Typical pioneer species rapidly invade the open area with numerous individuals. The speed of invasion depends on the seed stock and germination success. The dominance sequences observed at different ages are a consequence of differential survival abilities of individuals, rather than of continuous recruitment of new individuals. At the early stages of regeneration, very high densities were observed. The absolute numerical dominant was *Cecropia discolor* with 4180 stems ha^{-1}. At the end of the period studied, the densities of all surviving species in the 24-year old *rastrojo* were < 100 ind. ha^{-1}, none of them with obvious relative numerical dominance.

Considerable differences were observed between stem volume and numerical dominance sequences for the 12 species. *Cecropia discolor* was the numerically dominant species in the 3-year-old *rastrojo* and shared stem volume dominance with *Cecropia tessmannii*, both with values around 20 m^3 ha^{-1}. All the other species had values < 5 m^3 ha^{-1}.

Along the chronosequences, different species alternated in stem dominance. The highest stem volume values were in the 12-year *rastrojo*, with 256 m^3 ha^{-1}, the dominant species having values of up to 80 m^3 ha^{-1}. At the end of the 26-year regeneration period, all the stem volumes mentioned decreased, and only a few big trees of *Miconia minutiflora*, *Cecropia sciadophylla* and *Vismia macrophylla* survived, with values around 40 m^3 ha^{-1}. This moment could be considered as a turning point in the regeneration process: new species, different from the typical pioneers, begin to appear, but never with clear numerical or stem volume dominance (von Hildebrand and Walschburger 1985).

Chronosequences of dominant species in natural gaps and mature forest

The most frequent species in the *rastrojos* had very few individuals in natural gaps, where the dominants were *Clathrotropis macrocarpa* and *Macrolobium* sp., with 1380 and 930 trees ha^{-1} respectively. These two species showed no orderly sequence of temporal frequency dominance in the natural gaps studied – probably a consequence of their variable densities in the initial floristic composition of the gaps. This conclusion is supported by the high frequency variations showed by these two species in mature forest plots.

CONCLUSIONS

Slash-and-burn agriculture, considered generally as a small scale perturbation, shows direct influence on regeneration. Similarities were observed for the values of net primary productivity, decomposition rates and chemical characteristics of the soils between the *rastrojos*, natural gaps and mature (Table 13.1). Net primary productivity values, estimated as the dry litterfall weight ha^{-1}, showed similar values in the different vegetation stands after 12 years of regeneration; values ranged from 7–9 t ha^{-1}. Decomposition rates were also similar in the different sites, ranging between 63–77% of dry litter weight loss per year (von Hildebrand and Walschburger 1985). These comparisons lead to the conclusion that net productivity, decomposition rates and chemical properties of the soil attain similar values after 10 years of regeneration. The principal difference between natural gaps and *rastrojos* lies in the floristics of the regenerating stand.

During the first stages of secondary succession in *rastrojos*, seven typical pioneer species dominated both in numerical frequencies as well as in total stem volumes. These same species occurred only occasionally in natural gaps and mature forest. Their period of dominance covered principally the first 12 years of the regeneration, but their predominance did not necessarily inhibit other species, which may later become common in the mature forest. These secondary growth species may owe a considerable part of their survival success in this area of Amazonia to continuous Indian slash-and-burn agriculture throughout the last 2000 years, which may have greatly favoured their geographical dispersal.

The 24-year-old *rastrojo* is considered to be the turning point in regeneration. Numerous mature forest species begin to be conspicuous as a result of typical *rastrojo* species dying out and creating new gaps within the rastrojo. New successions may thus be initiated within the regeneration process, as proposed by Hallé *et al.* (1978). In sum, the role of *rastrojo* specialists in the regeneration process may be posed in terms of three questions, which are being addressed in continuing work in the Miriti-Parana region:

(1) Are these fast growing *rastrojo* species essential in guaranteeing the conservation of nutrients, through their incorporation into living biomass after the burning of a logged forest area – nutrients which otherwise would be easily washed out?

(2) Do these species delay or alter the regeneration process by temporarily inhibiting the growth of sun-loving juveniles of the mature forest?

(3) Do secondary growth areas such as *rastrojos*, with floristic composition patterns differing from those of natural gaps, finally alter the species composition of the mature forest on a long-term scale?

In conclusion, a small scale disturbance, like slash-and-burn agriculture, has direct effects on the floristic composition of secondary growth areas. This fact has clear implications for tropical rain forest management strategies. Any type and intensity of disturbance may change the dispersal and survival possibilities of different rain forest species. Thus, any new alternatives proposed for sustainable use of the rain forest, must be evaluated in the light of its consequences on the regeneration process. This is most important, especially if the goal is to maintain biological diversity on a long-term basis.

REFERENCES

Colombia. Asuntos Indígenas – Ministerio de Gobierno. (1988). *Política del Gobierno Nacional para la Protección y Desarrollo de los Indígenas y la Conservación Ecológica de la Cuenca Amazónica.* (Asuntos Indígenas – Ministerio de Gobierno: Bogotá)

Hallé, F., Oldeman, R.A.A. and Tomlinson, P.B. (1978). *Tropical Trees and Forests: an Architectural Analysis.* (Springer: New York)

PRORADAM (Proyecto Radargramétrico del Amazonas). (1979). *Estudio Radargramétrico de la Amazonia Colombiana.* 5 Vols. (IGAC: Bogotá)

Puerto Rastrojo Foundation. (1987). *Amazon Rain Forest Research Centre. Brochure.* (Puerto Rastrojo Foundation: Bogotá)

Von Hildebrand, P. and Walschburger, T. (1985). *Uso sostenido de la selva humeda tropical basado en el proceso de regeneración.* Informe preliminar. (Colciencias: Bogotá)

Walschburger, T. and Hildebrand, P.von. (1988). *Observaciones sobre la utilizacion estacional del bosque húmedo tropical por los indigenas del Rio Miriti (Amazonas, Colombia).* Colombia Amazonica 3(1). (Fundacion Puerto Rastrojo: Bogotá)

CHAPTER 14

PLANT DEMOGRAPHY AND THE MANAGEMENT OF TROPICAL FOREST RESOURCES: A CASE STUDY OF BROSIMUM ALICASTRUM *IN MEXICO*

C.M. Peters

ABSTRACT

Brosimum alicastrum *(Moraceae) is multipurpose forest tree whose seeds are extremely rich in protein. The species occurs naturally in almost pure stands in many regions of Mexico. To provide guidelines for the utilization and management of this important forest resource, a 3-year study of the growth, reproduction and population dynamics of* B. alicastrum *was conducted in Veracruz, Mexico. Detailed demographic data were collected for the population, and a matrix model was used to simulate forest dynamics over time. A sensitivity analysis was then performed to determine the maximum number of seeds which could be harvested sustainably from the forest. Data collection and matrix construction are examined, and the utility of plant demography in the management of tropical forest resources is discussed.*

INTRODUCTION

The change in numbers which a plant population exhibits over time is a direct result of birth and death processes. The population grows when the recruitment of new individuals exceeds the number of deaths, and the population declines when mortality is greater than the number of births. Population stability is achieved when the birth rate is exactly balanced by the death rate. These simple demographic relationships determine the sustainability of forest resource exploitation. Given an adequate understanding of the growth, survival and reproduction of a species, management systems can be developed to maximize the growth and productivity of the population. Detailed ecological information, however, is virtually non-existent for most of the plant resources of the tropics. With the exception of a few important timber species, the establishment, growth and reproductive biology of tropical trees with economic value have rarely been assessed. In consequence, the exploitation of most tropical forest resources has become a purely extractive activity with little regard for the continued regeneration of the species. Unfortunately, high mortality and low recruitment inevitably lead to population extinction.

An illustration of the utility of plant demography in the management of tropical forest resources is provided by the case of *Brosimum alicastrum* Sw. (Moraceae) in Mexico. This large forest tree produces edible fruit with protein-rich seeds, the leaves are an excellent source of browse, the wood is used in construction, and the fruits, leaves, latex and bark are all used medicinally (Peters and Pardo-Tejeda 1982). The species occurs naturally in high-density stands in several regions of Mexico (Gómez-Pompa 1973). From both an economic and ecological standpoint, *B. alicastrum* is an extremely important tropical tree. Before promoting the increased utilization of this species, we need to know how much fruit is produced by natural forests of the species and what is the maximum number of seeds which can be harvested sustainably from these forests.

To begin answering these questions, a 3-year study of *B. alicastrum* was initiated in co-operation with the Instituto Nacional de Investigaciones sobre Recursos Bioticos (INIREB) of Xalapa, Veracruz, Mexico. Individuals were marked and monitored within a permanent study site to quantify the number of individuals at each stage in the life cycle of the species, and to determine the magnitude and rate of flux from one stage to the next. Based on these data, a predictive model was then constructed to assess the impact of seed collections on the long-term stability of the forest.

DATA COLLECTION

The forest selected for study was located near the town of Papantla (20° 21' N, 97° 14' W) in the central part of the State of Veracruz, Mexico. A series of twenty-five 20 x 20 m plots was established in the forest and all *B. alicastrum* individuals taller than 1.0 m in each plot were measured for height and diameter and permanently tagged. The exact position of each individual in the plot was mapped. Canopy gaps, tree falls and stumps were also mapped. Seedlings were inventoried on a random sample of 1.0 m^2 plots (N = 100).

Size-specific rates of growth and mortality were determined biweekly over a 3-year period for a subsample of marked individuals of varying size. Height growth by the smaller individuals was measured directly, while vernier dendrometer bands were used to record the diameter increment of larger individuals. The survivorship of the seedling cohorts produced each year was monitored on sixteen 1.0 m^2 plots located randomly throughout the study area. Percentage germination and survival in these plots were recorded for 15 months.

Production of flowers and fruits was measured over two seasons by placing ten 0.5 m^2 circular litter traps under the crowns of eight reproductive adults of differing size. Size-specific fecundity rates were then multiplied by the number of reproductive adults in each size class to yield an estimate of the total annual production of fruit on the site. Pollination, fruit maturation and dispersal were examined in detail using special tree climbing equipment.

The change in population structure and size over time was simulated using a modified transition matrix (Lefkovitch 1965) based on size class data. By selectively varying the fecundity coefficients in the matrix model, a sensitivity analysis was performed to examine the demographic impact of annual seed collections.

RESULTS AND ANALYSES

Natural populations of *B. alicastrum* are extremely productive. The litter trap studies showed that the 51 adult trees on the study site produced fruit in excess of 6 t ha^{-1} yr^{-1}. All of the trees, however, did not contribute equally. Female trees, which produce the greatest amount of fruit, are concentrated in the smaller diameter classes; hermaphroditic trees, which are less productive, occur in the intermediate size classes, and all of the larger trees are male.

Apparently, adult *B. alicastrum* trees are sequential hermaphrodites (*sensu* Charnov and Bull 1977), and pass through female, hermaphroditic and male sexual states during the course of their development. After an individual reaches 10–15 cm diameter, it begins to produce female flowers and form fruit. If the tree is growing under a relatively open canopy, it will retain its female sexual expression as it increases in diameter. Female trees growing under suppressed conditions, however, gradually begin to produce male flowers and become functionally hermaphroditic. With further increases in diameter, female flower production declines until the tree begins to function strictly as a male.

To define a sustainable yield of seeds requires a more complex analysis. To be able to predict future changes in the population, we need to relate data on current population structure with information about the growth, survival and fecundity of individuals at different stages of the life cycle. The net result for the population at Papantla is shown in Table 14.1.

Table 14.1 Life table data for *B. alicastrum* on 1.0 ha study sites in Papantla, Veracruz. All fluxes are annual; growth rate units are cm yr^{-1} for all stages. Seedling and sapling classes based on height; juvenile and adult classes are based on diameter at breast height (DBH)

	Stage	*N*	*Survival*	*Growth %*	*Moving %*	*Remaining*	*Fecundity*
Seeds	S_0	247 607	0.742		1.000	0.000	
Seedlings	S_1	220 480	0.040	35.7	0.482	0.518	
Small Saplings	S_2	3510	0.150	17.8	0.287	0.713	
Large Saplings	S_3	142	0.823	25.2	0.094	0.906	
Juveniles	S_4	12	0.910	0.38	0.029	0.971	
Adults (10–30 cm)							
Female	S_5	9	1.000	0.33	0.014	0.986	2774
Hermaphrodite	S_6	1	1.000	0.42	0.021	0.979	1549
Adults (30–50 cm)							
Female	S_7	10	1.000	0.38	0.020	0.980	10 035
Hermaphrodite	S_8	5	1.000	0.67	0.033	0.967	548
Adults (50–70 cm)							
Female	S_9	4	0.933	0.49	0.000	1.000	25 384
Hermaphrodite	S_{10}	3	1.000	0.58	0.029	0.971	267
Male	S_{11}	7	1.000	0.58	0.029	0.971	
Adults (70–90 cm)							
Male	S_{12}	12	1.000	0.58	0.000	1.000	

The population has been grouped into nine size classes and 13 different life cycle stages (s_0–s_{12}). The pre-reproductive phase is represented by seeds (s_0), seedlings (s_1), saplings (s_2 and s_3) and juveniles (s_4); adults have been stratified into females (s_5, s_7 and s_9), hermaphrodites (s_6, s_8 and s_{10}) and males (s_{11} and s_{12}). It was felt necessary to subdivide adults into eight classes because of the differing fecundities of females and hermaphrodites within the same size class, and the importance of accurately predicting annual seed production.

The first column of Table 14.1 shows the number of individuals (N) at each stage or the structure of the *B. alicastrum* population. These data will later become the column vector in the matrix model. Survival and growth percentages are based on the repeated measurement of marked individuals over a 3-year period. It should be noted that male, hermaphrodite and female trees exhibit differing rates of diameter growth. The proportion of individuals moving from or remaining in a given stage during one time period was calculated using average growth rates and the actual size distribution within each class. For example, in the small sapling class (s_2), 1007 plants with heights greater than or equal to 82.2 cm were recorded. Given an average growth rate of 17.8 cm yr^{-1}, all of these individuals (28.7% of the class) would move out of stage s_2 in 1 year, while smaller plants (71.3% of the class) would remain in that stage until a future time period. Finally, the fecundity values shown represent 2-year averages calculated for female and hermaphroditic trees.

Arranging the life table data in a matrix format yields the transition matrix shown in Table 14.2. The top row of the matrix lists the size-specific fecundity values, the principal diagonal lists the proportion of individuals remaining in a given class, and the subdiagonal lists the proportion of individuals moving from one stage to another. The shift in the subdiagonal at s_5 takes into account the possibility of sex change from female to hermaphrodite. A female at s_5 may either remain female (transition from s_5 to s_7), or modify its sex expression to become a hermaphrodite (transition from s_5 to s_6). Equal probabilities (50%) were assigned to each alternative.

Multiplying the transition matrix by the column vector (N) yields the size and stage structure of the population at one time interval in the future. Given that the matrix is square with *s* rows and columns, repeated multiplication by the column vector will eventually produce the dominant latent root, or λ, of the matrix. As Leslie (1945), Lefkovitch (1965), Usher (1966) and other workers have shown, the dominant latent root of a transition matrix is equal to e^r, the net reproductive rate or finite rate of increase of a population, and, therefore, can be used to assess the stability or growth of a population. A λ greater than 1.0 indicates that the population is increasing in size, while a λ less than 1.0 shows that the population is

Table 14.2 Transition matrix constructed for *B. alicastrum* population. Flux probabilities have been rounded off to two digits; see text for explanation of matrix construction and parameter estimation

						Stage						
S_0	S_1	S_2	S_3	S_4	S_5	S_6	S_7	S_8	S_9	S_{10}	S_{11}	S_{12}
0	0	0	0	0	2774	1549	10035	548	25384	267	0	0
0.74	0.02	0	0	0	0	0	0	0	0	0	0	0
0	0.02	0.10	0	0	0	0	0	0	0	0	0	0
0	0	0.04	0.74	0	0	0	0	0	0	0	0	0
0	0	0	0.08	0.88	0	0	0	0	0	0	0	0
0	0	0	0	0.03	0.99	0	0	0	0	0	0	0
0	0	0	0	0	0.01	0.98	0	0	0	0	0	0
0	0	0	0	0	0.01	0	0.98	0	0	0	0	0
0	0	0	0	0	0	0.02	0	0.97	0	0	0	0
0	0	0	0	0	0	0	0.02	0	0.93	0	0	0
0	0	0	0	0	0	0	0	0.03	0	0.97	0	0
0	0	0	0	0	0	0	0	0	0	0.03	0.97	0
0	0	0	0	0	0	0	0	0	0	0	0.03	0.99

decreasing in size. A λ equal to 1.0 indicates that birth and death rates are exactly balanced so that the population remains stable.

Taking the exponential of the transition matrix shown in Table 14.2 yields a dominant latent root of 1.0635. Although this λ is reasonably close to the theoretical value of 1.0 expected for stable populations, the slight departure from unity indicates that the density of *B. alicastrum* on the Papantla site is continuing to increase. Not only are high-density aggregations of the species highly productive, they also are apparently self-maintaining over time. In the absence of exogenous disturbance, the population consistently maintains a level of seedling establishment which is more than sufficient to compensate for mortality in the larger size classes.

Given the balance between mortality and seedling establishment, what happens to the population if a large portion of the seed crop is collected each year? To simulate seed removals prior to germination, fecundity coefficients in the matrix were reduced in a stepwise fashion for all adult stages. Dominant latent roots were then calculated for each new matrix to determine the level of seed collection necessary to drive the λ value below 1.0. The results from this analysis are presented in Figure 14.1, which shows the dominant latent roots derived from matrices in which seed production parameters have been reduced from 10 to 99.9%. The dotted, horizontal line at 1.0 represents the λ value of a stable population. As indicated in the figure, latent roots greater than 1.0 are maintained under all levels of seed collection up to 98%. At greater intensities, λ values drop below 1.0, indicating that the level of regeneration is insufficient to balance natural mortality rates. Although it is highly unlikely that the collection of *B. alicastrum* seeds would ever be this intensive, the matrix projections suggest that less than 2% of the annual seed crop is necessary to keep the population at its existing density. It should be noted, however, that this result is based on the assumption that the seeds which remain on the site encounter the appropriate conditions for germination and establishment.

CONCLUSIONS

Perhaps the most important aspect of the experience with *B. alicastrum* in Mexico is the manner in which the development of this resource is being conducted by INIREB. Prior to increasing forest exploitation, ecological studies of the species were initiated to assess the potential sustainability of annual seed collections. This is a rather unusual ordering of priorities. More commonly, ecological analyses are employed only after exploitation has severely reduced the abundance of an important resource, the investigation, in effect, being an *a posteriori* assessment of what went wrong and what can be done about it. Regardless of the fact that moderate seed harvests

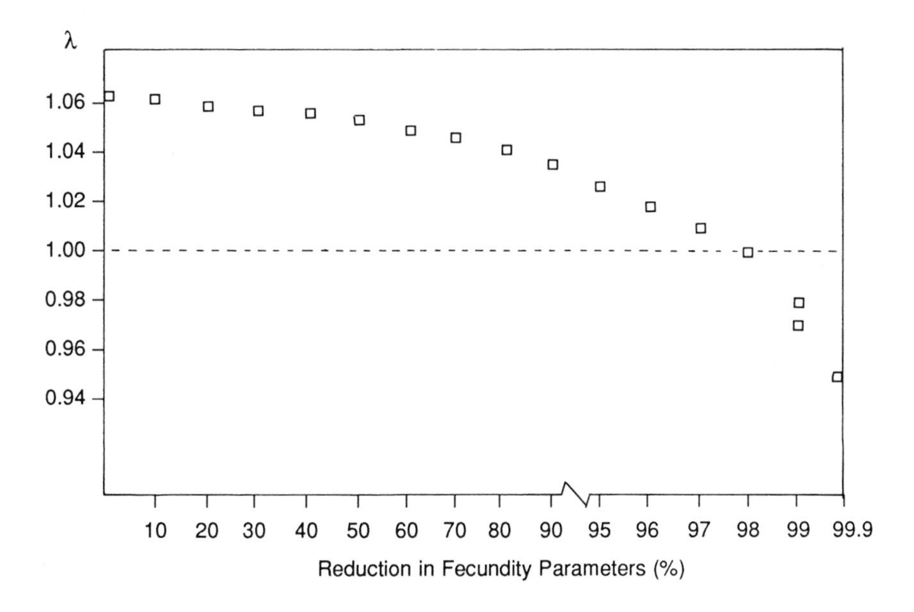

Fig. 14.1 Response of finite rate of increase (λ) of *B. alicastrum* population to reductions in annual seed production. Horizontal line at 1.0 indicates the λ value of a stable, self-maintaining population; see text for explanation of parameter adjustment and matrix operation

were ultimately shown to have little impact on the population dynamics of *B. alicastrum*, the logic of putting ecology before utilization is exemplary.

The results from the ecological study of *B. alicastrum* underscore the value of a demographic focus in the analysis of tropical forest resources. The abundance of a species may be controlled by selective pressures acting on germination, seedling establishment, pollination, fruit maturation or any other stage in the life cycle. Most surveys of useful plants, however, fail to take into account any of these parameters. Although tedious and time-consuming, repeated observation of the growth, mortality and reproduction of marked individuals of varying size or age is the only way to collect this type of information.

Transition matrices proved to be extremely useful in manipulating the demographic data collected for *B. alicastrum*. Matrix methods have been used in several demographic studies of trees (e.g. Hartshorn 1974, Enright and Ogden 1979, Pinero *et al.* 1984), but rarely with the objective of providing management guidelines for natural populations of economic plants. While the impacts of seed collections were examined in this study, transition matrices could be used with equal success to appraise the population consequences of timber exploitation, latex extraction or any other type of resource use. The form of the matrix and the specific data required would vary according to species and the intensity and type of exploitation.

In response to the accelerating destruction of tropical forests and the loss of many potentially valuable species, it is critical that we greatly increase our understanding of the ecology and management of tropical forest resources. Plant demography appears to offer a particularly effective means of achieving this goal.

ACKNOWLEDGEMENTS

I thank F.H. Bormann and P.B. Tomlinson for their excellent advice and encouragement throughout this study, and gratefully acknowledge the support and hospitality of the scientists and staff of the Instituto Nacional de Investigaciones sobre Recursos Bioticos. Special thanks to A. Gómez-Pompa for initially introducing me to *B. alicastrum* and for providing new insights on the management of tropical forest resources. This research was supported by a grant from the US Man and the Biosphere Program.

REFERENCES

Charnov, E.L. and Bull, J. (1977). When is sex environmentally determined? *Nature,* **266**, 828–30

Enright, N. and Ogden, J. (1979). Application of transition matrix models in forest dynamics: *Araucaria* in Papua New Guinea and *Nothofagus* in New Zealand. *Australian Journal of Ecology,* **4**, 3–23

Gómez-Pompa, A. (1973). Ecology of the vegetation of Veracruz. In Graham, A. (ed.) *Vegetation and Vegetational History of Northern Latin America*, pp. 73–148. (Elsevier: Amsterdam)

Hartshorn, G.S. (1975). A matrix model of tree population dynamics. In Golley,F. and Medina, E. (eds.) *Tropical Ecological Systems*, pp. 41–52. (Springer-Verlag: New York)

Lefkovitch, L.P. (1965). The study of population growth in organisms grouped by stages. *Biometrics,* **21**, 1–18

Leslie, P.H. (1945). On the use of matrices in certain population mathematics. *Biometrika,* **35**, 183–212

Peters, C.M. and Pardo-Tejeda, E. (1982). *Brosimum alicastrum*: Uses and potential in Mexico. *Economic Botany,* **36**, 167–75

Pinero, D., Martinez-Ramos, M. and Sarukhan, J. (1984). A population model of *Astrocaryum mexicanum* and a sensitivity analysis of its finite rate of increase. *Journal of Ecology,* **72**, 979–91

Usher, M.B. (1966). A matrix approach to the management of renewable resources, with special reference to selection forests. *Journal of Applied Ecology,* **3**, 355–67

CHAPTER 15

THE GROUND FLORA AND RAIN FOREST REGENERATION AT OMO FOREST RESERVE, NIGERIA

D.U.U. Okali and H.D. Onyeachusim

ABSTRACT

Regeneration processes, tree seed germination patterns, seedling features and the dynamics of change in the ground flora were studied in a relatively undisturbed forest and a Gmelina arborea *plantation at Omo Forest Reserve, Nigeria. Twenty-two of 25 tree species germinated rapidly (within 12 weeks of sowing); 78% of 113 species investigated germinated with cotyledons emerging above ground, and, among these, fleshy cotyledons were most common for large primary species; over 50% of the seedlings differed from adult forms in leaf arrangement and leaf type. Both the variation in germination type and in initial seedling features proved useful in the construction of a seedling identification key. Total species, and rates of recruitment and extinction of species in the ground flora, were higher in forests than plantations, but the density of plants and seasonal fluctuation in this density were higher in the plantations than in the forests. The practical value of this study is underlined.*

INTRODUCTION

Events at or near the forest floor are often missed in forestry studies of regeneration because assessments are commonly limited to established seedlings some 3 feet (about 91 cm) or more tall; moreover assessments often consider only 'economic' species. For these reasons, although regeneration has been studied extensively in the Nigerian rain forest (see Okali and Ola-Adams 1987), few of the studies (notably, Jones 1956, Lancaster 1960, Olatoye 1968) deal with tree seed germination or the early behaviour of tree seedlings at the forest floor. Germination patterns and initial seedling behaviour may, however, provide insights into establishment and regeneration strategies of forest trees (Ng 1976). Besides, knowledge of initial seedling features has practical value as an aid to early seedling identification in regeneration studies or forest management.

Here, we present early results of studies on forest tree seed germination patterns and ground flora dynamics from the Nigerian Rain Forest Project, under the Man and the Biosphere Programme, at Omo Forest Reserve. The studies were initiated in 1981 and are continuing. The preliminary data presented here are for germination patterns and initial seedling features of over 100 species looked at so far, and for seasonal changes in numbers of species and of individuals of selected species in the ground flora. Because the Nigerian Rain Forest Project is also concerned with changes accompanying conversion of rain forest to other uses, the observations on changes in the ground flora are being made for both natural forest and a *Gmelina* plantation.

OMO FOREST RESERVE

Omo Forest Reserve in Ogun State (area 130 500 ha, 6° 35'–7° 05'N and 4° 19'– 4° 15'E) was designated a Biosphere Reserve in 1977 and has since been the main base for investigations in the Nigerian Rain Forest Project. The reserve is one of the few remaining large blocks of high forest in Nigeria. The annual rainfall of about 1800 mm is bimodally distributed with peaks in June/July and September. The driest months are November to March which together receive less than 10% of the annual rainfall. The natural vegetation is mixed moist semi-evergreen rain forest according to the scheme of White (1983). Conversion of part of the reserve into plantations of indigenous and exotic species began in the mid-1960s and became intensified recently in favour of *Gmelina arborea* as a source of pulpwood for a projected pulp and paper mill nearby. The planting programme aimed to convert up to 11 300 ha of the 24 000 ha of plantable land in the reserve into *Gmelina* plantations by 1984, and annually to use 250 ha and 70 ha of the forest for maize and taungya farming respectively. At present, the reserve is a patchwork of a core area of undisturbed Strict Natural Reserve (460 ha), exploited and depleted forests, secondary forests of various ages around enclaves, and forest plantations, mostly of *Gmelina* and teak (*Tectona grandis*).

GERMINATION AND INITIAL FEATURES

We adopted the methods of Ng (1976) to examine germination patterns. So far, patterns have been examined for 113 species. For 25 of these, germination rates were also determined. Seeds were germinated in the glasshouse, in wooden boxes containing top soil from natural high forest, noting the dates for starting and completion of germination for each batch of seeds sown. The type of germination, whether epigeal, hypogeal, semi-hypogeal or durian (Ng 1976), and the initial seedling morphology were observed for the germinated seeds, as well as for a large number of readily identified wildlings collected from the forest.

Twenty-two (88%) of the 25 species sown germinated rapidly or simultaneously, in the sense that all the germination that could occur was completed within 84 days (Ng 1976); the remaining three species exhibited prolonged germination of the intermediate type, which means that only a portion of their viable seeds germinated within 84 days (Table 15.1). The slowest species, *Staudtia stipitata* (Myristicaceae), took 132 days to

Table 15.1 Germination rate for 25 Nigerian rain forest tree species.Successional category indicated in parenthesis

Species and (successional category)	*Gemination rate*
1. *Afzelia pachyloba* (LR)	Rapid
2. *Antiaris africana* (LP)	Rapid
3. *Blighia sapida* (SP)	Rapid
4. *Bosqueia angolensis* (SP)	Rapid
5. *Carapa procera* (SR)	Rapid
6. *Carapa* sp. (SR)	Intermediate
7. *Ceiba pentandra* (LP)	Rapid
8. *Cylicodiscus gabunensis* (LR)	Rapid
9. *Discoglypremna caloneura* (LP)	Rapid
10. *Diospyros dendo* (SR)	Rapid
11. *Distemonanthus benthamianus* (LR)	Rapid
12. *Entandrophragma angolense* (LR)	Rapid
13. *Entandrophragma cylindricum* (LR)	Rapid
14. *Hannoa klaineana* (LR)	Rapid
15. *Khaya ivorensis* (LP)	Rapid
16. *Pentaclethra macrophylla* (LR)	Rapid
17. *Piptadeniastrum africanum* (LR)	Rapid
18. *Scottellia coriacea* (LR)	Rapid
19. *Staudtia stipitata* (LR)	Intermediate
20. *Sterculia rhinopetala* (LP)	Rapid
21. *Strombosia pustulata* (LR)	Intermediate
22. *Terminalia ivorensis* (LP)	Rapid
23. *Terminalia superba* (LP)	Rapid
24. *Tetrapleura tetraptera* (LR)	Rapid
25. *Trichilia monadelpha* (SP)	Rapid

Thus: SP = small pioneers, LP = large pioneer, SR = small primary species, LR = large primary species (after Swaine and Hall, 1983). Germination completed within 84 days is rapid while that initiated before but prolonged beyond 84 days is intermediate.

complete its germination, so that, even for this species, the period of dormancy was less than 5 months. Germination rate (judged by the number of days from sowing to last germination) was very variable even among seed batches of the same species. In fact, some batches of seeds among the species classified as exhibiting prolonged germination completed their germination before 84 days. Thus the features – a short life span for seeds, differential dormancy and a preponderance of species that germinate within a short time, already noted for other rain forest areas – are also characteristic of Nigerian forest trees. It is noteworthy that the three species with a tendency to prolonged germination – *Carapa* sp. (Meliaceae), *Staudtia stipitata* and *Strombosia pustulata* (Olacaceae) – are primary forest species in the sense of Swaine and Hall (1983). Ng (1976) had also observed in Malaysian forests that prolonged germination was not necessarily a feature restricted to pioneer species as is sometimes reported, although it would be necessary to examine the germination behaviour of such well-known pioneers as *Musanga cecropioides* experimentally to resolve the issue. Whitmore (1983) reviews this question of longevity of seeds in relation to successional status of forest trees.

Germination was epigeal in the vast majority (78%) of the species observed. Hypogeal, semi-hypogeal and durian types of germination (see Ng 1976) occurred in 11%, 7% and 4% of the species respectively. To see if the type of germination was related to successional status or regeneration strategy, we classified the species into small pioneers, large pioneers, small primary species and large primary species, using the terminology of Swaine and Hall (1983, 1986). Trees that were capable of attaining a height of 30 m, thus forming part of the mature forest canopy, were classified as large. The data in Table 15.2 show that germination type, classified broadly into epigeal and non-epigeal, was not biased in favour of any successional group ($X^2 = 3.42$; $p > 0.25$). By contrast, data in Table 15.3 show that, among the 97 species in which the cotyledons emerged during germination (the epigeal and semi-hypogeal types), fleshy cotyledons were significantly more common in the large primary species than in the other successional categories, while foliaceous cotyledons occurred more frequently than would be expected from random distribution among the pioneer and small primary species ($X^2 = 13.14$; $p < 0.005$).

We have not yet determined seed sizes, but a large seed size coupled with abundant food reserves (which the fleshy cotyledons probably reflect) might be expected to enable large primary species to produce large seedlings that can persist in shade, ready to take advantage of gaps (Ng 1976). Conversely, foliaceous cotyledons common in pioneer species might be expected to support early photosynthesis by seedlings in the exposed situations in which these species normally germinate. We observed however that a good number of lower storey primary forest species, (e.g.

Table 15.2 Germination types in relation to successional categories in Nigerian rain forest trees. Observed numbers of species in each category. Expected numbers in parenthesis

	Successional category				
Germination type	*Small pioneer*	*Large pioneer*	*Small primary species*	*Large primary species*	*Totals*
Epigeal	32 (30.4)	22 (21.0)	9 (11.7)	25 (24.9)	88
Non-epigeal	7 (8.6)	5 (6.0)	6 (3.3)	7 (7.1)	25
Totals	39	27	15	32	113

$X^2 = 3.42$; $p > 0.25$

Table 15.3 Cotyledon types in relation to successional categories in Nigerian rain forest trees. Observed numbers of species in each category. Expected numbers in parenthesis

	Successional category				
Cotyledon type	*Small pioneers*	*Large pioneers*	*Small primary species*	*Large primary species*	*Totals*
Fleshy	16 (19.7)	13 (16.1)	3 (4.2)	26 (17.9)	58
Foliaceous	17 (13.3)	14 (10.9)	4 (2.8)	4 (12.1)	39
Totals	33	27	7	30	97

$X^2 = 13.14$; $p < 0.005$

Strombosia pustulata, *Picralima nitida* (Apocynaceae)), also have foliaceous cotyledons which persist, sometimes for several seasons, in the forest shade. Foliaceous cotyledons may thus not be solely an adaptation for taking advantage of exposed situations.

The differences between seedlings and adult trees in leaf features have their greatest value as an aid to identification. We have used seedling leaf features, together with germination characteristics, in constructing a provisional key for the identification of the 113 species studied so far, and it is the intention to add to this key as more species are studied. On the whole, 21 seedling forms differed from their parent plants in both leaf arrangement and leaf type, 34 differed only in leaf arrangement, while 10 differed only in leaf type.

The potential use of seedling differences for clarifying taxonomic problems is illustrated by the case of *Carapa* sp. in our sample. At present only one species, *Carapa procera*, is recorded for Nigeria, while the existence of a second species (*C. grandiflora*) is debated because of difficulties in distinguishing between adult forms. In our studies, the form of *Carapa* with small smooth seed produced seedlings with simple initial leaves, while the other form with large non-smooth seeds gave rise to seedlings with 1-pinnate initial leaves. Thus, there is the possibility of resolving the *Carapa* problem from further collection and study of initial seedling features.

GROUND FLORA POPULATION CHANGES

Seasonal changes in ground flora populations were monitored in 1981 and 1982 in both the forest and the *Gmelina* plantation, using ten 1 m^2 quadrats distributed approximately randomly in 50 x 50 m permanent plots in each forest type. All vascular plants < 100 cm tall were pegged at the start of the study and recounted 3–6 times. Seedling identification proved to be a major obstacle, particularly where seedlings were not close to their putative parents. Thus, up to 55% of the plant forms observed in the natural forest could not be determined.

Changes in species numbers

Over the period of observation, as many as 97 species/forms were recorded in the ten 1 m^2 study areas in the natural forest, as against 20 in the plantation. The greater richness of the ground flora in the natural forest is a direct reflection of its greater species diversity. In a separate assessment of floristic attributes, species diversity, computed as the reciprocal of Simpson's (1949) index, was found to be much higher for the forest (11.0) than for the plantation (2.6).

In general, seasonal fluctuations in total species, and in numbers of species surviving, recruited or dying out, were higher in the forest than in the plantation (Table 15.4). These seasonal changes were brought about by recruitment of new seedlings from seed germination and by vegetative reproduction mainly during the favourable wet season, and death of plants mainly during the dry season. Species recruitment and loss were higher in the forest than in the plantation, partly because of the greater diversity of species in the natural forest, but also because of the difference in the nature of the plants that dominated the ground flora in each forest type. In the forest, the majority of the species were woody (mainly tree) species,

Table 15.4 Seasonal variation in number of ground flora species in a 10 m^2 sample area in (1) forest and (2) *Gmelina* plantation at Omo Forest Reserve, Nigeria. February 1981 to August 1982

Sampling date	*No. of species*	*No. of species surviving*	*No. of species recruited*	*No. of species lost*
(1) Natural forest				
March 1981	43	38+	5	-
May 1981	58	40	18	3
July 1981	66	53	13	5
January 1982	50	46	4	20
March 1982	49	45	4	5
August 1982	47	36	11	13
Mean ±S.E.	52.2 ±3.4	43.0 ±2.6	9.2 ±3.2	9.2 ±3.2
(2) *Gmelina* plantation				
February 1981	10	10+	-	-
May 1981	12	10	2	0
July 1981	15	11	4	1
November 1981	15	14	1	1
January 1982	13	12	1	3
March 1982	13	12	1	1
August 1982	15	12	3	1
Mean ±S.E.	13.3 ±0.7	11.6 ±0.5	2.0 ±0.5	1.2 ±0.4

+ Survivors from previous growing season(s)

represented each by one or two seedlings, in most cases newly recruited and unestablished, with a high probability of dying within a short period from suppression under the forest shade or from predation.

By contrast, in the plantation, the majority of the species were of perennial weeds or shrubs each represented by many established, old and/or young individuals with a lower chance of all dying at the same time. Thus, 60 of the 97 species/forms recorded in the natural forest had only one or two individuals each, and 39 of these 60 were among the 46 species that were lost during the observation, whereas only seven of the 20 species in the plantation had only one or two individuals; four of these seven also were among the five species lost from the plantation during the study.

Changes in density

The density of plants in the ground flora fluctuated seasonally more widely in the plantation (18–59 plants m^{-2}) than in the forest (22–27 plants m^{-2}) (Table 15.5). Mass recruitment and death of seedlings and vegetative sprouts of such weeds as *Chromolaena odorata* Keay and Robinson (formerly *Eupatorium odoratum*) and of *Gmelina* seedlings in the more open plantation site largely account for this difference. Since about the same number of species, each represented by a few seedlings, was recruited and lost in the natural forest, there was little change in density of plants in the natural forest floor.

Behaviour of individual species

We took advantage of mast fruiting of *Sterculia rhinopetala*, a large pioneer forest tree, to monitor closely the fate of seeds and seedlings on the forest floor in 1981 and 1982. Firstly, we monitored seeds and seedlings in 1 m quadrats along two transects set at right angles to each other from the base of a selected mother tree. The quadrats were arranged systematically at 2 m intervals along each transect, beginning 1 m from the base of the tree. Second, we laid out plots of various sizes and numbers, depending on the abundance of seedlings, around three mother trees.

In the first set of observations (Table 15.6), seed density decreased with distance, being greatest around 1–3 m and least at 10 m, from the mother tree. Out of 368 seeds recorded in the total 10 m^2 area at the start of observations in March 1981, only 43 seedlings, of which only four were

Table 15.5 Seasonal changes in the total number of ground flora plants m^{-2} in a 10 m^2 sample area in forest and a *Gmelina* plantation at Omo Forest Reserve, Nigeria. February 1981 to August 1982

	No. of plants m^{-2} ± S.E.	
Sampling date	*Forest*	Gmelina *plantation*
February 1981	–	18 ± 3
March 1981	22 ± 5	21 ± 0
May 1981	26 ± 5	59 ± 29
July 1981	27 ± 5	27 ± 5
November 1981	–	33 ± 11
January 1982	23 ± 4	34 ± 12
March 1982	24 ± 5	35 ± 8
August 1982	28 ± 4	51 ± 15
Mean	24.4 ± 4.2	34.8 ± 5.0

Table 15.6 Seed and seedling distribution in relation to distance from mother tree in *Sterculia rhinopetala* in forest. Omo Forest Reserve, Nigeria

	Transect 1 quadrat					*Transect 2 quadrat*						
	1	2	3	4	5	1	2	3	4	5		
Quadrat distance from mother tree (m)	1	3	5	7	10	1	3	5	7	10	Totals	Mean m^{-2}
March 1981												
Number of seeds observed m^{-2}	93	100	6	3	1	115	47	0	3	0	368	36.8 ± 15.1
May 1981												
Number of dead seedlings	12	1	0	0	0	6	3	0	0	0	22	
Number of unsound living seedlings	5	0	0	0	0	10	2	0	0	0	17	
Number of healthy seedlings	1	0	0	1	0	1	0	0	1	0	4	
Total all seedlings	18	1	0	1	0	17	5	0	1	0	43	4.3 ± 2.3
July 1981												
Number of seedlings surviving	1	0	0	0	0	0	0	0	1	0	2	

sound, were detectable by May 1981. By July 1981, only two seedlings remained. In the second set of observations (Table 15.7), seedling densities were not high (ie. 1–3 seedlings m^{-2}) yet only a few seedlings survived between May 1981 and August 1982. Survival varied from 0–43% between locations and, of the total of 325 seedlings initially recorded, only 43 (13%) survived.

Seedling mortality occurred in all seasons but was most pronounced in the dry season (January–March). We observed that many seedlings germinating close to the mother tree could not root effectively in the mineral soil because of their inability to penetrate the thick layer of mother tree litter; the few seeds that were dispersed farther away from the parent location were better established.

Gmelina seedlings observed in two 30 m^2 plots behaved like *Sterculia* in the natural forest (Table 15.8), except that, for *Gmelina*, there was evidence of pronounced seedling loss during the favourable season for growth, March to August. We suspect that this loss may have been due to harvesting of seedlings for planting rather than predation by natural agents.

Table 15.7 Mortality rates over different periods among 325 seedlings of *Sterculia rhinopetala* in a total sample area of 168 m^2 in forest at Omo Forest Reserve, Nigeria

Period of assessment	*No.of seedlings at beginning of period*	*No. of seedlings at end of period*	% *Mortality per month* (28 days)
May–July 1981	325	222	10.6
Aug.1981–Jan.1982	222	106	8.6
Jan.–March 1982	106	63	17.8
March–Aug. 1982	63	43	5.9

Table 15.8 Mortality rates over different periods among 333 *Gmelina* arborea seedlings surveyed in two 30 m^2 plots in a 10–11 year-old *Gmelina* plantation at Omo Forest Reserve, Nigeria, February 1981 to August 1982

Period of assessment	*No.of seedlings at beginning of period*	*No.of seedlings at end of period*	% *Mortality per month* (28 days)
May– July 1981	333	227	10.6
Aug–Nov. 1981	227	152	7.6
Nov.1981–Jan.1982	152	119	9.7
Jan.–March 1982	119	112	2.8
March–Aug. 1982	112	45	10.9

SUMMARY AND CONCLUSIONS

The majority of Nigerian forest tree seeds tested, like tree seeds in other rain forest regions, germinated rapidly. The few that, by contrast, showed a tendency to dormancy were not pioneers. The ability of seeds to remain dormant for long periods is, however, necessary for a species to contribute significantly to the soil seed-bank. Thus, the results given raise doubt that seeds of such well-known pioneers as *Antiaris africana* (Moraceae), *Blighia sapida* (Sapindaceae), *Bosqueia angolensis* (Moraceae) and *Ceiba pentandra* (Bombacaceae) can be stored for much longer than 84 days in a soil seed-bank. Germination was epigeal in 78% of 113 species observed and this did not appear to be related to successional status, whereas the possession of fleshy cotyledons by large primary species or foliaceous cotyledons by pioneers appeared to be part of the syndromes of these successional categories. The number of ground flora species and the fluctuations in this number were greater in natural forest than in a *Gmelina* plantation because of the greater diversity of species in the natural forest. By contrast ground flora plant density was higher and fluctuated more widely in the plantation than in the natural forest, presumably because of exposure and the preponderance of such weeds as *Chromolaena odorata* (Compositae), with mass production and death of seedlings.

A provisional key (not presented here) for the identification of tree seedlings is the main gain of practical importance from this study. Precise identification is essential for better understanding of life histories and probable causes of regeneration failures, while being of immediate practical value to the forester applying or assessing the effects of management treatments.

REFERENCES

Jones, E.W. (1956). Ecological studies on the rain forest of Southern Nigeria. IV. The plateau forest of the Okomu Forest Reserve. *Journal of Ecology,* **44**, 83–117

Lancaster, P.C. (1960). Investigation into natural regeneration in Omo Forest Reserve. A report of trial methods of obtaining natural regeneration in lowland rain forest of Southern Nigeria (Inv. 208). *Nigerian Forestry Information Bulletin* (*New Series*), **13**. (Federal Printer: Lagos)

Ng, F.S.P. (1976). Strategies of establishment in Malayan forest trees. In Tomlinson,P.B. and Zimmermann, M.H. (eds.) *Tropical Trees as Living Systems*, pp. 129–62. (Cambridge University Press: Cambridge)

Okali, D.U.U. and Ola-Adams, B.A. (1987). Tree population changes in treated rain forest at Omo Forest Reserve, south-western Nigeria, *Journal of Tropical Ecology,* **3**(**4**), 291–313

Olatoye, S.T. (1968). Seed storage problems in Nigeria. Paper presented at IXth Commonwealth Forestry Conference, New Delhi, India.

Simpson, E.H. (1949). Measurement of diversity. *Nature,* **163**, 688

Swaine, M.D. and Hall, J.B. (1983). Early succession on cleared forest land in Ghana. *Journal of Ecology,* **71**, 601–27

Swaine, M.D. and Hall, J.B. (1986). Forest structure and dynamics. In Lawson, G.W. (ed.) *Plant Ecology in West Africa: Systems and Processes*, pp. 45–90 (John Wiley: New York, Chichester)

White, F. (1983). *The Vegetation of Africa.* Natural Resources Research Series 20. (Unesco: Paris)

Whitmore, T.C. (1983). Secondary succession from seed in tropical rain forests. *Forestry Abstracts,* **44**, 767–79

CHAPTER 16

COMPARATIVE DYNAMICS OF TROPICAL RAIN FOREST REGENERATION IN FRENCH GUYANA

G. Maury-Lechon

ABSTRACT

Experimental forest management and associated research have been carried out at three sites in French Guyana, including a 25 ha clear-felled area. Research related to forest regeneration has touched on such aspects as reproductive ecology, the effects of hunting on seed dispersal, the structure and species composition of regrowth, organic matter inputs to the soil, changes in hydrological and erosion regimes under different management treatments. Suggested implications include the crucial role of litter in revegetation of compacted and charred soils (including the differential effect of Cecropia *litter), the manipulation of dispersers to favour the establishment of desirable plants, and the use of knowledge on the morphological, biological, eco-physiological and ecological characteristics of the economically interesting plant species to optimize biological performance.*

RESEARCH OBJECTIVES AND STUDY SITES

In 1975 was launched the so-called 'Green programme' for the development of French Guyana. It envisaged the intensive exploitation of the rain forest for pulp industry, and the conversion of deforested zones into pastures, orchards and tree plantations of eucalypts and pines (Guiraud 1979, Sarrailh 1980). Multidisciplinary studies were undertaken as part of the ECEREX project (acronym derived from EC-Ecology, ER-Erosion, EX-Experiment), with a view to assessing the consequences of management practices on the future evolution of the forest. The three main goals were:

(1) to describe and define the natural ecosystem and more especially its composition and regeneration,

(2) to elaborate management practices compatible with the dynamics and other characteristics of the ecosystem, which would permit the rational exploitation of the forest and the establishment of simplified systems, and

(3) to study the effects of modifications in factors affecting production (soil evolution, water balance, genetic reserves, etc.).

Two main study sites were chosen initially: one in the central part of French Guyana in 'Arataye' primary forest, which acted as the 'control' forest (Maury-Lechon and Poncy 1986); the other 'primary forest' in the coastal zone at Saint Elie, in which 25 ha were clear-felled, and 10 watersheds managed in different ways within the so-called 'Arbocel operation' (Sarrailh 1980). Subsequently, a third study site was added – an 80-year-old secondary forest at Cabassou in the coastal zone (Charles-Dominique *et al.* 1981).

The three sites are located within the same climatic zone, characterized by a 8-month rainy season from December to July, interrupted by a short dry season in February–March, with a long dry season of 4 months from August to November. Topography corresponds to a succession of low hills on altered ferralitic soils, on the Bonidoro schists series at St. Elie and on a granitic base in the Arataye forest. Annual rainfall at St. Elie is about 3300 mm (Roche 1982). Some 92.8% of rainfall passes through the foliage and reaches the soil, 0.6% drips along the boles, and 6.6% is intercepted by foliage (Ducrey and Finkelstein 1983). Severe water shortages may occur during the dry season. The two main drainage types (free vertical drainage and blocked vertical drainage) do not appear markedly to influence species composition, though a few taxa do grow preferentially under certain hydrological conditions (Lescure 1981, Lescure *et al.* 1983). Drainage does however seem to restrict diameter increase of trees in plots with blocked vertical drainage. Vegetation biomass (dry weight) varies from 263–372 t ha^{-1} according to individual plots, with an average value of 318 t ha^{-1} (±17%), and a litter production of 7.9 t ha^{-1} yr^{-1} (Lescure *et al.* 1983, Puig 1979). The annual diameter increase is equivalent to 4429 kg ha^{-1} (Prévost *et al.* 1981).

Initial results of the ECEREX project were presented at a symposium held in Cayenne in March 1983 (ORSTOM 1983), and an overall synthesis has been prepared under the aegis of the Technical Centre for Tropical Forestry (Sarrailh 1990). The following paragraphs provide insights to some of the information that has been obtained on the comparative dynamics of regeneration, and its possible implications to rain forest management.

REGROWTH AND REPRODUCTIVE BIOLOGY

Detailed studies have been carried out on the structure of the natural regeneration at the various study sites, and have provided information on such aspects as the relative importance of various woody families and the total numbers of stems of all plant species (Maury-Lechon 1982*a*). Of the 158 054 stems recorded in one 1 ha area, 97% had a diameter < 1 cm, with 630 individuals, or 0.4% of the total, having a diameter ≥ 5 cm and a basal area of 38.25 m^2. Trees represented 43% of the stems present, and trees plus treelets 82%.

The breeding rhythm of all dicotyledonous trees > 2 cm diameter at breast height (dbh) in the French Guyana rain forest is closely linked to climatic seasonality (Sabatier 1985). Flowering is most abundant during the latter part of the dry season (October–November), with 76% of the total number of species in flower at that time in 1980 compared with 49% during the wettest period of the year. Comparable figures for dry and wet seasons in 1981 were 80% and 45% respectively. In contrast, fruiting takes place mainly in the rainy season of March–April: 86% and 88% of species were in fruit in the rainy seasons of 1980 and 1981, with 49% and 41% the relevant dry season figures for the 2 years. Seasonality was particularly marked for species with fleshy zoochorous fruits, but seasonality also existed for those species with wind dispersal or autochorous processes. In 56% of species, germination of ripe seeds occurred within 1 month.

MAN-VERTEBRATE EFFECTS ON DISPERSAL AND REGENERATION

Considerable differences were recorded within the coastal and interior forests, and in the means of diaspore dispersal. At each of the two main sites, more than 50% of the total tree population of dbh ≥ 20 cm and 30 cm is made up of a small number of families: three in Saint Elie, five in Arataye (Maury-Lechon and Poncy 1986). However, great differences appear when the main tree families are considered. In the Arataye forest, Burseraceae predominate (17%). In this family, ripe fruits with mature seeds often expose a bright and contrasting set of colours (the seeds with white arils hanging from the bright carmine placenta axis and the open valves of the pericarp), attracting birds and mammals that are eaten by man. Humans have not lived in this forest for at least a century.

In the St. Elie forest, Lecythidaceae is the dominant family (23%). Fruits are generally greenish or vine coloured, and mainly attract bats (Greenhall 1965), which are not eaten by man. A few people do utilize this forest zone, living there or just coming to hunt, mainly for game.

In the Cabassou forest, the two main families are the Sapindaceae and Myristicaceae, the fruits of which have bright and contrasting colours, as in Burseraceae. Here, however, the size of the fruits is on average smaller than in Arataye forest. Thus, there are just eight species in a 8.5 ha area at Cabassou, whose seed volume is greater than 3000 mm^3, whereas, in the Arataye forest, there are at least 30 species in this seed category in a 6 ha area. Also for closely related species such as *Virola melinonii* (= *V. michelii*), the length of seeds is only 1.8 cm in Cabassou (Charles-Dominique *et al.* 1981), compared to 2.5 cm or more at Arataye for *V.* cf *melinonii* (Maury-Lechon and Poncy 1986). Being smaller, the diaspores are dispersed by smaller vertebrates that are not eaten by humans. As in St. Elie, the fauna of Cabassou forest represents a source of game for local people.

One conclusion is that hunting suppresses the populations of most large vertebrates which disperse coloured and large diaspores (fruits or seeds). This action modifies the forest structure. In the St. Elie forest, 83% of species have zoochorous dispersal (Sabatier 1985), while 93% species in the Arbocel regrowth are so dispersed (de Foresta *et al.* 1984). Medium to small size dispersal agents such as Chiroptera, which are not eaten by humans, disperse greenish diaspores of large to small dimensions – large diaspores being considered those having a volume >3000 mm^3, medium 40–3000 mm^3, and small <40 mm^3 (Charles-Dominique *et al.* 1981). Small birds and mammals, also not eaten by humans, disperse several types of propagules of reduced size, generally those of the pioneer regrowth.

The hunting of edible vertebrates does not seem to disturb greatly the secondary regrowth (in contrast to primary forest regeneration). Effectively, in Arbocel regrowth, 30% of pioneer species are dispersed by Chiroptera and 60% by small birds (de Foresta *et al.* 1984, Tostain 1986). All the principal vertebrate dispersers and all the main pioneer species of the regrowth in the St. Elie region are present in the corresponding primary forests (de Foresta *et al.* 1984, Maury-Lechon and Poncy 1986).

These findings have several implications for forest management in French Guyana, particularly the management of regeneration. First, hunting results in an accumulation of ripe fruits and seeds under the fruiting trees, which, on germination, have little possibility of survival under the canopy of the mother tree. Second, hunting leads to increasing rarity of species dispersed by edible vertebrates and the clumping of species dispersed by bats (e.g. Chrysobalanaceae, Clusiaceae, Lecythidaceae). Third, a study of *Carapa procera* and Acouchi (its main predator in the Arataye forest) has shown how a wild rodent intervenes as a 'seeding technician' (Maury-Lechon and Poncy 1986) and has led to the suggestion that natural dispersal techniques such as this might be encouraged within silvicultural schemes, for example through the introduction of such dispersal agents into suitable forest areas.

NATURAL REGENERATION ON A CLEAR-CUT AREA

Regeneration has been studied on a 25 ha clear-cut area known as Arbocel. The structure and species composition of secondary regrowth has been studied 3.5 years after clear-cutting and burning, and regrowth dynamics have been shown to be driven by the interactions of six major sets of factors (Maury-Lechon 1982*b*, 1982*c*, Maury-Lechon *et al.* 1986): fire; drainage (itself linked to topography and erosion); proportion of soil surface covered by abandoned logs; soil seed-bank present after experimental exploitation and fire; seed transport into and within the clear-cut area after exploitation and fire; proximity of forest borders.

The revegetation of bare ground on the logging trails varied considerably (Maury-Lechon *et al.* 1986), ranging from 2 years on the edges of the logging trails (to where soil had been moved), to 3 years on the periphery of zones compacted by mechanical vehicles, to 6 years in the more compacted parts. Three-and-a-half years after clear-cutting, 94% of the soil surface of the 25 ha plot had been covered by vegetation, some 4% of the total corresponding to the bare ground of logging trails. Six years after felling, the remaining 4% area was being colonized and the only patches of bare soil corresponded to the more severely burnt zones. On very charred soils, the first seedlings only appeared at the end of the 6th year, and even then were strictly confined to the periphery of areas where new litter had been accumulating for 3–4 years.

The supply of litter from adjacent regrowth is determinant in the revegetation of very compacted and charred soils, and areas where the top soil has been removed by machines. Studies have elucidated the role of litter in the population dynamics of soil organisms and in soil development (Betsch *et al.* 1980). In unburnt areas, soil populations develop in a reasonably uninterrupted way, independent of season, while the litter populations are affected by climatic variations. In the charred areas, the soil itself was subject to these variations, and recolonization of the soil was slow.

Comparative analyses have shown that *Cecropia* (Cecropiaceae) litter is more effective than that of *Vismia* (Clusiaceae) in promoting soil restoration, in terms for example of Collembola diversity, micro-arthropod biomass, organic matter content, water retention capacity (Betsch and Betsch-Pinot 1983). *Cecropia* species, though considered to be of no economic value, thus play a fundamental ecological role (Maury Lechon-1979, Maury-Lechon *et al.* 1986). They produce the first shade and litter, which enable more delicate species to germinate and establish on the bare ground of the clear-cut areas. They also encourage the commercial species which grow in their understorey (e.g. *Goupia glabra*, *Laetia procera*) to produce straight boles, in their search for light under *Cecropia* crowns. The mortality of these two valuable species is also related to the covering crown of *Ceropia* during their first 6–7 years of growth.

Pluvio-lessivates contribute 1.5–2.7 t ha^{-1} yr^{-1} of organic matter to the soil in the primary control forest, with microbial measures of $16–47.10^6$ germs g^{-1} during the rainy season and 24.10^6 germs g^{-1} during the dry season (Kilberthus 1983). The results highlight the intensity of microbial activity and suggest that leaf breakdown has already started before leaf fall, which might explain the rapidity of their ultimate decomposition within the soil litter.

AGRICULTURAL AND FORESTRY MANAGEMENT TRIALS

Treatments following forest exploitation and watershed clearance have included transformation into fertilized pastures and pine and eucalypt plantations (Sarrailh 1980). Seedlings of *Pinus caribaea* var. *hondurensis* (Poptum-Guatemala origin) planted at 6 months reached a height of 1.54 m after 14 months and 4.23 m after 2 years. In 1 ha experimental trials, growth was 10% faster in areas with free vertical drainage compared to blocked vertical drainage (Ayphassorho 1983). For *Eucalyptus urophylla* (Flores origin), there was a 40% difference in growth on free compared to blocked vertical drainage.

Detailed studies have been carried out on changes in hydrological and erosion regimes under different treatments. Among the results reported by Fritsch (1983) are an increase in surface run-off following clearance (from 11% to 27% after 25 mm rainfall). The most marked effects on run-off water and erosion rates occurred during the clearing operation itself. For example, soil erosion rates reached 45 t ha^{-1} yr^{-1} during clearance but decreased to 0.2–0.7 t ha^{-1} yr^{-1} only one year after the plantation of grass. In respect to run-off water in an area of blocked vertical damage, proportions changed from 15% of total rainfall prior to clearance, to 50–64% during the conversion to pasture, and 29–32% 3 years after pasture establishment (Sarrailh 1980). Especially on soils with blocked vertical drainage, clearance with large machines can have a traumatic effect on tree growth, as reported by Ayphassorho (1983) following measurements on height variation among 1585 young pines and 914 eucalypts.

CONCLUSIONS AND FUTURE RESEARCH DIRECTIONS

Experimental forest management and associated research within the ECEREX programme in French Guyana have highlighted the large effect of vegetation removal and disturbance of soils and macrofauna on the composition and structure of natural secondary regrowth and on plantation development. Fire and the compaction, removal and erosion of soil markedly affect soil biology and water dynamics – to the extent that fire combined with soil removal may preclude seed germination and seedling establishment for up to 6 years (Betsch and Betsch-Pinot 1983, Maury-Lechon *et al.* 1986) and strongly delay the growth of planted seedlings (Ayphassorho 1983, Finkelstein 1983). Fires and the use of machines should be tightly controlled and if possible avoided, more especially during the rainy season and just after ground clearance.

The specific composition and general structure of the natural regrowth can be broadly foreseen from the preliminary analyses of the soil seed-bank, of the seed rain in naturally open zones (e.g. natural gaps) and the potential animal dispersers of available seeds (Prévost 1983). Thus of the 19 most common species in the Arbocel regrowth, eight were present in the soil seed-bank, nine were dispersed by mammals (eight of these by bats) and 12 by birds (Charles-Dominique 1986). Small vertebrates also play an important role in the dispersal of small sized seeds of pioneer vegetation. The disperser fauna may be manipulated to favour the establishment of desirable plants. This may be particularly important in rehabilitating large surface areas of degraded land, for example through the introduction of bats and *Cecropia* fruiting trees.

It should be possible to use knowledge on the morphological, biological, eco-physiological and ecological characteristics of the economically interesting plant species to optimize biological performance (for example, in terms of flowering and pollination biology, conserving artificially the reproductive diaspores or very young seedlings, inoculating symbionts, etc.). Another possibility is to create new types of landscapes based on the optimal complementarity of plant and animal species within simplified ecosystems in the proximity of human populations.

The work in French Guyana has emphasized that natural regeneration in large openings, in this area at least, depends largely upon sexual reproduction – on seed production first, and on seed germination and seedling survival second. Research is required on the pollen and seed stages, which are critical to species survival. Studies are especially called for on the shapes and mechanisms of the ontogenesis of the juvenile stages of plants, from the ripe fruit to the autotrophic seedling. Practical implications include the possibility of shortening the duration of juvenile development and promoting the growth of plant species on very infertile soils (e.g. through the experimental inoculation of symbionts).

REFERENCES

Ayphassorho, H. (1983). Première contribution à l'étude des interactions sol-végétation sur les bassins versants G (pins) et H (eucalyptus) du dispositif ECEREX. In *Compte-rendu Journées ECEREX*, pp. 351–61. (ORSTOM: Cayenne)

Betsch, J.M. and Betsch-Pinot, M.Ch. (1983). Recolonisation d'une coupe papetière par les microarthropodes du sol, en particulier les collemboles en forêt dense humide subéquatoriale (Guyane française). In Lebrun, Ph., André, H.M., Medts, A. de, Gregoire-Wibo, G. and Wauthy, G. (eds.) *Proceedings of VIII International Colloquium of Soil Zoology* (*Louvain*), pp. 519–33

Betsch, J.M., Kilberthus, G., Proth, J., Betsch-Pinot, M.Ch., Couteaux, M.M., Vannier, G. and Verdier, B. (1980). Effets à court terme de la déforestation à grande échelle de la forêt dense humide en Guyane française sur la microfaune et la microflore du sol. In Dindal, D. (ed.) *Proceedings of VII International Colloquium of Soil Biology* (*Washington*), pp. 472–90

Charles-Dominique, P. (1986). Inter-relations between frugivorous vertebrates and pioneer plants: *Cecropia*, birds and bats in French Guyana. In Estrada, A. and Fleming, T.H. (eds.) *Frugivores and Seed Dispersal,* pp. 119–35. (Junk: Dordrecht)

Charles-Dominique, P., Atramentowicz, M., Charles-Dominique, M., Gerard, H., Hladik, C.M. and Prévost, M.F. (1981). Les mammifères frugivores arboricoles nocturnes d'une forêt guyanaise: inter-relations plantes- animaux. *Revue d'Ecologie* (*Terre et Vie*), **35**, 341–435

Ducrey, M. and Finkelstein, D. (1983). Contribution à l'étude de l'interception des précipitations en forêt tropicale humide de Guyane. In *Compte-rendu Journées ECEREX*, pp. 305–26. (ORSTOM: Cayenne)

Finkelstein, D. (1983). Influence des propriétés hydriques et de la disponibilité en eau du sol sur la croissance du pin des Caraïbes. In *Compte-rendu Journées ECEREX*, 345–50. (ORSTOM: Cayenne)

Foresta, H. de, Charles-Dominique, P., Erard, Ch. and Prévost, M.F. (1984). Zoochorie et premiers stades de la régéneration naturelle après coupe en forêt guyanaise. *Revue d' Ecologie* (*Terre et Vie*), **39**, 369–400

Fritsch, J.M. (1983). *Modifications du comportement hydrologique et de l'érosion en bassins versants sous l'effet des aménagements.* (ORSTOM: Cayenne)

Greenhall, A.M. (1965). Sapucaia nut dispersal by greater spear-nosed bats in Trinidad. *Caribbean Journal of Science,* **5**, 169–71

Guiraud, A. (1979). Objectifs et méthodologie. *Bulletin Liaison ECEREX,* **1**, 3–4. (ORSTOM: Cayenne)

Kilberthus, G. (1983). Microflore et activités microbiennes en forêt dense humide guyanaise. In *Compte-rendu Journées ECEREX.* pp. 375–80. (ORSTOM: Cayenne)

Lescure, J.P. (1981). La végétation et la flore dans la région de la piste de Saint Elie. *Bulletin Liaison ECEREX,* **3**, 4–24. (ORSTOM: Cayenne)

Lescure, J.P., Puig, H., Riera, B., Leclerc, D., Beekman, A. and Beneteau, A. (1983.) La phytomasse éipigée d'une forêt dense en Guyane française. *Acta Oecologica,* **4**(**3**), 237–51

Maury-Lechon, G. (1979). Plantules et régénération forestière en Guyane française: premières constatations sur une coupe à blanc de 25 ha. *Bulletin. Société botanique de France,* **126**, *Actualitiés botaniques* (3), 165–71

Maury-Lechon, G. (1982*a*). Régénération forestière en Guyane française. III. Plantules et jeunes en forêt témoin. *Bulletin Liaison ECEREX,* **6**, 119–35. (ORSTOM: Cayenne)

Maury-Lechon, G. (1982*b*). Régénération forestière en Guyane française: recrû sur 25 ha de coupe papetière en forêt dense humide (ARBOCEL). *Bois et Forets de Tropiques,* **197**, 3–21

Maury-Lechon, G. (1982*c*). Régénération forestière sur 25 ha de coupe papetière en forêt dense humide de Guyane française. *Comptes rendu de l'Académie des sciences de Paris,* **294** (III), 975–8

Maury-Lechon, G., Betsch, J.M. and Betsch-Pinot, M.Ch. (1986). Dynamiques comparées de la végétation et de la pédofaune dans un recrû en zone forestière tropicale (Guyane francaise). *Mémoires du Museum nationale d'Histoire naturelle. Entretiens du Museum, série A, Zoologie,* **132**, 243–55

Maury-Lechon, and Poncy, G. O. (1986). Dynamique forestière sur 6 ha de forêt dense humide de Guyane française, à partir de quelques espèces de forêt primaire et de cicatrisation. *Mémoires du Museum nationale d'Histoire naturelle. Entretiens du Museum, série A, Zoologie*, **132**, 211–42

ORSTOM. (1983). Le Projet ECEREX (Guyane). Analyse de l'écosystème forestier tropical humide et des modifications apportées par l'homme. Journées de Cayenne du 4–8 mars 1983, organisées par GERDAD (CTFT), INRA, Museum Nationale d'Histoire Naturelle, ORSTOM. (ORSTOM: Cayenne)

Prévost, M.F. (1983). Les fruits et les graines des espèces végétales pionnières de Guyane française. *Revue d'Ecologie* (*Terre et Vie*), **38**, 121–41

Prévost, M.F. and Puig, H. (1981). Accroissement diamétral des arbres en Guyane. *Bulletin liason* ECEREX, **1**, 93–106 (ORSTOM: Cayenne)

Puig, H. (1979). Production de litière en forêt guyanaise: résultats préliminaires. *Bulletin. Societé d'Histoire naturelle de Toulouse,* **115 (3–4)**, 338–46

Roche, M.A. (1982). Comportements hydrologiques comparés et érosion de l'écosystème forestier amazonien à ECEREX en Guyane. *Cahiers. ORSTOM, Série Hydrologie,* **19(2)**, 81–114

Sabatier, D. (1985). Soisonnalité et determinisme du pic de fructification en forêt guyanaise. *Revue d'Ecologie* (*Terre et Vie*), **49**, 289–320

Sarrailh, J.M. (1980). L'écosystème forestier guyanais. Etude écologique de son évolution sous l'effet des transformations en vue de sa mise en valeur. *Bois et Forêts des Tropiques,* **189**, 31–6

Sarrailh, J.M. (Ed). (1990). Mise en valeur de l'écosystème forestier guyanais. Opération ECEREX. INRA, Paris (in press).

Tostain, O. (1986). Etude d'une succession terrestre en milieu tropical: les relations entre la physionomie végétale et la structure du peuplement avien en mangrove guyanaise. *Revue d'Ecologie* (*Terre et Vie*), **41**, 315–42

CHAPTER 17

REGENERATION AFTER DISTURBANCE IN A LOWLAND MIXED DIPTEROCARP FOREST IN EAST KALIMANTAN, INDONESIA

S.Riswan and K. Kartawinata

ABSTRACT

A study on regeneration after clear-cutting and burning was carried out in a primary lowland forest in Lempake, East Kalimantan, Indonesia. Two (0.5 ha each) contiguous clear-cut plots were established. One was burnt, the other unburnt, and observations were carried out over a period of 78 weeks. In the early stage of succession, seedlings played a more important role than resprouts. The number of species, percentage of cover and frequency of seedlings and resprouts, as well as the number of primary forest species, were greater in the unburnt than in the burnt plot. The dominant species in the unburnt and burnt plots differed, even though the plots were adjacent. It is suspected that recovery in the unburnt plot could be attributed mainly to the undisturbed seed-bank in the soil, as indicated by a high number of primary forest species developed therefrom. The natality period in the unburnt plot was longer than in the burnt plot, but was followed by high mortality, particularly of light demanding species.

INTRODUCTION

In East Kalimantan (Indonesia), most secondary forests were for many years primarily the product of shifting cultivation, in which clearance of patches of primary forest was followed by burning and subsequent abandonment after one or more cultivations. More recently, during the last 2 decades, mechanized selective logging practices have led to the formation of millions of hectares of depleted primary forests, in which secondary forest species develop in the gaps created by tree felling and tractor operations.

Studies on secondary succession can play an important role in understanding basic biological problems. Succession involves interactions between many community processes, such as nutrient cycling, productivity, rate of decomposition, activities of micro-organisms and biotic interactions in the whole system. Richards (1955) claimed that study of secondary succession by ecologists is one of the essential requirements for a solution to the problem of establishing a rational system of land-use in the tropics. To date, secondary succession studies in the tropics have been limited to a few areas, and there is a paucity of knowledge concerning succession in these regions (Richards 1952, Whitmore 1975, 1983). Understanding of succession has mostly relied on a reconstruction of a time sequence from simultaneous studies on patches of different ages, and very little is based on experimental studies (Whitmore 1983).

The present contribution summarizes results of observations on vegetation recovery processes in experimental plots in primary dipterocarp forest after clear-cutting and burning. Detailed results of the study are presented in Riswan (1982), while information on the broader MAB pilot project in East Kalimantan is given in Kartawinata *et al.* (1981) and Kartawinata and Vayda (1984).

STUDY AREA

The study area is located within the 300 ha teaching and research forest of the Mulawarman University at Lempake, about 12 km north east of Samarinda, the capital city of the Province of East Kalimantan in Indonesia. The site includes primary, secondary and logged-over forests. The Lempake area consists of sedimentary rocks formed during the Upper Miocene period (Anon. 1965). Topography is undulating to somewhat hilly, with an altitude range of 40–80 m. The soils belong to the red-yellow podzolic group (Hardjono 1967), roughly equivalent to ultisols.

The climate is humid, belonging to the rainfall type A with a Q value (ratio between the mean number of dry months and wet months) of 7.4 (Schmidt and Ferguson 1951). Data from 76 years of records at the rainfall station at Sungai Kunjang near Samarinda, show that the average rainfall is 1964 mm, with the monthly values ranging from 110 mm in August to 240 mm in December (Anon. 1982, Berlage 1949). The dry season occurs from July to September and the rainy season between October to June. Mean monthly temperature ranges from 26.5° C in January to 28° C in July (Berlage 1949).

The primary lowland mixed dipterocarp forest (MDF) of the area is rich in species. In one 1.6 ha plot, Riswan (1982) recorded 209 tree species with dbh greater than 10 cm. Twelve were dipterocarp species, of which the most common were *Shorea polyandra*, *S. parvifolia* and *Hopea rudiformis*. The prevalent non-dipterocarp species were *Eusideroxylon zwageri* (Bornean ironwood), *Baccaurea macrocarpa*, *Cleistanthus myrianthus* and *Pentace laxiflora*. In one 0.8 ha plot of the 35-year old secondary MDF, 121 tree species with dbh > 10 cm were found. The common tree species were *Macaranga gigantea*, *M. conifera*, *M. pruinosa* and *Nauclea orientalis*. *Hopea rudiformis* was the only dipterocarp species occurring there.

METHODS

Two experimental plots of 0.5 ha each were established in the primary MDF. The vegetation in these plots was clear-cut. In one plot, the felled vegetation was burnt. In the other, it was removed and the plot was left unburnt. In each plot, a strip of 100 x 1 m was established, and divided into 100 sub-plots of 1 x 1 m each. All seedlings and resprouts growing in all sub-plots were recorded every 6 weeks during the first 6 months, and then every 6 months during subsequent 1.5 years. Seedlings and resprouts were grouped into three life-forms: tree, shrub (including lianas or woody climbers) and herb (i.e. grasses, vines and ferns). Density, frequency and cover of trees and shrubs were recorded, but only frequency and cover were registered for herbs. We used the definition of Whitmore (1975, 1978) and our experience in field observations elsewhere in grouping plants into pioneer/secondary and primary forest species. The control plots were established in primary forest and 35-year old secondary forest located adjacent to the experimental plots. The primary and old secondary forest plots were 1.6 ha and 0.8 ha respectively. Data for control plots have been reported in Riswan (1982, 1987).

RESULTS AND DISCUSSION

Species composition

Seventy-eight weeks after disturbance: the number of species of seedlings and resprouts was greater in the unburnt plot; the common species of trees, shrubs and herbs in the unburnt plot differed from those in the burnt plot; and the ratio of primary forest to secondary forest seedlings was higher in the unburnt than in the burnt plot.

Unburnt plot

After 78 weeks, the number of species of seedlings and resprouts was 105 and 57 respectively. Of 105 species of seedlings, 64 were tree species (28 pioneer species). The number of species and frequency of tree seedlings were lower than those at the 1-year phase, but the cover value was higher. The common tree species were *Macaranga pruinosa*, *M. conifera*, *M. hypoleuca*, *M. gigantea*, *Endospermum diadenum*, *Ficus arfakensis*, *Hopea rudiformis* and *Callicarpa pentandra*. The dipterocarps were represented by *Hopea rudiformis* and *Shorea parvifolia* (which appeared after 6 weeks), as well as *S. leprosula* (which appeared after 6 months). The last two species have been reported as species with pioneer characteristics, being tolerant to light and fast-growing (Wyatt-Smith 1955). The high frequency of *Hopea rudiformis* is of interest. It is normally considered as a primary forest species, but like *S. leprosula* and *S. parvifolia*, it seems to behave more like secondary forest species. *Anthocephalus chinensis* was absent, although it was common during the period of 6–24 weeks, suggesting that it failed to compete with other pioneer tree species for light.

The number of species and percentage of cover of shrub seedlings increased slightly, but the number and frequency of seedlings decreased. The common species were *Melastoma affine* (shrub) and *Maesa polyantha*, *Uncaria setiloba*, *U. glabrata*, *Embelia* sp. and *Grewia* sp. (woody climbers). The common herbs were *Curculigo glabrescens*, *Paspalum conjugatum*, *Globba pendula*, *Dryopteris* sp. and *Imperata cylindrica*. *Paspalum conjugatum* was still abundant although it was dying, because of shading. It would eventually disappear when the forest floor becomes completely shaded.

Imperata cylindrica grass (alang-alang) appeared after 12 weeks, but could not spread rapidly because of the shade created by the pioneer tree canopy. The spread of this species would be enhanced by clear-cutting and burning, since the rhizomes remain alive and are resistant to fire. A patch of primary forest can be taken over by this grass within 3 years after repeated clear-cutting and burning. This is one of the reasons why shifting cultivation practices with short fallow periods have led to the formation of extensive alang-alang fields in the tropics, particularly in South-east Asia.

One and a half years after clear-cutting, the resprouts of primary forest trees showed much better growth, since they apparently could withstand full shading. They increased in species number, percentage cover and frequency. The number of species of tree resprouts was particularly high (42 species). The most common species was *Eusideroxylon zwageri*, followed by *Cleistanthus myrianthus*, *Claoxylon* sp., *Trioma malaccensis*, *Nauclea subdita* and *Pentace laxiflora*. The number of species and frequency of shrubs decreased, but the percentage of cover increased slightly. The common species were still similar to those at 1 year, i.e. *Dinochloa scandens* and *Spatholobus* sp. The herbs were particularly dominated by *Stachyphrynium jagorianum* and *Hallopegia blumei*.

Burnt plot

In this plot, the total number of seedling species was constant. The number of species of tree and shrub seedlings increased, but that of the herbs decreased. The frequency of seedlings declined, but the percentage of cover increased. The resprouts decreased in number of species and frequency, but the percentage of cover increased.

The number of species of tree seedlings (55) was less than that in the unburnt plot, but the number of pioneer tree species was higher (32). The most common tree seedlings were *Anthocephalus chinensis* followed by *Endospermum diadenum*, *Macaranga* species (*M. gigantea*, *M. hypoleuca* and *M. pruinosa*), *Trema orientalis*, *Ficus arfakensis* and *Callicarpa pentrandra*. Compared with the unburnt plot, the composition of tree species differed, although the plots were adjacent. *Anthocephalus chinensis* and *Endospermum diadenum* were the most common, followed by *Macaranga* spp. *Trema orientalis* grew well. Dipterocarps were represented by one seedling of *Shorea parvifolia*.

The seedlings of shrubs were still increasing in species number and percentage of cover, but decreasing in frequency and number of seedlings. The number of species of pioneer shrubs (woody climbers) was much higher than that of the primary forest shrubs. The common shrubs were woody climbers, such as *Maesa polyantha*, *Uncaria ferrea*, *U. setiloba*, *U. glabrata* and *Embelia* sp.

The species composition of shrubs differed in burnt and unburnt areas, indicating that fire may determine species composition. Burning at an earlier stage killed both primary and secondary forest seeds in the forest floor. Thus, fire determined the direction and rate of the forest succession, as noted by Ahlgren and Ahlgren (1960) in temperate forest areas.

Herb species were also different in the burnt compared with unburnt plot. The most common species were ferns, *Nephrolepis biserrata* and *Lygodium microphyllum*, followed by a herb, *Curculigo glabrescens*. The grass *Paspalum conjugatum* died completely under the shade of pioneer trees. *Imperata cylindrica* was present, but not common, and would not spread further provided that the succession is left undisturbed.

The resprouts showed a decline in species number and frequency, but increased in the percentage of cover. The species composition was not different from that in 1-year burnt and unburnt areas. At this stage the common tree species were *Eusideroxylon zwageri*, *Dichapetalum gelonoides* and *Monocarpia marginalis*. Coppicing may hasten the vegetation recovery process, since most of the resprouts were primary forest species. An interesting species is the Bornean ironwood (*Eusideroxylon zwageri*), which was the dominant tree resprout in the burnt and unburnt plots. A similar phenomenon was also observed in a dipterocarp forest stand in East Kalimantan after the extensive forest fires of 1983 (Riswan and Yusuf 1986), but its ability to recover rapidly after disturbance through resprouting might be considered a special strategy which compensates for its slow growth rate.

Vegetation recovery

The early stages of secondary succession in both experimental plots in mixed dipterocarp forest (MDF) showed that the initial seedlings play a major role in vegetation recovery. Resprouts of the original forest vegetation seem unimportant. After 1.5 years, there were twice as many primary forest tree species in the unburnt plot but the numbers of secondary forest trees were similar in the two plots. In the unburnt plot, primary species number soon exceeded secondary forest species. In the burnt plot, there were always more secondary species and the number of individual primary forest seedlings and resprouts was very low. At the same time, the species composition of the dominant trees was quite different, even though the plots were contiguous. However, after 1.5 years, both had the same similarity value (14%) as the 35 year-old secondary MDF.

The number of individual tree seedlings of both primary and secondary forest species was very much reduced in the burnt plot, especially in the early stages. Since there were twice as many species of primary forest trees developing in the unburnt plot, this suggests the presence of a seed-bank in the soil. In this case, however, the number of secondary species was not different, although they differed in species composition. Numbers of individuals were much higher in the unburnt plot; this suggests that while the secondary forest species have large numbers regenerating, and in general have a high species number, it is small differences in the number of primary forest species which may influence the rate of regeneration to "primary" forest.

CONCLUSION

It appears that floristic composition of the forest depends upon natural regeneration and the rate of recovery from disturbance. The new composition of species after disturbance may depend on the following factors: kind of disturbance (clear-cutting with burning or without burning); remnants in individual plots which survive following disturbance (resprouts); survival of seed-bank in the forest floor, unaffected by disturbance; immigrant species (seed-rain), which arrive after disturbance. The nature of the disturbance strongly determines species composition in MDF. Even though the burnt and unburnt plots in MDF were contiguous, the major dominant species were different after 1.5 years.

REFERENCES

Ahlgren, I.F. and Algren, C.E.. (1960). Ecological effects of forest fires. *Botanical Revue,* **46**, 304–10

Anon. (1965). *Peta geologi Indonesia*. (Direktorat Geologi: Bandung)

Anon. (1982). Rainfall records, East Kalimantan. Report, analysis summaries and histograms. Transmigration Area Development Project, Samarinda, Kalimantan Timur.

Berlage, H.P., Jr. (1949). *Regenval in Indonesia.* Verhandelingen No.37. Koninklijk Magnetish en Meterologisch Observatorium te Batavia.

Hardjono. (1967). *Uraian Satuan Peta Tanah Eksplorasi Kalimantan Timur baegian Selatan.* (Lembaga Penelitian Tanah: Bogor)

Kartawinata, K., Adisoemarto, S., Riswan, S. and Vayda, A.P. (1981). The impact of man on a tropical forest in Indonesia. *Ambio,* **10(2–3)**, 115–19

Kartawinata, K., and Vayda, A.P. (1984). Forest conversion in East Kalimantan, Indonesia: the activities and impact of timber companies, shifting cultivators, migrant pepper-farmers and others. In Castri, F. di, Baker, F.W. and Hadley, M. (eds.) *Ecology in Practice.* Vol. **1**. *Ecosystem Management*, pp. 98–126. (Tycooly: Dublin, and Unesco: Paris)

Richards, P.W. (1952). *The Tropical Rain Forest. An Ecological Study.* (Cambridge University Press: London)

Richards, P.W. (1955). The secondary succession in the tropical rain forest. *Science Progress (London),* **43**, 45–57

Riswan, S. (1982). *Ecological Studies in Primary, Secondary and Experimentally Cleared Mixed Dipterocarp Forest and Kerangas Forest in East Kalimantan, Indonesia.* Ph.D. Thesis. (University of Aberdeen: Aberdeen)

Riswan, S. (1987). Structure and floristic composition of Mixed Dipterocarp Forest at Lempake, East Kalimantan. In Kostermans, A.J.G.H. (ed.) *Proceedings of the Third International Round Table Conference on Dipterocarps*, pp. 435–57. (Unesco: Jakarta)

Riswan, S. and Yusuf, R. (1986). Effects of forest fires on trees in the lowland dipterocarp forest, East Kalimantan, Indonesia. In *Proceedings of the Symposium on Forest Regeneration in South-east Asia*, pp. 155–63. BIOTROP Special Publication No.25. (BIOTROP: Bogor)

Schmidt, F.H. and Ferguson, J.H.A. (1951). *Rainfall Types Based on Wet and Dry Period Ratios for Indonesia with Western New Guinea.* Verhandelingen No.**12**. (Djawatan Meteorologi and Geofisika: Djakarta)

Whitmore, T.C. (1975). *Tropical Rain Forests of the Far East.* (Clarendon Press: Oxford)

Whitmore, T.C. (1978). Gaps in the forest canopy. In Tomlinson, P.B. and Zimmermann, M.H. (eds.) *Tropical Trees as Living Systems*, pp. 639–55. (Cambridge University Press: Cambridge)

Whitmore, T.C. (1983). Secondary succession from seed in tropical rain forests. *Forestry Abstracts,* **44(12)**, 767–79

Wyatt-Smith, J. (1955). Changes in composition in early natural plant succession. *Malayan Forester,* **18(1)**, 44–9

CHAPTER 18

RECOVERY OF FOREST VEGETATION FOLLOWING SLASH-AND-BURN AGRICULTURE IN THE UPPER RIO NEGRO

J.G. Saldarriaga and C. Uhl

ABSTRACT

Changes in species composition, forest structure and biomass have been studied at 24 tropical forest sites along the Upper Rio Negro region of Colombia and Venezuela. Stands were selected from the tierra firme *forests (non-flooded) to represent a chronosequence of succession following slash-and-burn agricultural practices. After abandonment, the number of species increases from early successional to mature forests. The species composition of the mature forests depends on a small fraction of primary species that survive from early stages of succession and on the introduction of many primary species at later stages of succession. Small areas disturbed by slash-and-burn agriculture recover their original species composition, but the time required varies, depending on the intensity and frequency of disturbance in the area. On a large scale, the forest is a mosaic of different aged patches and structural characteristics, with high variability among stands, depending on soils, micro-relief, species composition, and disturbance dynamics. Approximately 140–200 years is required for an abandoned farm to attain the biomass values comparable to those of a mature forest. Recovery is thus five to seven times longer in the Upper Rio Negro than in other tropical areas in South America.*

INTRODUCTION

This study was conducted in a remote area of the Upper Rio Negro region of Colombia and Venezuela where the dominant vegetation is tropical rain forest. These forests are a mosaic of successional and mature forest stands interspersed with many abandoned farm sites of known history. Twenty-four stands were selected from *tierra firme* forests to represent a chronosequence of succession from abandoned farms to mature forests following slash-and-burn agricultural practices. The objectives of the study were to document changes in species composition, vegetation structure, and woody biomass. The study is the longest known quantitative record of successional development for the Amazon region.

METHODS

Study site

The research sites are situated near the confluence of the Canal del Casiquiare and the Rio Negro in Colombia and Venezuela (1° 56' N, 67° 03' W) at an elevation of 119 m. The climate is equatorial with 3500 mm of annual rainfall (Heuveldop 1980). There is an average of >200 mm of rainfall for each month of the year, except for May through July when the average is >400 mm.

The vegetation is classified as tropical, moist forest (Ewel and Madriz 1968), and the landscape is generally flat with smooth, rolling hills rising 25–50 m above the river. There are three main vegetation types. *Tierra firme* (non-flooded) forests occur on the sides and tops of hills, and these forests are the only ones that are being cut and burned for agriculture (Uhl and Murphy 1981). *Caatinga* is a non-flooded forest that occurs on white-sand podosols, and *igapo* is a seasonally flooded forest occurring on hydromorphic gley soils.

Site selection and plot establishment

Twenty-four sites of *tierra firme* forests were studied to determine changes in species composition, forest structure, and biomass following slash-and-burn agriculture. Twenty-three of the sites were between 9 years old and mature forests. In these sites, three 10 x 30 m plots were established (Saldarriaga *et al.* 1988). The other site has been studied since the farm was abandoned in 1981 with one plot 30 x 50 m (Uhl 1987). These stands were located on four mature forest sites and on 20 farms abandoned 7–80 years ago.

The age of fallows was established by interviewing long-time residents of the region and by taking into account significant historic events, e.g. the 'rubber boom', the migration of the people from San Carlos to the interior of the forests during the Funes administration, and the immigration of Europeans after the Second World War. Most of the selected sites were owned either by the informants or their relatives.

RESULTS

Species diversity and dominance

Forbs (particularly *Eupatorium cerasifolium* (Sch. Bip.) Baker and *Phyllanthus* sp.) and grasses (notably *Andropogon bicornis* L., *Panicum pilosum* Sw., and *Paspalum decumbens* Sw.) dominate the plots during the first year following abandonment. After the first year, there were 17 tree species in the plot (0.15 ha), and 12 of these were pioneer species. While pioneer species richness remained fixed for the first 5 years, primary forest species did accumulate slowly (Fig. 18.1). After 5 years, there were 35 tree species present and 24 of these were primary forest species. Most of these (79%) had established from seeds; the remainder were sprouts. Plant diversity indices (Simpson's and Shannon-Weaver) increased with time except for a slight decline in the 3rd year. This decline coincided with the peak of *Vismia* dominance.

In comparing 10-year-old fallows and mature forests, the number of species was found to be lower in the younger stands, ranging from 33 to 96 for trees ≥ 1 cm diameter at breast height (dbh) and from 5 to 30 for trees ≥ 10 cm dbh (values are totals calculated from three 10 x 30 m plots).

The Shannon-Weaver index ranged from 3.32 to 5.90 for successional stands and from 4.75 to 5.45 for mature forests. These data show little if any trend in species diversity from young to old stands. Simpson's index indicates that dominance is high for the 10-year-old stands, declines somewhat at year 20, and remains more or less constant thereafter.

The species were grouped according to their relative importance value (RIV). In 10-year-old-stands, two to five species accounted for 60% of the RIV. Two species, *Vismia japurensis* and *Bellucia grossularioides*, had the highest RIV in three of the four 10-year-old stands. In 20-year-old successional stands, four to eight species made up 60% of the RIV. Two species from early stages of succession, *Vismia japurensis* and *Vismia lauriformis*, were still among the dominant species at two of the four stands. *Humiria balsamifera* and *Eperua purpurea* were the dominant species in density and basal area in three of the four 30–40 year-old stands. *Humiria balsamifera* accounted for almost 40% of the RIV in stand XII. The RIVs for 60-year-old stands ranged from 13.97 to 35.90 for the dominant species. The most visibly dominant species was *Eperua purpurea*. The RIVs of the most dominant species for the 80-year-old stands ranged from 13.40 to 22.68. Three species, *Bellucia grossularioides*, *Alchornea* sp. and *Protium* sp. were among the dominant ones found in three of the four stands. Five to 12 species made up 60% of the RIVs for the mature forest. Two species, *Eperua purpurea* and *Swartzia schomburgkii*, had values > 18% for number of stems and basal area.

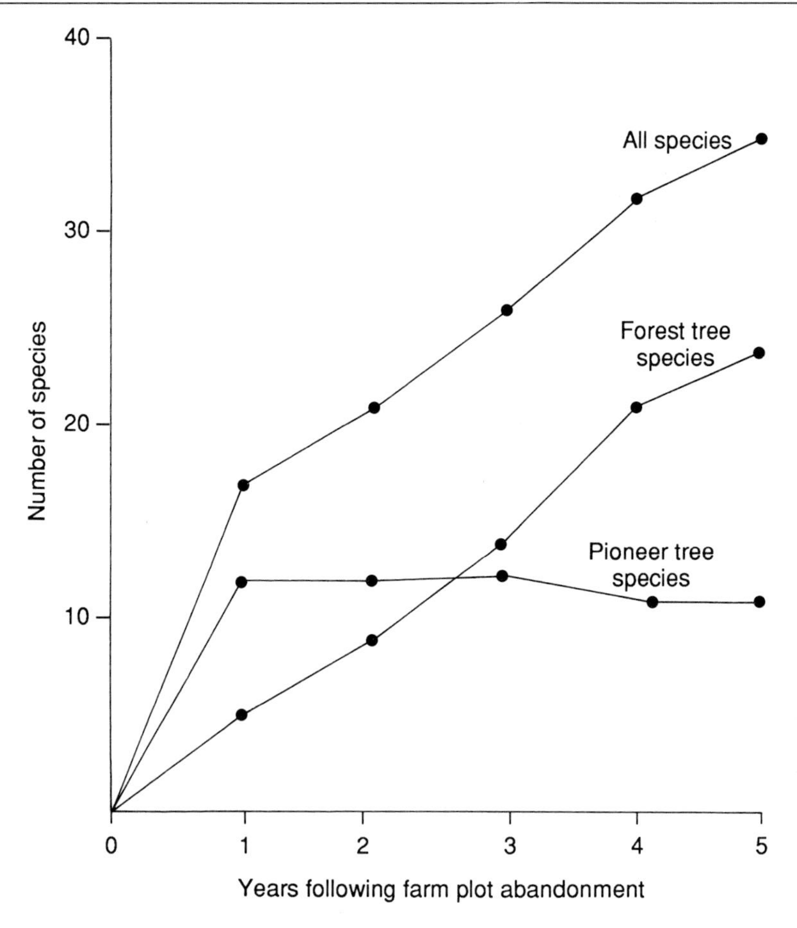

Fig. 18.1 The number of primary forest and pioneer tree species present in the 0.15-ha study plot near San Carlos de Rio Negro, Venezuela during the first 5 years of succession

Structure

Pioneer trees were widely spaced and generally 2–3 m tall after 1 year. During the 2nd and 3rd years, the pioneer, *Vismia*, grew to form a partially closed canopy at 8 m. Meanwhile, the grass-forb understorey was replaced by a dense layer of melastomataceous shrubs with a maximum height of 2 m. By the 5th year, there were 26 trees that were 10 cm or greater in dbh in the study plot (0.15 ha).

Tree density increased steadily during the first four years of succession and then declined in the 5th year (Fig. 18.2). *Vismia* represented over 50% of the stems for the first 5 years. With the decline of *Vismia*, other species,

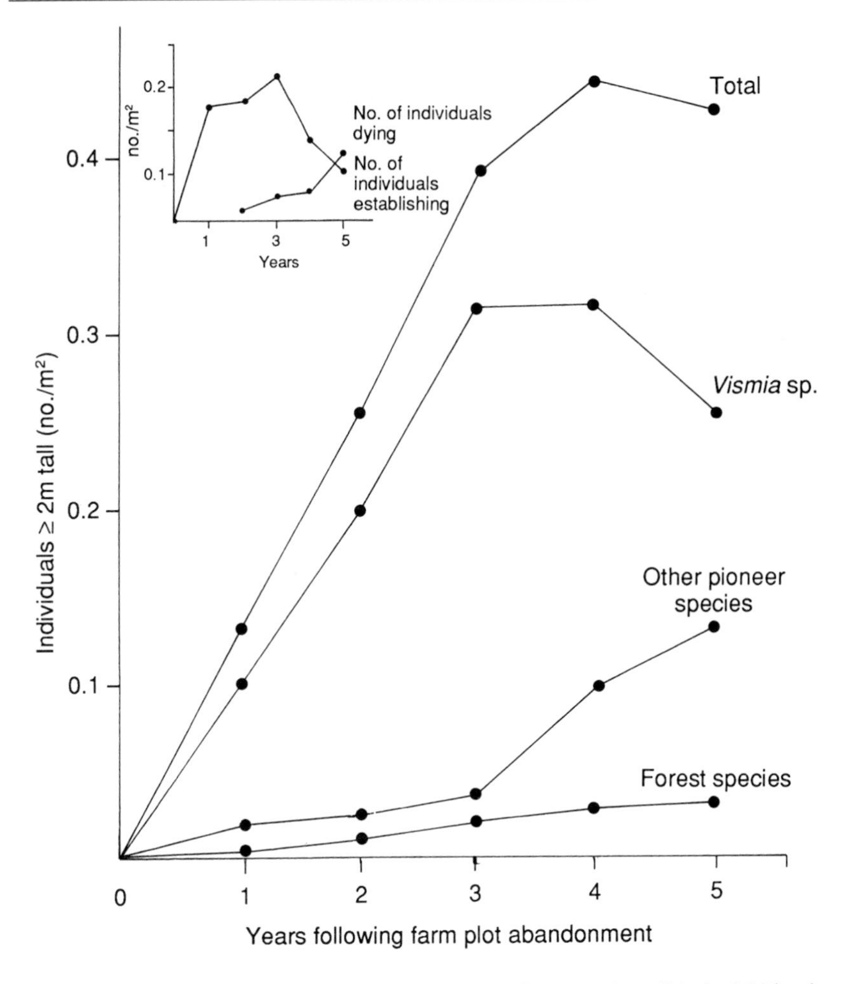

Fig. 18.2 Changes in density, establishment, and mortality for trees ≥ 2 m tall in the 0.15-ha plot near San Carlos de Rio Negro, Venezuela during the first 5 years of succession

such as *Bellucia grossularioides* (L.) Triana, *Miconia dispar* Bentham, and *M. myriantha* Bentham, began to establish in significant numbers, and, by the 5th year, these three species represented 29% of all stems.

After the first 9 years, the highest densities of trees were found in the two smallest dbh classes (dbh < 1 cm and 1 ≤ dbh ≤ 5 cm). The total number of stems, including sprouts, in these two classes ranged from 6776 to 20 180, representing 76–95% of the total number of stems on each stand. For trees 20–40 cm dbh, stems were concentrated in the old second-growth and mature forests. Trees with a dbh of ≥ 60 cm were found only in the four mature forest stands.

The number of stems was also grouped by height. The highest densities of trees were found among the three smallest height classes (1 to < 2 m, 2 to < 4 m, and 4 to 9 m height). The total number of trees, including sprouts, in these three groups, ranged from 7454 to 20 404, representing 84–95% of the total number of individuals on each stand. Trees taller than 26 m were missing in most of the successional stands except in two stands 60- and 80-year-old, but they were present in all mature forests.

The total basal area of living trees varied from 11.12 to 36.95 m^2 ha^{-1} for 9 year-old successional stands to mature forests. Trees from 5 to 20 cm dbh had large basal areas in some of the stands younger than 40 years and small basal areas in the 60-year-old group. Trees from 20 to 40 cm dbh were concentrated in the older successional stands and mature forests. Trees with a dbh larger than 60 cm were found only in the mature forests. Their basal area values ranged from 3.30 to 12.95 m^2 ha^{-1} and represented from 9 to 35% of the total basal area.

The number of trees with broken crowns varied from 0 to 150 ha^{-1} for stands 9 years and older. The largest numbers of damaged trees, 131 and 150, were found in two 80-year-old stands. In the mature forests, the number of damaged trees ranged from 12 to 43. Standing dead trees were found in the successional stands as well as in the mature forests. The largest number of dead trees was 82 and was found in an 80-year-old stand. Mature forests had 22 to 44 dead trees.

Biomass

Regression equations for estimating biomass of all trees measured in each stand were developed (i.e. Uhl 1987, for the first 5 years of succession, and Saldarriaga *et al.* 1988 for successional stands older than 9 years). During the first 5 years, above-ground biomass accumulated at an average of 6.77 t ha^{-1} yr^{-1}. After 5 years of succession, the total living biomass (including roots) was 38.1 t ha^{-1} which is equivalent to 12% of the biomass present at a near-mature forest. The average above-ground biomass ranged from 58 t ha^{-1} for trees in 10-year-old stands to 110 t ha^{-1} at 40-year-old stands. Biomass continues increasing to 150 t ha^{-1} at 60- to 80-year-old stands. The largest average value, 255 t ha^{-1}, was achieved by a mature forest (Fig. 18.3).

Above-ground biomass values for all successional stands were fitted to a linear and logarithmic regression to obtain coefficients for estimating the time required for an abandoned farm to attain values similar to those of a mature forest. Using the mean above-ground living biomass of 255 t ha^{-1} for a mature forest as the dependent variable, the time required for the above-ground biomass of a successional forest to attain the above-ground biomass of a mature forest is 144 years using the linear model, and 189 years using the logarithmic model (Fig. 18.4).

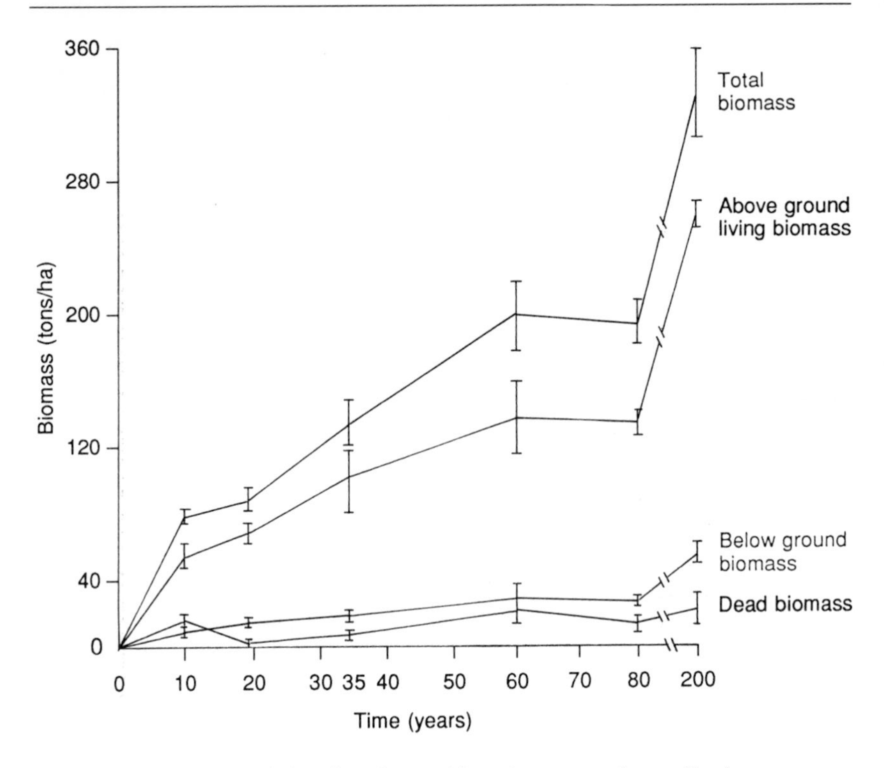

Fig. 18.3 Biomass accumulations from 9-year-old stands to mature forests. Numbers are means with one standard error, n = 4, except age class 60 with n = 3

Above-ground dead biomass was estimated using a logarithmic regression equation (Saldarriaga *et al.* 1988). Average dead biomass increased through succession, varying between 17 t ha^{-1} at 10-year-old stands to 26 t ha^{-1} at mature forests (Fig. 18.3). However, the lowest dead biomass value is 1 t ha^{-1} found at 20-year-old stands. Dead biomass is low on younger stands because of the size of the trees and because the soft-wooded species that dominate the early stages of succession decay quickly.

Root biomass data from six stands were used to develop regression equations for estimating root biomass for 23 stands (Saldarriaga *et al.* 1988). Below-ground root biomass values obtained in the field varied from 7 to 42 t ha^{-1}. The average below-ground biomass ranged from 10 to 30 t ha^{-1} for successional stands 10–80 years old. The maximum average value, 58 t ha^{-1}, was reached at mature forest stands (Fig. 18.3).

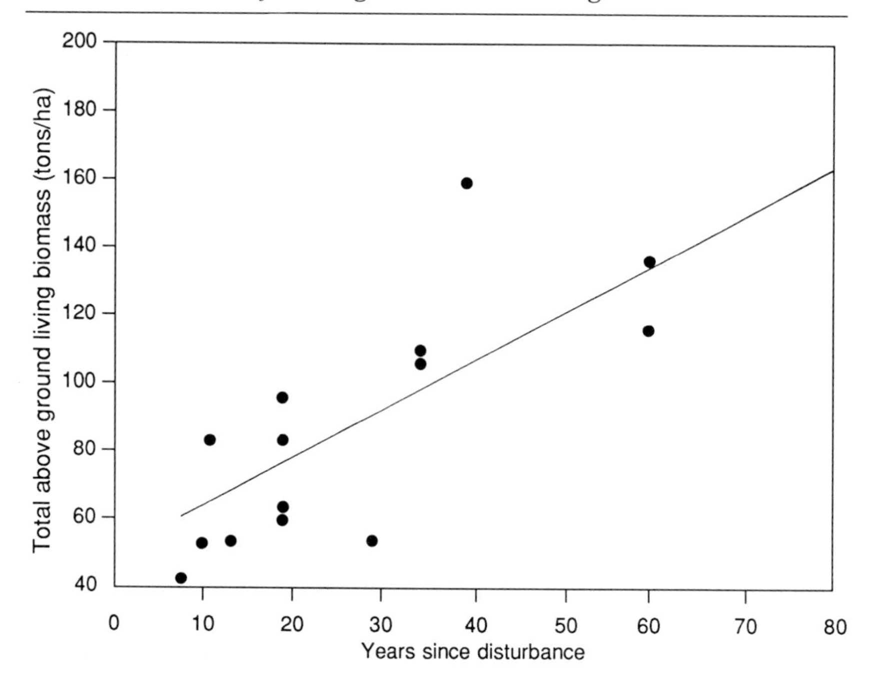

Fig. 18.4 Relationship between total above-ground living biomass and time in the Upper Rio Negro region of Colombia and Venezuela. The regression is as follows: $Y = 48.73 + 1.43X$ ($n = 19$, $r^2 = 0.67$, $p < 0.0001$, CV = 25.50%)

DISCUSSION

The early years of plant succession following slash-and-burn agriculture at San Carlos are characterized by the predominance of grasses and forbs, followed by a period of *Vismia japurensis* and *V. lauriformis* dominance. *Vismia* spp. usually persist for 10 to 20 years on abandoned farms, but, after 5 years, their mortality exceeds establishment. As *Vismia* spp. die, more long-lived, melastomataceous trees invade the site. This group appears to form a distinct third seral stage. Meanwhile, primary forest species slowly establish and grow in the shade cast by the pioneer canopy. The primary species will become an important part of the plant community in 50 years (Saldarriaga 1985).

The number of tree species increases from early successional stands to mature forests. The number of species from early stages decreases with stand age. During the first 10 to 20 years of succession, pioneer species are dominant, but a high percentage of these species die during this interval. At 30 to 40 years, the early successional species are replaced by groups of other fast-growing and more persistent species such as *Vochysia* sp., *Alchornea* sp., and *Jacardanda copaia*. These species become dominant in number of stems and basal area for the next 50 years, and, in some cases, may continue to be important in the mature forests. Major species changes take place in 40–80 year-old stands when significant events occur, such as senescence of previously dominant species, the death of subsequent falling trees, and a reduction in the pioneer species. The size of the gaps produced by the death of the previously dominant trees determines which species next occupy the site. For example, small gaps contain samplings of mature forest species, but large gaps contain pioneer species such as *Cecropia* sp. and *Vismia* spp.

In this study, the maximum number of tree species was attained in the four mature stands. Only a few species from early successional stages were found on these mature sites. These results agree with the hypothesis that species diversity increases through time (Budowski 1965, Margalef 1968, Odum 1969). Species diversity in these forests is not an indicator of stability.

These forests are characterized by numerous small trees. The distribution of the number of stems by dbh class is represented by a J-shaped curve, with trees often absent in the larger diameter classes. Similar stem distributions were observed in Amazon *tierra firme* forests at the Rio Xingu, Brazil (Campbell *et al.* 1986).

The mean mature forest basal area value of 34.8 m^2 ha^{-1} is similar to the 34.28 m^2 ha^{-1} reported by Jordan and Uhl (1978) for another *tierra firme* forest near San Carlos de Rio Negro. However, the basal area values found in the Upper Rio Negro region are considerably less than the 40–60 m^2 ha^{-1} values for lowland wet forests found in Costa Rica (Holdridge 1972).

Above-ground living biomass increases from 34 t ha^{-1} at 5 years to a mean value of 58 t ha^{-1} at 9–14 years. Above-ground living biomass continues increasing, reaching a mean of 150 t ha^{-1} in stands 60–80 years old. During this time above-ground biomass is 50–60% of the maximum which is reached in the mature forest.

In the first 5 years, dead biomass was greater than the net accumulation of living biomass (Uhl 1987). Mean dead biomass values decline from 97 t ha^{-1} at 3–4 years to 18 t ha^{-1} at 9–14 years. Dead biomass showed high variability at 20 years with the lowest value of 1.05 t ha^{-1} increasing to 23 t ha^{-1} at 60 years and remaining constant until 80 years.

Fallen trees begin to appear at 20 years. In stands 60–80 years old, gaps are common as a result of wind throw and senescence of relatively short-lived species that are canopy dominants. When tree-falls occur over a short period of time, living biomass can be reduced to levels of successional stands 30–40 years old or younger. By contrast, the stand will remain essentially unchanged when the canopy trees are gradually reduced because of senescence, suppression, or competition.

The rate of recovery of floristic composition, structure, and biomass following disturbance is relatively slow. The maximum species number for the 60–80 year-old successional stands approaches the value for a mature forest, whereas the basal area is *c*.60% of the maximum value at mature stages. Approximately 140–200 years is required for an abandoned farm to attain biomass comparable to a mature forest. The results of this study indicate that recovery is five to seven times longer in the Upper Rio Negro than in other tropical areas in South America.

REFERENCES

Budowski, G. (1965). Distribution of tropical American rain forest species in the light of successional processes. *Turrialba,* **15**, 40–2

Campbell, D.G., Daly, D.C. and Prance, G.T. (1986). Qualitative ecological inventory of tierra firme and Varzea tropical forest on the Rio Xingu, Brazilian Amazon. *Brittonia,* **38**, 369–93

Ewel, J. and Madriz, A. (1968). *Zonas de vida de Venezuela.* (Ministerio de Agricultura y Cria: Caracas)

Heuveldop, J. (1980). Bioklima von San Carlos de Rio Negro, Venezuela. *Amazoniana,* **7**, 7–17

Holdridge, L.R. (1972). *Forest Environments in Tropical Life Zones: A Pilot Study.* (Pergamon Press: New York)

Jordan, C.F. and Uhl, C.. (1978). Biomass of a "tierra firme" forest of the Amazon Basin. *Oecologia Plantarum,* **13**, 387–400

Margalef, D.A. (1968). *Perspectives in Ecological Theory.* (University of Chicago Press: Chicago)

Odum, E.P. 1969. The strategy of ecosystem development. *Science,* **164**, 262–70

Saldarriaga, J.G. (1985). *Forest Succession in the upper Rio Negro of Colombia and Venezuela.* Ph.D. Thesis. (University of Tennessee: Knoxville)

Saldarrriaga, J.G., West D.C., Tharp, M.L. and Uhl, C. (1988). Long-term chronosequence of forest succession in the upper Rio Negro of Colombia and Venezuela. *Journal of Ecology,* **76**, 938–58

Uhl, C. (1987). Factors controlling succession following slash-and-burn agriculture in Amazonia. *Journal of Ecology,* **75**, 377–407

Uhl, C. and Murphy, P.G. (1981). Composition, structure and regeneration of a *tierra firme* forest in the Amazon Basin of Venezuela. *Tropical Ecology,* **22**, 220–37

CHAPTER 19

REGENERATION FOLLOWING PULPWOOD LOGGING IN LOWLAND RAIN FOREST IN PAPUA NEW GUINEA

S. Saulei and D. Lamb

ABSTRACT

Clear-felling to produce pulpwood commenced in Papua New Guinea in 1973. The operation produces a patchwork of logged areas mostly of a maximum radius < 400 m. Regrowth is rapid and 10-year-old regrowth is c.20 m tall. Many of the early secondary species originate from seed stored in the topsoil. Vegetative regrowth is also common, especially amongst primary forest tree species. Eleven years after logging, 0.12 ha regrowth plots have about 60 tree species, some 60 % of which are secondary forest species. By comparison, unlogged plots of the same size have about 64 trees species of which 30% are secondary species. Early patterns of regeneration after logging suggest the forests are recovering both structurally and floristically. How long this will take is not known. Several implications for forest management can be suggested.

(1) There is considerable scope for improving the rate of forest recovery after logging.

(2) Logging methods that reduce the level of soil disturbance would improve early seedling density.

(3) Smaller logging areas would improve the rate of seed dispersal.

(4) More scattered residual trees would improve structural diversity.

INTRODUCTION

Papua New Guinea (PNG) is estimated to have 14 x 10^6 ha of accessible and commercially productive forest. On a simple per capita basis, it is better endowed than its neighbours in South-east Asia (for example Papua New Guinea has 583 m^3 ha^{-1} $capita^{-1}$ of timber while Indonesia has 97 m^3 ha^{-1} capita). Management of these forests for timber production has, however, been difficult. There are several reasons for this. First, little is known of the ecological requirements of the main commercial species. Second, the volumes per hectare of these species are usually small. Third, even commercial species have been difficult to sell internationally because they are comparatively unknown. Finally, overriding all these difficulties, is the fact that all the forest land of Papua New Guinea is owned by the traditional land-owning clans and not by the government so there is no dedicated national forest estate.

This situation has some considerable social advantages for the citizens of PNG but it and the other constraints has led the government to develop a two-pronged approached to forest development. The first approach was to exploit the natural forest by purchasing rights to harvest timber from the traditional owners. These rights were then allocated to a logging company which could harvest the merchantable species selectively. Following logging, the site would be returned to the traditional owners.

The second approach was to create a system of plantations of commercially valuable species strategically located near ports or centres of high density population. Land for these plantations had to be acquired from traditional owners, but the amounts of land were smaller than needed for a forest managed by some form of natural regeneration because of the greater productivity of plantations. Most landowners in Papua New Guinea are extremely unwilling to sell land, but, by the late 1960s, the government had made some progress in this direction.

In 1971, it seemed a third alternative had appeared when the Honshu Paper Company of Japan reached agreement with the then colonial administration of PNG to commence clear-felling of lowland rain forest to produce mixed-species wood pulp. From the outset of the project, the intention was that part of the logged area should be used to establish plantations of fast growing tree species suitable for wood pulp. The remainder of the land was to be used for various agricultural projects, or to be allowed to regenerate naturally back to forest. The project has been controversial, and further background is given in De'Ath (1980), Lamb (1977, 1980, 1990), Seddon (1984) and Webb (1977). We confine ourselves here to reviewing some aspects of the process of natural regeneration occurring on the clear-felled areas and to considering some management implications arising from the project.

THE SITE AND THE FOREST

The logging operation is centred on the Gogol Valley, which lies 30 km south west of Madang on the north coast of Papua New Guinea (5° 0' S, 145° 45' E). The two main logging areas (the Gogol and Naru blocks) together cover 66 000 ha and are made up of broad alluvial valleys separated by a series of low hills. In the valleys, the soils are loams and silts, but most soils are silty clays or clays in the hills. Little profile differentiation has occurred, and most soils are of at least moderate fertility. The climate in the area is typical of much of the lowlands of Papua New Guinea, with a mean monthly maximum temperature of 32° C and a mean monthly minimum temperature of 23° C with little seasonal variation. Rainfall totals 2985 mm, falling mainly between December and April. The driest months are July to September with *c.* 100 mm per month. There are reports that extended dry periods have occurred in the past, sufficient to have allowed forest fires to occur in parts of the Gogol Valley (Johns 1986).

The forests of the Gogol Valley are owned by 260 clans, and, at the time the logging operation commenced, the population was 2336. Prior to the logging, these people practised shifting cultivation. Subsequently, a number have obtained employment with the logging company and many have established cash crops on cleared land. None-the-less, shifting cultivation remains the main form of subsistence. The social diversity resulting from this pattern of land ownership means that any larger-scale land-use planning is extremely difficult.

Most of the Gogol Valley is covered by lowland rain forest, but there are areas of *Imperata cylindrica* grassland (3%) and secondary regrowth forest (16%). These have probably originated from past shifting cultivation, landslips, or perhaps from a drought induced wildfire. The forest is 30–40 m tall and has an estimated 100 m^3 ha^{-1} of timber. About 27 species make up 70% of this volume. There are probably over 140 tree species in the area (Lamb 1977).

LOGGING

Logging is carried out using crawler tractors. All tree species may be felled, except *Ficus* species which are unsuitable. Other species, especially fruit trees, may be retained if the owners mark them for retention. Each year, up to 3000 ha of forest may be logged, and it is anticipated that, when logging in the natural forest is completed, about 55% or 36 300 ha of the original 66 000 ha will have been felled. The unlogged area is made up of forest on steeper hills, small village reserves, larger 'benchmark' reserves, and grassland or secondary regrowth resulting from earlier disturbances.

The area felled each year is usually broken into a number of smaller logging areas, commonly seven or eight. These separate logging areas (Table 19.1) are commonly < 400 m radius, although several larger areas have also been cleared. Some areas are separated by large expanses of unlogged forest. In other cases separation is maintained by narrow (up to 60 m wide) buffer strips along streams. The overall effect of logging has thus been to leave a patchwork of cleared areas, mostly concentrated along the floodplain of the Gogol River and its tributaries. At least this was the pattern at the time when many of the studies reported in this account were carried out. In more recent years, logging has tended to be more aggregated in any one year and less care has been taken with buffer strips.

Table 19.1 Frequency of non-contiguous logging areas of various size classes in the Gogol Valley 1974–1981 (n = 62)

	Maximum distance within logging area to unlogged boundary (m)							
	< 200	201–300	301–400	401–500	501–600	601–700	701–800	> 800
Number	20	13	13	7	0	2	4	2
Percentage	32	11	21	11	0	3	6	3

Source: Lamb, unpublished data

REGENERATION

Regeneration after logging is rapid. By 3 months, most areas are covered by plant growth, and most of this growth is at least 1 m tall by 6 months. By 10 years, regrowth forests on well drained sites are usually about 20 m tall. The recovery of tree species richness has been examined in a sequence of 0.12 ha plots covering the first 11 years of succession. These show a gradual increase with time in the number of tree species present, reaching a total of 50–60 species at 11 years (Table 19.2). Altogether, the 10 regrowth plots contained 132 species. The majority (56%) of the species in the oldest regrowth plot were secondary species. By comparison, a similar sized plot in nearby unlogged forest had 64 tree species of which only 19 species (or 30%) were secondary species. These data suggest that floristic recovery is underway following logging, but that it is still incomplete.

Most of the initial revegetation appears to originate from seed stored in the topsoil, and from vegetative regrowth; there seems to be only a minor contribution from seedlings on the forest floor or other advanced growth. Soils in the Gogol Valley have seed banks containing some 410 seeds m^{-2} and 32 species, most of them secondary forest species. After logging, many

Table 19.2 Number of primary and secondary forest tree species in 0.12 ha plots in regrowth of various ages

	Age of regrowth (years)									
	1.5	2.5	3.5	4.8	5.8	6.5	7.8	8.8	9.8	10.8
Secondary species										
seedlings, vegetative	26	31	19	27	34	19	36	25	33	28
growth plus residual trees	1	4	4	1	3	1	3	11	1	6
Primary species										
seedlings, vegetative	0	0	1	0	11	2	5	8	5	6
growth plus residual trees	5	3	18	11	20	1	11	11	7	21
Total	32	38	42	39	68	23	55	55	46	61

Source: Saulei 1985

germinate and the size and diversity of the soil seed store declines. Thus, the undisturbed forest soil has 298 seeds of tree and shrub species prior to logging, but only 86 seeds of tree and shrub species were recorded 18 months after logging (Saulei 1985). Many of the species lost from the seed store are dominant members of the regrowth community.

Vegetative regrowth occurs from coppice and root suckers. Saulei (1984, 1985) has found about 50% of all species regenerating in the wet season could do so vegetatively. The proportion is slightly higher in the dry season. Vegetative regrowth is thus a powerful mechanism for maintaining species richness following logging. It is especially significant because most of the coppicing species are primary forest species rather than secondary forest species. Included amongst the coppicing species are the commercially important sawlog species *Intsia bijuga*, *I. palembanica* and *Pometia pinnata*, all of which form a significant proportion of the original forest (Lamb 1977).

Overall, most individual plants in the early post-logging regrowth are secondary species originating from the previous soil seed store, but vegetative regrowth ensures that primary forest species are also represented in the new succession.

The actual density of trees establishing on logged sites, whether from seed or from vegetative growth, is strongly affected by the extent to which the soil is disturbed during logging. Where little topsoil disturbance occurs, tree density is high, but the density declines where substantial disturbance occurs (Fig. 19.1).

With time, other species have been dispersed into the logged areas. As mentioned earlier, the logged areas form gaps of various sizes. In some cases, the dispersal distances are relatively short, but, in others, they are quite long. Many species have no particular dispersal mode other than,

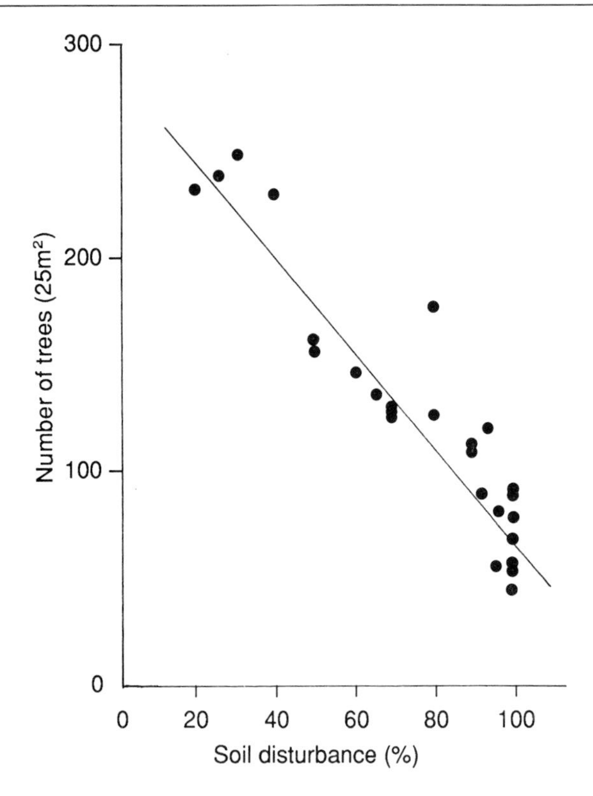

Fig. 19.1 Density of 2-year-old seedlings in 25 m² plots in relation to soil disturbance (% of area disturbed)

perhaps, gravity (Table 19.3). Such species will probably be under-represented in large gaps for some considerable time except where vegetative regrowth allows them to persist. There is also, however, a substantial proportion of species with seed or fruit dispersed by birds and bats (about 34%). These species are therefore potential colonists of the large gaps.

Seed dispersal by birds may, in fact, take some time to occur because many bird species normally responsible for seed dispersal in undisturbed forest are adversely affected by logging. The effect of clear-felling on birds with various feeding preferences is illustrated in Table 19.4. Frugivores, the bird group most important in seed dispersal, are amongst those most strongly affected by logging. Some of these may later return when further succession increases the structural complexity of the regrowth forests. In the meantime, it seems that scattered residual trees (many of which are *Ficus*) act as focal points for bird activity within logged areas, and it appears that seedlings of some bird-dispersed species may establish around these trees. There are at least four frugivorous bat species present within

Table 19.3 Putative dispersal modes of 141 tree species in the Gogol Valley

Percentage of total				
Wind	*Bird*	*Bat*	*Water*	*No special mode*
17	35	34	9	50

Source: Lamb, unpublished data

Table 19. 4 Number of common birds (i.e. seen perching on at least 10 occasions) with particular feeding preferences in the canopy of undisturbed forest and regrowth forest

	Number of species		
Bird feeding type	*Primarily in undisturbed forest*	*Primarily in regrowth forest* (4.5 years old)	*Both habitats*
Obligate frugivore	9	2	5
Frugivore	15	7	16
Nectivore	5	3	6
Granivore	4	0	1
Insectivore	8	8	11
Mixed (herbivore and insectivore)	5	2	10
Carnivore	0	0	2
Total	22	13	22

Source: Driscoll 1984

the Gogol Valley, but little is known of their role in dispersing seed within logged areas or, indeed, within the Gogol Valley in general.

NUTRIENT LOSSES

Pulpwood logging inevitably imposes a heavier nutrient drain on forest ecosystems than selective logging because more of the forest biomass is removed, and there is a greater potential for topsoil erosion and nutrient leaching following complete canopy removal. The amounts of nutrients removed in pulp logs from the Gogol Valley (Table 19.5) are small in comparison with the estimated nutrient reserves remaining in the upper soil profile, based on levels of 'available' nutrients and an arbitrarily chosen soil depth. The exception is phosphorus, for which the soil reserves are comparatively lower, especially in the hill forest sites. These soil reserves are an underestimate of the actual amounts of nutrients remaining in the ecosystem after logging because they do not include nutrients contained in

Table 19.5 Nutrients removed in logs (plus bark) in relation to estimated * soil nutrient reserves (0–50 cm)

	Nutrient content (kg ha^{-1})				
	N	*P*	*K*	*Ca*	*Mg*
Floodplain forest					
Logs	124	39	160	520	34
Soil	7460	45	1399	32365	11404
Hill forest					
Logs	108	33	140	452	29
Soil	8040	13	1657	44640	7811

* available phosphorus (using Olsen method), exchangeable cations (using ammonium acetate), total nitrogen (using Kjeldahl method). All plant analyses are total.

logging debris (e.g. leaves, branches). Thus, the actual reserves are greater than they might seem.

Some nutrients are probably lost from many sites by topsoil erosion. Some leaching may also occur. In both cases, the losses are likely to be small and short-lived. Revegetation soon prevents much erosion (Lamb and Beibi 1977) and the root systems sustained by coppicing, as well as the roots from new regrowth, probably intercept and immobilize much of the nutrients being released from decomposing organic matter. It therefore seems unlikely that the post-logging succession is limited by nutritional factors.

LONGER-TERM TRENDS

Evidence from the early patterns of regeneration after logging suggests that the forests are recovering both structurally and floristically. How long this recovery will take is not known. As already mentioned, there is evidence that many areas of the Gogol Valley have been disturbed in various ways in the past, and that much of the apparently undisturbed forest is, in fact, old regrowth forest. Saulei (1985) has compared one area thought to be 55 years with another where there was no evidence of disturbance. Total tree species richness was similar, although the regrowth forest still had a greater proportion of secondary species than the more mature forest. Structurally, the two forests were dissimilar, with the older one having slightly taller trees and a higher proportion of trees with larger girths. The recovery process clearly had still some way to go.

White (1975) has suggested that natural phenomena such as earthquakes, floods, volcanic eruptions, fires, etc. may effectively prevent much of the lowland forest in Papua New Guinea from maturing much beyond 150–200 years. It is unlikely that a recovery period as long as this will occur in the Gogol Valley. The logging road network means that previously isolated clans now have road access to markets, and there has been a gradual spread of cash crops since logging began. It is also the government's aim to establish a forest plantation on part of the area. The longer-term situation on the logged lands is therefore likely to be a heterogeneous mix of cash cropping, plantations, and various-aged regrowth forest. Some intact remnants of older forest will probably remain on steeper topography.

SOME MANAGEMENT IMPLICATIONS

Although, ecologically speaking, the post-logging succession is very young, there are several implications for forest managers in Papua New Guinea, and perhaps elsewhere, arising from the Gogol logging operation.

First, pulpwood logging reduces management flexibility. Although it has helped finance 'development' in the region and increased opportunities for agriculture and plantation forestry, it has, at the same time, reduced the scope for sawlog production and nature conservation.

Second, successive pulpwood rotations may be impractical. It has been suggested that there should be a second pulpwood logging after, say, 30 years, as is taking place at the Bajo Calima pulpwood logging operation in Colombia (Ladrach 1985). But apart from having a lower proportion of primary forest species in any resulting wood chip mix, and hence a lower pulp yield, the option is likely to cause a heavy nutrient drain from the site. When this loss of nutrients would begin to affect recovery processes and timber yields is difficult to say. There is little doubt, however, that few tropical forest soils could sustain such an operation for long.

Third, there is considerable scope for improving the rate of recovery of the forest after pulpwood logging. A logging method that reduced the level of soil disturbance would improve early seedling density. Ladrach (1985) reported excellent regeneration after clear-felling at Bajo Calima because extraction here is by a high-lead logging technique. Likewise, smaller sized logging areas would improve the rate of seed dispersal and more scattered residual trees would improve structural diversity of the logged areas, and hence benefit birds able to disperse seeds.

REFERENCES

De'Ath, C. (1980). *The Throwaway People: Social Impact of the Gogol Timber Project, Madang Province.* Monograph **13**. Institute of Applied Social and Economic Research, Boroko, Papua New Guinea.

Driscoll, P.V. (1984). *The Effects of Logging on Bird Populations in Lowland New Guinea Rain Forest.* Ph.D. Thesis. (University of Queensland: Brisbane)

Johns, R. (1986). The instability of the tropical ecosystem in New Guinea. *Blumea,* **31**, 341–71

Ladrach, W.E. (1985). *Forest Research in the Bajo Calima Concession: Ninth Annual Report.* (Celulosa y Papel de Columbia S.A.: Cali, Colombia)

Lamb, D. (1977). Conservation and management of tropical rain forest: a dilemma of development in Papua New Guinea. *Environmental Conservation,* **4**, 121–9

Lamb, D. (1980). Some ecological and social consequences of logging rain forests in Papua New Guinea. In Furtado, J.I. (ed.) *Tropical Ecology and Development. Proceedings of Fifth International Symposium of Tropical Ecology, Malaysia*, pp. 55–63. (International Society of Tropical Ecology: Kuala Lumpur)

Lamb, D. (1990) *Exploiting the Tropical Rain Forest. An Account of Pulpwood Logging in Papua New Guinea.* Man and the Biosphere Series 3. (Unesco: Paris and Parthenon Publishing: Carnforth)

Lamb, D. and Beibi, F. (1977). *The Effect of Logging Operations on Stream Turbidity Levels in the Rain Forests of the Gogol Valley*. Tropical Forest Research Note, **36**. (Dept. Forests: Papua New Guinea)

Saulei, S. (1984). Natural regeneration following clear-fell logging operations in the Gogol Valley, Papua New Guinea. *Ambio,* **13**, 5–6, 351–4

Saulei, S. (1985). *The Recovery of Tropical Lowland Rain Forest after Clear Fell Logging in the Gogol Valley, Papua New Guinea*. Ph.D. Thesis. (University of Aberdeen: Aberdeen)

Seddon, G. (1984). Logging in the Gogol Valley, Papua New Guinea. *Ambio,* **13**, 5–6, 345–50

Webb, L.J. (1977.) *Ecological Considerations and Safeguards in the Modern Use of Tropical Lowland Rain Forests as a Source of Pulpwood: Example, the Madang Area, PNG.* (Office of Environment and Conservation, Department of Natural Resources: Papua New Guinea)

White, K. (1975). *The Effect of Natural Phenomena on the Forest Environment.* Presidential Address to Papua New Guinea Scientific Society. 26 March 1975. (Department of Forests: Port Moresby)

CHAPTER 20

RAIN FOREST ECOSYSTEM FUNCTION AND ITS MANAGEMENT IN NORTH-EAST INDIA

P.S. Ramakrishnan

ABSTRACT

Shifting agriculture (known as jhum in India) is one of the most important factors in the conversion of tropical rain forests world-wide, and is the chief land-use of the hill regions of the north-eastern part of India. Jhum is based on sound scientific principles, though distortions that have crept in due to the shortened cycle have made this system untenable in its present form. Nonetheless, better understanding of the processes that operate in traditional ecological systems, such as jhum, can provide insights useful for improved land-use and resource management, including restoration of degraded sites. Such thinking has shaped a series of investigations on ecosystem dynamics in north-eastern India. Studies on secondary succession in this region have included measures of biomass and productivity, weed potential, allocation strategies of weeds (including changing patterns in C_3/C_4 strategies), tree growth strategies and leaf dynamics. Research findings have a number of implications for rain forest management, including the manipulation of species mixtures in time and space in order to 'condense' the successional sequence.

INTRODUCTION

The relationships between man and forests have always been closely interlinked in traditional societies. Nowhere is this as pronounced as in the tribal societies in north-east India (Ramakrishnan 1985*a*). Shifting agriculture (commonly called jhum) is the chief land-use of the hill regions of the north-east. This highly complex form of land-use involves mixed croppings of as many as 30 to 35 crop species, is extremely energy efficient, and is based on efficient recycling of resources. However, the rapid shortening of the jhum cycle (the length of the intervening fallow phase between two successive croppings from the same plot) due to increased population pressure and decreased land availability has resulted in the take-over of vast areas of forest land by weeds, often exotic, and has resulted in large scale 'desertification' of the landscape, in spite of high rainfall.

While we have done extensive studies on land-use and animal husbandry, relating these to the village ecosystem function under varied socio-economic and socio-cultural backgrounds of the tribal communities (Mishra and Ramakrishnan 1982), the present paper attempts to summarize the complex interactions between land-use in forested areas, animal husbandry and village functions as related to forest ecology and management in the humid tropics.

AGRICULTURE AND ANIMAL HUSBANDRY

On the basis of economic analysis of shifting agriculture under different cycles, we have shown that a 10-year jhum cycle is economically efficient (Table 20.1). If energy efficiency is considered to be a measure of ecological efficiency, shifting agriculture systems have higher output/input ratios compared to terrace cultivation (Table 20.2). Even though the energy under a 10 year cycle was not different from that under a 5-year cycle, the energy output was markedly higher for the former. When a correction factor of 1/30, 1/10 or 1/5 was used, a 10-year cycle had a higher energy output compared to other cycles. This comparison is significant because the energy input is largely in the form of labour, which is contributed by the family unit itself, at no monetary cost to the farmer. Valley cultivation, another land-use system, is both economically and ecologically efficient (Toky and Ramakrishnan 1981*a*, 1982), but is limited by topography.

Swine husbandry is (as elsewhere) an integral part of the jhum system (Mishra and Ramakrishnan 1982), mainly because of its low maintenance costs, and because it is based on efficient recycling of resources with a reasonable level of energy efficiency. Agricultural produce that is unfit for human consumption is used as animal feed in addition to crop residues and grazing (Mishra and Ramakrishnan 1982). This is an efficient way of storing excess food energy.

PERTURBATION EFFECTS

After slash-and-burn for jhum, the ecosystem functions are upset (Toky and Ramakrishnan 1981*b*). Apart from the loss of nutrients through ash blow-off,

Table 20.1 Monetary input-output (Rs ha^{-1} yr^{-1}) into jhum, terrace and valley agro-ecosystems (after Toky and Ramakrishnan 1981*a*)

	Jhum (yr)				*Valley*
	30	*10*	*5*	*Terrace*	*Crops I and II*
Input	2616	1830	896	2542 (4544)	4843
Output	5586	3354	1690	3658	5565
Net gain (loss)	2970	1524	794	1116 (-886)	722
Output/input	213	1.83	1.88	1.43 (0.80)	1.14

Table 20.2 Energy ratios in agricultural systems – jhum, terrace and valley cultivation (after Toky and Ramakrishnan 1982)

	Energy (MJ ha^{-1} yr^{-1})		
Agricultural system	*Input*	*Output*	*Output/input ratio*
Jhum – 30-yr cycle	1665	56766	34.1
Jhum – 10-yr cycle	1181	56601	47.9
Jhum – 5-yr cycle	510	23858	46.7
Terrace	6509 (8003)	43602	6.7 (5.4)
Valley – I and II crops	2843	50596	17.8

significant losses also occur due to changes in hydrology and related run-off and leaching. Carbon and nitrogen are lost through volatilization. We have shown that the losses are accelerated under short jhum cycles because of the poor physical quality of the soil. Further, the chemistry of the ash and the losses were related to the type of vegetation slashed and burnt under different jhum cycles.

The rapid decline in the losses through hydrology, even in the very first year of fallow regrowth, was related to the initial weed growth during secondary succession (Table 20.3). Indeed, weeds play an important conservation role even during the cropping phase itself, through the 'non-weed' concept as perceived by the jhum farmer (Ramakrishnan 1984). Thus, the jhum farmer does not do total weeding. He knows how intense the weeding should be so that the weed population does not interfere with the crops and yet the beneficial effects are manifest. Apart from providing cover to the soil and checking erosion losses, weeds contribute to recycling of nutrients because they are put back into the jhum plots (Table 20.4). Apart from the role of weeds in disease control and reducing pest attacks,

Table 20.3 Nutrient losses (kg ha^{-1} yr^{-1}) through run-off and percolation water under 10-yr jhum cycle agroecosystem and a 5-yr fallow (after Toky and Ramakrishnan 1981*b*, Mishra and Ramakrishnan 1983*a*)

	10-year cycle jhum plot		*5-year fallow*	
	Run-off	*Percolation*	*Run-off*	*Percolation*
Low elevation jhum:				
Nitrate nitrogen	4.2	10.7	0.8	1.1
Available phosphate	1.3	0.1	0.1	0.02
Potassium	91.2	21.1	0.9	0.5
High elevation jhum:				
Nitrate nitrogen	1.7	0.5	1.0	0.9
Available phosphate	0.9	0.1	N.D	N.D
Potassium	80.1	25.8	19.6	N.D

N.D – not detectable

Table 20.4 Use pattern of biomass by the Garos at lower elevations in Meghalaya (after Swamy and Ramakrishnan, unpublished)

Jhum cycle (years)	*Weeding*	*Total weed biomass produced* (kg ha^{-1})	*Weed biomass ploughed back* (kg ha^{-1})	*Weed biomass retained* (kg ha^{-1})
20	1	9357	7110 (76.0)	2247 (24.0)
	2	10180	8257 (81.1)	1923 (18.9)
10	1	8085	6189 (76.5)	1896 (23.5)
	2	8608	6930 (80.5)	1678 (19.5)
5	1	4378	3160 (72.2)	1218 (27.8)
	2	5780	4815 (83.3)	965 (16.7)
	3	4242	3592 (84.7)	650 (15.3)

Percentage values given in parentheses

many are used as food. The skill of the jhum farmer lies in his ability to distinguish the 'weed' state from the 'non-weed' state.

SECONDARY SUCCESSION

Biomass and productivity

When the jhum cycle is longer than 10 years, the natural process of secondary succession operates, and we have considered the structural attributes of rain forest development (Toky and Ramakrishnan 1983*a*). An initial weed stage up to 5–6 years of fallow regrowth, is replaced by bamboo (*Dendrocalamus hamiltonii*). Beyond 25 years, a mixed broadleaved forest develops, replacing bamboo. The succession is accompanied by increased species diversity (Fig. 20.1), reduced dominance and increased above ground net primary productivity which reached 1.8 kg m^{-2} yr^{-1} in a 20 year-old fallow, with similar patterns for above ground biomass and litterfall, with 15 kg m^{-2} and 1 kg m^{-2} respectively in a 20-year fallow.

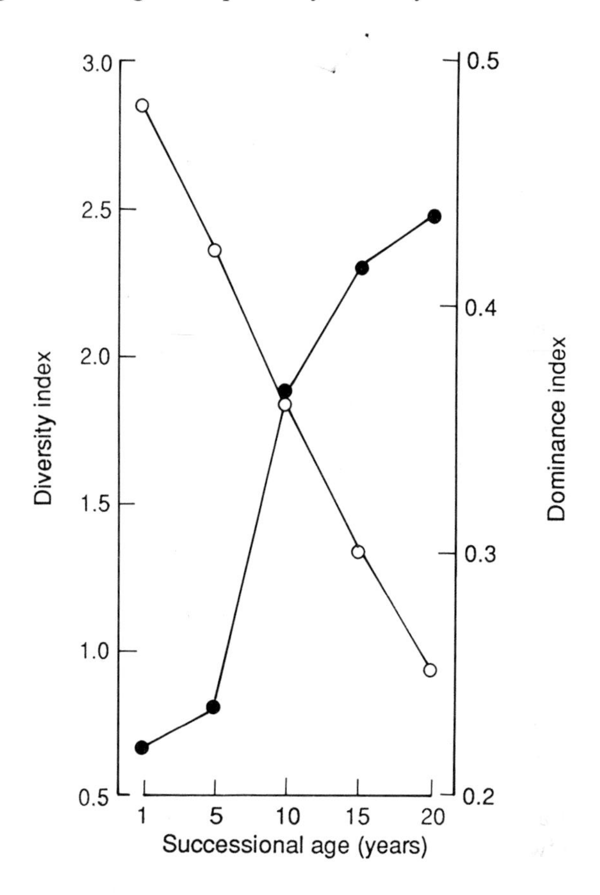

Fig. 20.1 Species diversity (•) and species dominance (○) in successional communities up to 20 years in Meghalaya, north-eastern India (after Toky and Ramakrishnan 1983*a*)

Weed potential

Imposition of a short jhum of 4–5 years, which we have studied through extensive demographic analysis of the weed population during the cropping phase and in the early successional environments, results in increased weed potential of the site (Saxena and Ramakrishnan 1984*a*). Fire adaptation of weeds is important. Thus *Imperata cylindrica*, which normally reproduces through an extensive below ground rhizome, resorts to seed reproduction only under constant disturbance such as through fire. Through a series of studies (Saxena and Ramakrishnan 1982, 1983) on the biomass and allocation strategies of weeds, we have apportioned the relative importance of sexual versus vegetative reproduction of both annual and perennial weeds. We have also followed the changing patterns in the C_3/C_4 strategies of herbaceous communities through succession (Saxena and Ramakrishnan 1984b). C_4 populations are generally commoner in the early phase of succession. In a highly heterogeneous early successional environment, C_4 species, which have a high nutrient-use efficiency, occupy nutrient-poor microsites, whereas C_3 species occur in nutrient-rich sites. Such a micro-distribution pattern permits coexistence of species through mutual avoidance.

Soil fertility and nutrient cycling

Depletion in soil carbon, nitrogen and phosphorus occurs during the cropping phase, and generally continues up to a 5-year-old fallow period. Since *Dendrocalamus hamiltonii* is an accumulator of potassium, depletion of this element in the soil continues up to about 20 years of fallow regrowth. Similarly, the low soil calcium levels in a 50-year-old fallow is related to its storage in the trunks and branches of dicotyledonous trees (Ramakrishnan and Toky 1981). Nutrient recovery in the soil normally starts with a 10-year fallow and continues in older fallows, except for variations for phosphorus (Fig. 20.2) and calcium.

In the above-ground living biomass, nutrients increase linearly up to 20 years (Toky and Ramakrishnan 1983*b*). The annual rate of accumulation is maximal after 10–20 years for nitrogen, 10–15 years for potassium, and during the first 5 years for phosphorus. The enrichment quotient (the weight of an element in the vegetation divided by its rate of uptake) is higher for phosphorus and potassium than for nitrogen, calcium, and magnesium, indicating the rapid accumulation of the former two elements in the standing biomass. This is because, during the early phases of succession, herbaceous species rapidly take up phosphorus and bamboo (*Dendrocalamus hamiltonii*) rapidly accumulates potassium. The annual return of elements through litterfall increases with the age of the fallow during the first 20 years.

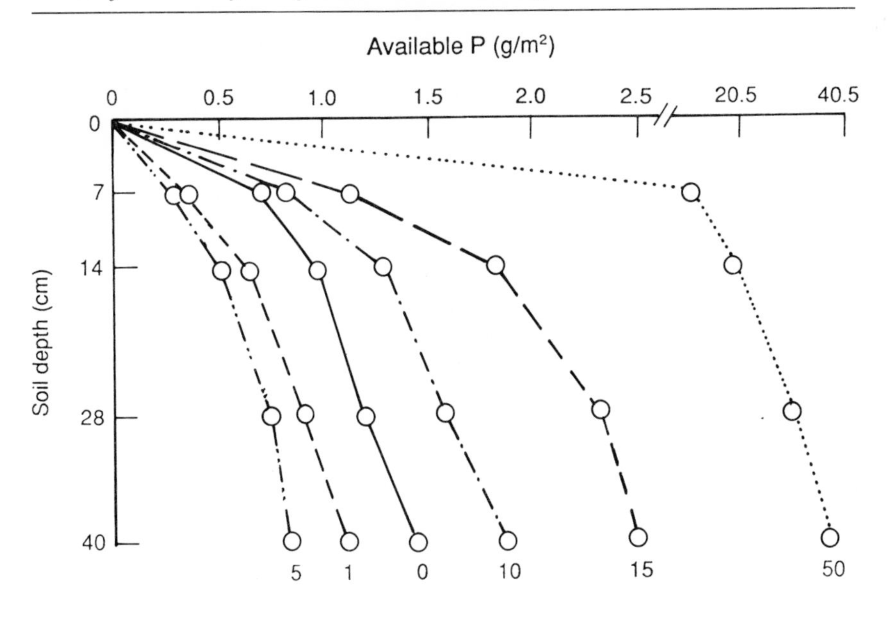

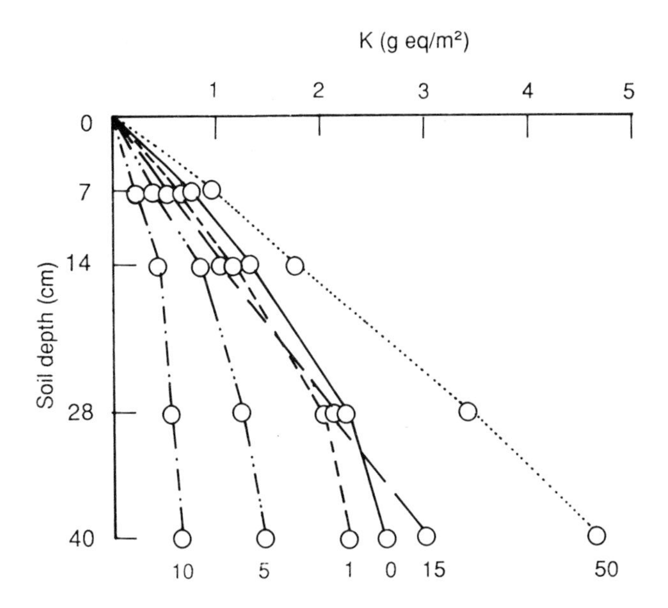

Fig. 20.2 Changes in cumulative quantity of available phosphorus (top) and potassium (bottom) within a soil column of 40 cm depth after fallows of various ages (after Ramakrishnan and Toky 1981)

TREE GROWTH STRATEGIES AND FOREST MANAGEMENT

Tree architecture

Tree architecture based on morphometric analysis of trees was related to the adaptive strategy over a successional gradient (Boojh and Ramakrishnan 1982*a*, 1982*b*, Shukla and Ramakrishnan 1984*a*, 1984*b*, 1986). The early successional trees, with faster extension and radial growth over a longer period and sylleptic branch production, contrast with the slower extension growth and proleptic branch production of late successionals. Self-pruning of the first order branches is better developed in the early than late successional species, resulting in rapid upward growth of the canopy and the greater extension of the first order branches compared with second order branches. The result is an excurrent crown form with a multilayered canopy and little overlapping of leaves. By comparison, increased production and extension of second order branches in late successional trees results in the development of a decurrent crown form where the leaves are placed more peripherally. The multilayered leaf placement of early successional species is further facilitated by less forked axes and increase in first order branch angle at lower canopy positions, in contrast to late successional species.

Leaf dynamics

The early successional species have higher leaf turnover rates, reduced leaf longevity, more uniform production and fall of leaves, and a larger leaf population of younger age classes, compared to late successional species. All these differential attributes enable early successional species to exploit high light regimes as opposed to a conservative strategy of late successional species.

Rain forest management

For a mixed plantation forestry programme, the choice of species mixtures is important. Architectural designs of the canopy that could exploit high light regimes would complement designs that could make adequate growth under the lower light regimes of the fast growing light demanders. Similarly, root architectural design should be such that the tree species that could exploit upper layers of the soil would avoid competition with those that would exploit the deeper soil layers. Early successional tree species that are opportunistic ones, with an 'exploitive' growth strategy, would be complementary to late successional trees with a 'conservative' growth strategy (Ramakrishnan *et al.* 1982).

Early and late successional shrubs follow similar differential growth strategies to the tree species (U. Baruah, and P.S. Ramakrishnan, unpublished), so both shrubs and trees could be used for a well conceived reconstruction of vegetation of degraded sites. By manipulation of species in the mixture in time and space, it is possible to reduce the time span of the normal process of succession from a few hundred years to a few decades.

DEVELOPMENT STRATEGIES

Why not terracing?

The significance of jhum as a land-use in the humid tropics cannot be appreciated without a brief mention of the alternatives for the tribal farmer. Terracing for sedentary agriculture is one alternative that has failed to find his acceptance (Ramakrishnan 1985*a*). Apart from the fact that infiltration losses – particularly of nitrogen and phosphorus – are considerable (Mishra and Ramakrishnan 1983*a*), the fertility depletion is heavy, as observed for a 2nd year of terrace cultivation (Mishra and Ramakrishnan 1983*b*). In fact, both physical and chemical qualities of the soil are so badly altered that the farmer has often to leave the terrace plots after 6–8 years of continuous cropping, as the land is totally degraded by then. Under high monsoonic rainfall conditions, maintenance costs for terraces are heavy. In addition, high inputs of inorganic fertilizers that are difficult to obtain make this option very expensive. Under constant cultivation of terrace plots, weed potential is exaggerated. To top it all, social considerations related to land tenure do not permit terracing, as the land belongs to the community rather than to individuals.

Is jhum unavoidable?

There are many reasons to suggest that jhum is an advanced form of agro-ecosystem very closely adapted to the humid tropics (Ramakrishnan 1984). Considering that a 10-year cycle is the minimum length for jhum, both ecologically and economically, it is imperative that the pressure on the land for jhum is taken off by alternative land-use development. One of the chief reasons why all alternative land-use strategies have failed is because man has not been able to come up with any other technology that could effectively replace the efficient way in which the natural process of secondary forest succession is able to build up soil fertility. Apart from the fact that a minimum of 10 years is required after cropping for fertility to recover, this minimum cycle also helps in the reduction of weed intensity under jhum. After 5–6 years of secondary succession, weeds are suppressed by larger shrubs and trees through reduction in light availability at the ground level (Kushwaha *et al.* 1981, Ramakrishnan and Mishra 1981).

Integrated rural development and forest management

With greater emphasis on valley cultivation by introducing more modern agricultural practices, including improved varieties of crops of shorter duration of growth, one could have more crop harvests in a year from the same site. Further reduction of pressure on land for jhum could be effected through plantation/horticultural economy developed co-operatively. Apart from a wide variety of tropical and temperate fruit trees, the region is also good for coffee, tea and rubber.

Jhum itself could be made to give better returns by varying the species composition in the crop mixture. Emphasis on potatoes at higher elevations, as compared to rice at lower elevations, has resulted in a many-fold increase in economic yield (Table 20.5). This increase is in spite of low soil fertility in the more acid soils at higher elevations.

Having achieved a jhum cycle of 10 years or more, and even under shorter cycles, innovation could be made in the jhum system itself. Nitrogen fixing species could be introduced during both cropping and fallow phases, e.g. *Alnus nepalensis*. By having a shelterbelt of forest and fruit trees along the jhum plot boundary, it should be possible to check some of the losses from the jhum system through wind-blow of ash after the slash-and-burn, and also check losses through hydrology (Toky and Ramakrishnan 1981*b*). Agro-forestry and social forestry systems within the village boundaries would meet the food, fodder and fuelwood needs of the rural community. Fast growing early successional species are particularly important.

Table 20.5 Monetary input-output analysis (Rs ha^{-1} yr^{-1}) of jhum under a 10-yr cycle at lower and higher elevations of Meghalaya (after Toky and Ramakrishnan 1981*a*, Mishra and Ramakrishnan 1981)

	Low elevation jhum	*High elevation jhum*
Input	1830	3842
Output	3354	14171
Net gain	1524	10329
Output/input	1.83	3.9

Animal husbandry systems, particularly swine husbandry and poultry developed on modern lines with improved breeds, could not only meet the protein needs of the people in the region, but provide better monetary returns through export outside the village boundaries. These innovations could be combined with rural technology for semi-processing of some of the products before export. Conservation of resources could also be effected by promoting more efficient energy use, such as through improved cooking stoves or through biogas technology.

Such an integrated approach to development villages would take the pressure off the land under forest outside the village boundary. Not only would existing forests be protected through such a management strategy, but forests regenerated through 'condensed succession' in degraded lands would also be conserved. Therefore, rain forest management demands linkage of forest management to village development. Such an ecologically based development plan has wider applications for similar areas elsewhere in the developing world (Ramakrishnan 1985*b*).

The major deficiency to date in the developmental strategy design for the tribal peoples of the humid tropics of north-eastern India has been that scientists, planners and administrators have tried to impose, from outside, a value system that they considered to be good for the people, without trying to understand the processes that operate in the traditional ecological systems. Any future development strategy should be based on an understanding of traditional society; if it is based on a value system which people can appreciate and identify with, that would ensure people's participation and acceptance. Nowhere else is natural forest ecosystem management so closely interlinked with rural development as in the tropics.

REFERENCES

Boojh, R. and Ramakrishnan, P.S. (1982*a*). Growth strategy of trees related to successional status. I. Architecture and extension growth. *Forest Ecology and Management,* **4**, 355–74

Boojh, R. and Ramakrishnan, P.S. (1982*b*). Growth strategy of trees related to successional status. II. Leaf dynamics. *Forest Ecology and Management,* **4**, 375–86

Kushwaha, S.P.S., Ramakrishnan, P.S. and Tripathi, R.S. (1981). Population dynamics of *Eupatorium odoratum* in successional environments following slash and burn agriculture. *Journal of Applied Ecology,* **18**, 529–35

Mishra, B.K. and Ramakrishnan, P.S. (1981). The economic yield and energy efficiency of hill agroecosystem at higher elevations of Meghalaya in northeastern India. *Acta Oecologica/Oecologia Applicata,* **2**, 369–89

Mishra, B.K. and Ramakrishnan, P.S. (1982). Energy flow through a village ecosystem with slash and burn agriculture in northeastern India. *Agricultural Systems,* **9**, 57–72

Mishra, B.K. and Ramakrishnan, P.S. (1983*a*). Secondary succession subsequent to slash and burn agriculture at higher elevations of northeastern India. I. Species diversity, biomass and litter production. *Acta Oecologica/Oecologia Applicata,* **4**, 97–107

Mishra, B.K. and Ramakrishnan, P.S. (1983*b*). Slash and burn agriculture at higher elevation in northeastern India. II. Soil fertility changes. *Agriculture, Ecosystem and Environment,* **9**, 83–96

Ramakrishnan, P.S. (1984). The science behind rotational bush-fallow agriculture systems (Jhum).

Proceedings of the Indian Academy of Sciences (*Plant Science*), **93**, 400
Ramakrishnan, P.S. (1985*a*). Tribal man in the humid tropics of the northeast. *Man in India,* **65**, 1–32
Ramakrishnan, P.S. (1985*b*). Jhum cultivation – Prospects of developing countries. *Science, Technology and Development* (*COSTED*), **9**, 1–3
Ramakrishnan, P.S. and Mishra, B.K. (1981). Population dynamics of *Eupatorium adenophorum* Spreng. during secondary succession after slash and burn agriculture (Jhum) in northeastern India. *Weed Research,* **22**, 77–84
Ramakrishnan, P.S. and Toky, O.P. (1981). Soil nutrient status of hill agro-systems and recovery pattern after slash and burn agriculture (Jhum) in northeastern India. *Plant and Soil,* **60**, 41–64
Ramakrishnan, P.S., Shukla, R.P. and Boojh, R. (1982). Growth strategies of trees and their application to forest management. *Current Science,* **51**, 448–55
Saxena, K.G. and Ramakrishnan, P.S. (1982). Reproductive efficiency of secondary successional herbaceous populations subsequent to slash and burn of sub-tropical humid forests in northeastern India. *Proceedings of the Indian Academy of Sciences (Plant Science),* **91**, 61–8
Saxena, K.G. and Ramakrishnan, P.S. (1983). Growth and allocation strategies of some perennial weeds of slash and burn agriculture (Jhum) in northeastern India. *Canadian Journal of Botany,* **61**, 1300–6
Saxena, K.G. and Ramakrishnan, P.S. (1984*a*). Herbaceous vegetation development and weed potential in slash and burn agriculture (Jhum) in northeastern India. *Weed Research,* **24**, 135–42
Saxena, K.G. and Ramakrishnan, P.S. (1984*b*). C_3/C_4 species distributions among successional herbs following slash and burn in northeastern India. *Acta Oecologia/Oecologia Plantaria,* **5**, 335–46
Shukla, R.P. and Ramakrishnan, P.S. (1984*a*). Leaf dynamics of tropical trees related to successional status. *New Phytologist,* **97**, 697–706
Shukla, R.P. and Ramakrishnan, P.S. (1984*b*). Biomass allocation strategies and productivity of tropical trees related to successional status. *Forest Ecology and Management,* **9**, 315–24
Shukla, R.P. and Ramakrishnan, P.S. (1986). Architecture and growth strategies of tropical trees in relation to successional status. *Journal of Ecology,* **74**, 33–46
Toky, O.P. and Ramakrishnan, P.S. (1981*a*). Cropping and yields in agricultural systems of the northeastern hill region of India. *Agro-Ecosystems,* **7**, 11–25
Toky, O.P. and Ramakrishnan, P.S. (1981*b*). Run-off and infiltration losses related to shifting agriculture in northeastern India. *Environmental Conservation,* **8**, 313–21
Toky, O.P. and Ramakrishnan, P.S. (1982). A comparative study of the energy budget of hill agro-ecosystems with emphasis on the slash and burn system (Jhum) at lower elevations of northeastern India. *Agricultural Systems,* **9**, 143–54
Toky, O.P. and Ramakrishnan, P.S. (1983*a*). Secondary succession following slash and burn agriculture in northeastern India. I. Biomass, litterfall and productivity. *Journal of Ecology,* **71**, 735–45
Toky, O.P. and Ramakrishnan, P.S. (1983*b*). Secondary succession following slash and burn agriculture in northeastern India. II. Nutrient cycling. *Journal of Ecology,* **71**, 747–57

CHAPTER 21

LEARNING FROM TRADITIONAL ECOLOGICAL KNOWLEDGE: INSIGHTS FROM MAYAN SILVICULTURE

A. Gómez-Pompa

ABSTRACT

The presence of tropical forests rich in useful trees as dominant species in the Maya area is used as the starting point to reconstruct an hypothetical silvicultural system that the ancient Maya people had for the managing and construction of different types of man-made ecosystems. The reconstruction is based on a series of isolated silvicultural and agricultural techniques that the present-day Maya use in different regions of the area. The ancient Mayan silviculture raises some serious questions about recent trends in the use of the land and resources in tropical regions, and suggests a new research approach for conservation and development that may help to improve the management of forest resources in some tropical lowland areas for the benefit of the local people.

THE MAYAN CULTURE AND USE OF NATURAL RESOURCES

The Mayan culture was one of the few successful cultures that developed and flourished in a tropical forest environment. Much is known about the accomplishments of the Maya (e.g. Coe 1984, Hammond 1982). Though there remain many unresolved problems related to the subsistence systems, land-use and conservation practices of the Maya, some advances have been made in recent years (e.g. Flannery 1982, Gómez-Pompa and Golley 1981, Pohl 1985; see also Gómez-Pompa 1987 for an expanded description with literature sources). It is now known that the Maya had a very efficient and complex shifting agricultural system that many of their descendants still practise very skilfully. We know that the ancient Maya used terraces very extensively, and that they had extensive hydraulic systems with channels and raised fields. They probably used intensive agriculture similar to the surviving chinampas system of the valley of Mexico. We know that the present-day Maya have very rich and diverse forest gardens, and we assume that they were also extensive in the past.

The Maya knew and used a great diversity of wild plants (Sosa *et al.* 1985). A deep knowledge of the native fauna is suspected from the richness of names applied to that fauna (Hartig 1979). The Maya had, and still have, a profound knowledge of their soils. The classification they made is much better than any known classification of soils for the area (Beltrán 1959). Their decisions on management were based on soil attributes, a method still followed by many present-day farmers. The same applies to their knowledge of vegetation. Their classification of vegetation is based on "ecological" insights from the successional process, particularly from the age of the fallow (Fig. 21.1). It also draws on knowledge of the past management of vegetation (through species indicators), and on the agricultural potentiality of the site based on past yields and soil type (Flores and Ucan Ek 1983).

We believe that the old Maya managed their forest ecosystems (Barrera *et al.* 1977), but we know almost nothing about how they did it. This subject is of great interest because today we are still struggling to find suitable methods for managing tropical forests. Different opinions have been expressed concerning the old Mayan forest management and conservation methods. On the one hand, it has been claimed that the Maya destroyed their forests, and it has even been suggested that the classic collapse of their civilization was caused by the loss of soils produced by shifting agriculture, deforestation, erosion, and siltation of lakes. In contrast, other findings suggest that the Maya protected and probably managed their forests as sources of many plant and animal products. Botanists exploring the region many years ago noted the abundance of useful tropical trees in the Mayan ruins (Lundell 1937) and suggested that the old Maya had something to do with such abundance.

Moreover, there is evidence that one tree species, the *Ramón* or *osh* tree (*Brosimum alicastrum*) may have played a central role in Mayan subsistence (Puleston 1982, Peters 1983) as a complement to, or substitute for, corn, especially in dry years. This species is still used and is widely cultivated by the Maya at the present time. The use of the *osh* tree is not a unique feature of the Maya. The species has been found to be widely used for similar purposes in several other tropical cultures in Mexico and Central America (see also contribution by Peters in this volume). A closely related species, *Brosimum utile*, with its seven varieties, is widely used in northern South America (Berg 1972). In addition to the *osh* tree, the Maya had a great variety of other tropical fruits, including *Acrocomia mexicana*, *Casimiroa edulis* and *Theobroma cacao* (see Gómez-Pompa 1987 for a more complete listing).

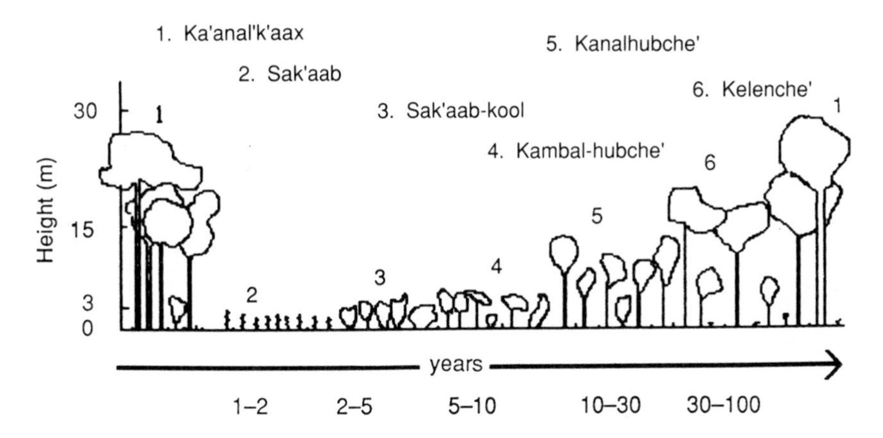

Fig. 21.1 Vegetation classification based on successional process.1. Ka'anal'k'aax. Old tropical forest (30 or more years old). 2. Sak'aab (or Sak'ab). 2nd year *milpa*. 3. Sak'aab-kool. Recently abandoned *milpa*. Early succession. 3–6. Hubche'. Secondary vegetation. 4. Kambal-hubche'. 5–10-year-old succession. 5. Kanalhubche'. 10–15-year-old succession. 6. Kelenche'. 15–30-year-old succession.

It is assumed that all of these trees were present in the native flora. These species were, and are (until recently), very abundant in the different natural ecosystems of the region, and many of them are dominants in a number of communities. If this is indeed so, then it would seem to refute the assertion that the old Maya cut down all these valuable forests to plant the annual crops used in shifting (*milpa*) cultivation. I do not think that was the case, and evidence is available that supports the idea that the Maya probably not only protected these forests but planted them for their future use (Gómez-Pompa *et al.*1987).

SILVICULTURAL METHODS, NATURAL FORESTS AND CONSERVATION

The question is "How did they do it?" From a series of isolated clues that have been accumulating in the past years of research on the old and modern Maya, a new concept of the Mayan silviculture has emerged. This Mayan silviculture has a series of methods and techniques, many of which still exist and are practised in different parts of the area. These isolated techniques (Table 21.1) are not all practised in any one place, but all occurred in the Maya area, and, for this reason, I assume that they were integrated at times in the past. They explain the presence of useful "natural" forests in the zone and their role in Mayan subsistence.

Table 21.1 Silvicultural techniques of the Maya (after Gómez-Pompa 1987)

Cenotes
Introduction of useful trees

Dooryard (Orchard) gardens
Germination of seeds in *caanchés* (elevated soil beds constructed with wood and organic soil)
Tree planting, contributing to greater abundance and diversity of tree species in gardens and producing shade, firewood, useful flowers, fruits, seeds, and green forage. Many of the most common trees are the same species found in "natural" vegetation (such as *Brosimum*, *Manilkara*, *Calocarpum*, *Cordia*, *Sabal*, etc.). In addition, introductions such as lemon, orange and other citrus fruits.

"Natural" forest ecosystems
Conservation of forest patches, selection of useful trees
Introduction of useful species

"Pet Kot"
Circular wall of stones, within which patches of useful plants concentrated
Selection of forested sites
Selection and protection of useful wild trees
Introduction of useful trees

Raised fields
Trees in borders of fields
Tree plantations (*cacao?*)

Trees in shifting agriculture
Selection and protection of trees on site chosen for shifting cultivation (best individuals protected, remain standing). Selection for usefulness probably included many properties in addition to food values (e.g. hardness of wood, toxicity of bark and wood). Religion also played a role in selection and protection of tree species (e.g. *Ceiba* species)
Coppicing of selected species in slash at about 50 cm height, leaving stumps ready to take advantage of fallow when area abandoned after 2–3 years
Tree planting before fallow, including shade trees (mainly legumes) and (at a later date) *cacao* and coffee

"Tolché"
Different sizes and forms of forested belts surrounding the *milpa* – a key factor in the regeneration process of the fallow

Tree plantations
Fruit trees
Cacao plantations with shade legume trees

Other
Living fences
Trees in urban and religious centres
Sacred groves
Trees in terraces?

The Mayan silviculture consisted of a series of activities of protection, cultivation, selection, and introduction of trees in their *milpas*, fallows, plantations, natural forests, houses, living fences, *cenotes*, and urban centres. Selection for usefulness probably included many properties in addition to food values, such as hardness of the wood and toxicity of bark and wood. Religion also played a role in the selection and protection of tree species; that is the case, to the present, of several species of *Ceiba*.

In the slash process of Mayan *milpa* cultivation, the farmer does an additional and more massive selection. He will identify some useful species, mainly fast-growing secondary ones that he will slash at a height of 50 cm or so (a sort of pruning and coppicing), leaving the stumps ready to take advantage of the fallow that will occur when the area is abandoned after 2 to 3 years.

All the various silvicultural activities and techniques assume the existence of some kind of natural ecosystems from which they could draw the species that they needed from time to time. It is clear that the first colonizers of the Maya area, whoever they were, found a rich and diverse mosaic of ecosystems in which they lived and from which they derived their subsistence. They managed and used them for an unknown length of time, starting a selection process that the Maya have continued up to the present.

In order to accomplish this hypothetical silviculture, there must have been a biological conservation strategy entailing a system of resource management that included intensive crop cultivation in raised fields, the creation of artificial forests, and the conservation of some natural ecosystems. They had many intermediate production systems with high biological diversity. This probably helps to explain why no evidence is available that mass extinctions of species occurred due to past actions, or that species diversity or richness was diminished by Mayan activities, in spite of the fact that the Maya area was a highly populated and intensively used area in the past. The proof of this can be found in the flora of the Maya area (Sosa *et al.* 1985), rich in endemic species, from the humid areas of the Lacandone rain forest to the drier tropical deciduous forests and swamps of the Yucatan Peninsula. It is important to mention the special richness of the secondary successional flora, and this may be another gift that the old Maya gave us that is worthy of research (Gómez-Pompa 1971).

The regeneration of the ecosystems of the Maya area after successive abandonments (the last one after the Conquest), was possible only because of the existence of seed-banks in managed and protected "natural" ecosystems in the area (Gómez-Pompa *et al.*1972), and of land-uses that did not cause irreversible damage to the soils. It is clear that Mayan silviculture played an important role in the success of Mayan society and also in the biological and ecological conservation of the area and its resources. We still do not know the cause of the Mayan collapse, but we have enough evidence to discard the hypothesis of poor management of soils and deforestation. Most of the techniques that the Maya used to manage their forests are not unique; they have been found scattered in many other traditional cultures, not only of the New World, but also in the Old World. This is not surprising, since we know that efficient subsistence techniques and useful species spread very rapidly, and it is not improbable that they might also have been discovered independently.

LESSONS FROM THE PAST

It is clear that there are more questions than answers in relation to the hypothetical silvicultural system of the old Maya. But, from the available information, there are several conclusions that emerge that may be of importance for the future conservation and management of tropical forests. From the research side, each abandoned *milpa* can be considered as an empirical experiment in directed succession.

Shifting agriculture, as practised by the Maya, can feed more people than we had assumed, while conserving a biological diversity for future use. It should be seen as a starting point for future permanent agriculture and silviculture in the lowland tropics, and it should not be seen as only a destructive technology. Secondary successions (fallows) in the lowland tropics can be managed in "the Mayan way" to produce a combination of useful species for multiple purposes.

Biological diversity can be conserved, even in densely populated tropical lowland areas, if appropriate resource use practices are followed. Regenerating rain forests from heavily used areas is possible if germplasm pools are conserved. Small forest patches (natural or managed) can help to maintain a high level of diversity in the lowland tropics. They should be considered as areas that can contribute to the conservation of biological diversity. Artificial forest gardens for biological conservation can be designed and created by man if he wishes to do so, in order to preserve species that he chooses to preserve.

Many tropical traditional cultures that still exist today have a great knowledge of their environment and resources. This knowledge is a human heritage that we should not lose. It has been valuable in the past, and it could be of utmost importance for the present and the future. Present-day agriculture is the result of an accumulated folk knowledge obtained from hungrily watching thousands of generations of food plants. When studying a tropical area that has been inhabited in the past, we should play close attention to the distributional patterns of abundance of species, since these are likely to be the result of man's actions.

A final point concerns the population density found in the lowland tropical areas of south-eastern Mexico today. This should not be considered as a problem. Rather, the problem lies in certain management practices, incorrect technology, and absence of broad ecological considerations in land-use planning. The present-day activities of dubious agricultural projects, extensive grazing and timber mining on the same sites where the old Maya had temples, towns, intensive hydraulic agriculture, permanent agriculture, and artificial and "natural" forests, should cause us to doubt our wisdom and our congruency.

ACKNOWLEDGEMENTS

I thank the Regents of the University of California for permission to base this contribution on a longer article which appeared in *Mexican Studies/ Estudios Mexicanos* 3 (1), 1–17 (1987).

REFERENCES

Barrera, A.M., Gómez-Pompa, A. and Vázquez, C. (1977). El manejo de las selvas por los Mayas: sus implicaciones silvícolas agrícolas. *Biotica,* **2**, 47–60

Beltrán, E. (1959). (ed.) *Los recursos naturales del sureste y su aprovechamiento.* (Ediciones INMERNAR: México, D.F)

Berg, C.C. (1972). *Olmediae, Brosimeae.* Flora Neotropica Monographs **7**. (Organization for Flora Neotropica, New York Botanical Garden: New York.)

Coe, M.D. (1984). *The Maya.* (Thames and Hudson: London)

Flannery, K.V. (1982). (ed.) *Maya Subsistence: Studies in Memory of Dennis E. Puleston.* (Academic Press: New York)

Flores, S. and Ucan, E. Ek. (1983). Nombres usados por los Mayas para designar a la vegetación. *Cuadernos de Divulgación INIREB*, **10**, 1–33

Gómez-Pompa, A. (1971). Posible papel de la vegetación secundaria en la evolución de la flora tropical. *Biotropica,* **3**, 125–35

Gómez-Pompa, A. (1987). On Maya silviculture. *Mexican Studies/Estudios Mexicanos,* **3(1)**, 1–17

Gómez-Pompa, A., Flores, E. and Sosa, V. (1987). The "pet kot": A man-made tropical forest of the Maya. *Interciencia,* **12(1)**, 10–15

Gómez-Pompa, A. and Golley, F.B. (eds.) (1981). *Estrategias del uso del suelo y sus recursos por las culturas mesoamericanas y su applicación para satisfacer las demandas actuales.* Memorias del Simposio CONACYT-NSF. Biótica **5(1,2)**. (INIREB: Xalapa)

Gómez-Pompa, A., Vázquez, C. and Guevara, S. (1972). The tropical rain forest: A non-renewal resource. *Science,* **177**, 762–5

Hammond, N. (1982). The exploration of the Maya World. *American Scientist,* **70**, 482–95

Hartig, H.M. (1979). *Las Aves de Yucatán. Fondo Editorial de Yucatán*, Cuaderno 4. (Mérida: México)

Lundell, C.L. (1937). *The Vegetation of Petén.* Carnegie Institute, Washington, D.C. 478, 10

Peters, C.M. (1983). Observations on Maya subsistence and the ecology of a tropical tree. *American Antiquity,* **48**, 610–15

Pohl, M. (1985). *Prehistoric Lowland Maya Environment and Subsistence Economy.* (Harvard University Press: Cambridge)

Puleston, D.E. (1982). The role of Ramón in Maya subsistence. In. Flannery, K.V (ed.) *Maya Subsistence: Studies in Memory of Dennis E. Puleston*, pp. 353–66. (Academic Press: New York)

Sosa, V., Flores, J.S., Rico-Gray, V., Lira, R. and Ortiz, J.J. (1985). *Lista Florística y Sinonimia Maya.* Etnoflora Yucatanense, Fascículo 1. (INIREB: México)

CHAPTER 22

MANAGEMENT OF SECONDARY VEGETATION FOR ARTIFICIAL CREATION OF USEFUL RAIN FOREST IN UXPANAPA, VERACRUZ, MEXICO – AN INTERMEDIATE ALTERNATIVE BETWEEN TRANSFORMATION AND MODIFICATION

S. del Amo R.

ABSTRACT

Since 1981, studies have been carried out in Veracruz, Mexico, on the management of secondary vegetation of corn fields, 2-year-old orchards and 9- and 11-year-old secondary forest. Elucidation of recovery processes has aimed at identifying ways and means of managing and accelerating the succession towards more useful and productive stages. Clearance of some plants and enrichment with others are among the experimental treatments which have sought to take advantage of knowledge about traditional systems of horticulture (Mayas), multiple use of land (Lacandones), and use of different ecological habitats (Chontales, Nahuas). Data include rates of biomass accumulation during the fallow, and chemical interactions between and among the naturally occurring and introduced plant species. Management insights generated by the study include the use of the acahual *as a substrate for introducing valued species with different ecological strategies into the successional sequence, and the imitation of the "natural" structure of the various successional stages.*

INTRODUCTION

Since pre-hispanic times, human impact on the Mexican tropical rain forest has intensified and increased with the growing population, resulting in widespread secondary forests (Gómez-Pompa 1979). The species of secondary communities are particularly important in tropical lowlands because they are abundant, adaptable to alteration, and potentially useful for present and future generations. At present, about 95% of the tropical rain forest in Mexico has disappeared. Secondary vegetation already covers 6.4 million hectares, and is spreading daily. For this reason, it is important to manage secondary vegetation and to incorporate it into the productive sector.

A great variety of management systems has been developed over the centuries by different ethnic groups in Mexico. These include horticultural practices (Mayas), multiple land-use systems (Lacandones), uses of simultaneous ecological areas (Chontales), management of sustainable resources in time and space, and selection of common tropical food plants.

This project is intended to provide a background of information about alternatives for the establishment of a variety of agro-forestry systems in the Mexican tropics. The main objective is to lead the secondary succession towards stages with species of economic value, assisting the process by clearing and removing some undesirable plants, and enriching with more desirable ones.

The experiment was begun in 1981 in the Valley of Uxpanapa, in south-east Veracruz, almost on the border of Oaxaca in the Ejido Agustin Melgar. It forms part of the coastal plain of the Gulf of Mexico, located between 17° 17' and 17° 21' north and 94° 05' and 94° 95' west at an altitude of 200–300 m above sea-level. The climate is hot and humid with rains in summer and an average annual temperature of 25° C. Rainfall is 2000–3000 mm per annum with a dry season from February to May, and a humid season from June to October.

APPROACHES AND MANIPULATIONS

Three different successional stages were chosen. The first is the diversified *milpa*, the second is 9 years old, and the third is about 11 years old.

In the *milpa*, one hectare of *acahual* was prepared in the traditional way (slash-and-burn) with four different treatments: a single corn field, and three different multiple crops with a mixture of annual and perennial species. The results indicate that the total production was twice as high in the multiple cropping. Sweet potato was an efficient species for weed control and it was evident that the polyculture was an efficient and successful system for continuous production. The *milpa* was also used to compare the dynamics of weeds in a monoculture and in a polyculture, utilizing the Sörensen index for diversity and the Motyka index to compare biomass. The results indicate that the polyculture system exerts greater weed control, especially of grasses.

The orchard system develops through gradual change from *milpa*. After 3 years of good results in the management of multiple crops, productivity was ensured for a longer period by gradual change to an orchard. Patches for annual crops were fertilized with legume mulches. The next step in the orchard structure was the introduction of more species of tropical fruit trees, forest trees and dominant secondary species.

In a younger successional stage (about 9 years old), management includes thinning, enrichment and addition of organic matter. The objectives are to establish the optimal conditions for growth and yield of introduced species, and to evaluate the changes in the succession produced by the disturbance of canopy and organic matter. The experiment was set up in late 1985 in an *acahual* with three different structures and floristic communities. It comprised nine plots and three treatments: clearing and leaving organic matter to decompose in situ, clearing and removing organic matter, clearing the understorey only and leaving organic matter to decompose. The species chosen for this study are given in Table 22.1.

The three communities differ in structure and floristic composition but have the same age and are located together. In one of them, *Cecropia obtusifolia* is the dominant tree and *Aegiphylla monstrosa* and *Heliconia* spp. are dominant shrubs. In another community, *Gmelina arborea* and *Cecropia obtusifolia* are the dominant trees and *Heliconia* spp. the dominant shrub. The third community is characterized by *Cestrum racemosum* and *Heliocarpus appendiculatus* in the canopy and has more primary species than the others (e.g. *Quararibea funebris*, *Guarea chichon* and *Dialium guianense*). The lower layers are more complex, and *Olmeca reflexa* and *Heliconia* spp. are dominants. Each treatment has two zones: one without any introduced species in order to observe the natural secondary succession and to compare the recovery rate in different initial conditions; the other one with the enrichment of introduced species of different life cycles in order to establish the best conditions for their growth. The parameters evaluated in this experiment were rate of growth, biomass, floristic potential and differences in the floristic composition in both areas (Fig. 22.1).

Table 22.1 Species used to enrich the younger successional stage

Shade tolerant:	*Theobroma cacao* *Chamaedora tepejilote*
Shade intolerant:	*Schizolobium parayba* *Cordia megalantha* *Anacardium occidentale*
Understorey species:	*Dioscorea composita*

In an older successional stage (about 11 years), three different thinnings have been made according to the canopy layers. The objective is to produce a model of growth of shade tolerant species and shade intolerant ones. The model consists of a number of submodels which describe the light climate, the interception of light by individual tree crowns, the photosynthetic response of the leaves, the assimilation and allocation of fixed CO_2, and their subsequent utilization for growth. From these submodels, the contribution of the different individuals trees to the stand growth and to the successional process has been evaluated. The submodels are interdependent and the outputs from the submodel will be used as inputs for another. The equations used within the model are either theoretical with reference to the underlying physical or physiological processes, or empirical based on observations and measurements.

Three different thinnings have been made, i.e. cleaning up, thinning codominants below 10 cm and cleaning understorey. Four different tree species have been introduced within the thinnings – *Brosimum alicastrum*, *Swietenia macrophylla*, *Cedrela odorata* and *Cordia alliodora*. These represent a range between the "primary" shade tolerant species and "secondary" shade intolerant tree species. Four repetitions of the species position are carried out in each plot. With this study, we will be able to determine which areas of the model are the weakest and require further research, and we may be able to suggest some management techniques (based on the simulation) directed towards the optimum use of these ecosystems, and the "regeneration" of the tropical rain forest.

EVALUATION OF SOME BIOLOGICAL INTERACTIONS WITH WILD SPECIES

One of the secondary objectives of this project is to incorporate within the management systems some of the wild common species that are available. *Pteridium aquilinum* (bracken) is an aggressive fern that is a pioneer in the abandoned fields. Bioassays were carried out to detect the allelopathic effect of green fronds of *P. aquilinum* on the germination and radicle growth of eight species of cultivated plants and five weed species. The results indicate that the growth of four of the five weed species was markedly inhibited by macerated fronds. The aqueous extract had no significant effect on either the cultivated or weed species. Mustard and tomato appeared to be the most susceptible species to the methanolic extract of the fronds (Nava *et al.* 1987).

The effect on the growth of four species of phytopathogenic fungi and four phytopathogenic bacteria was also tested. Fungi were tested with the frond's aqueous, methanolic and ethanolic extracts. Bacteria were tested

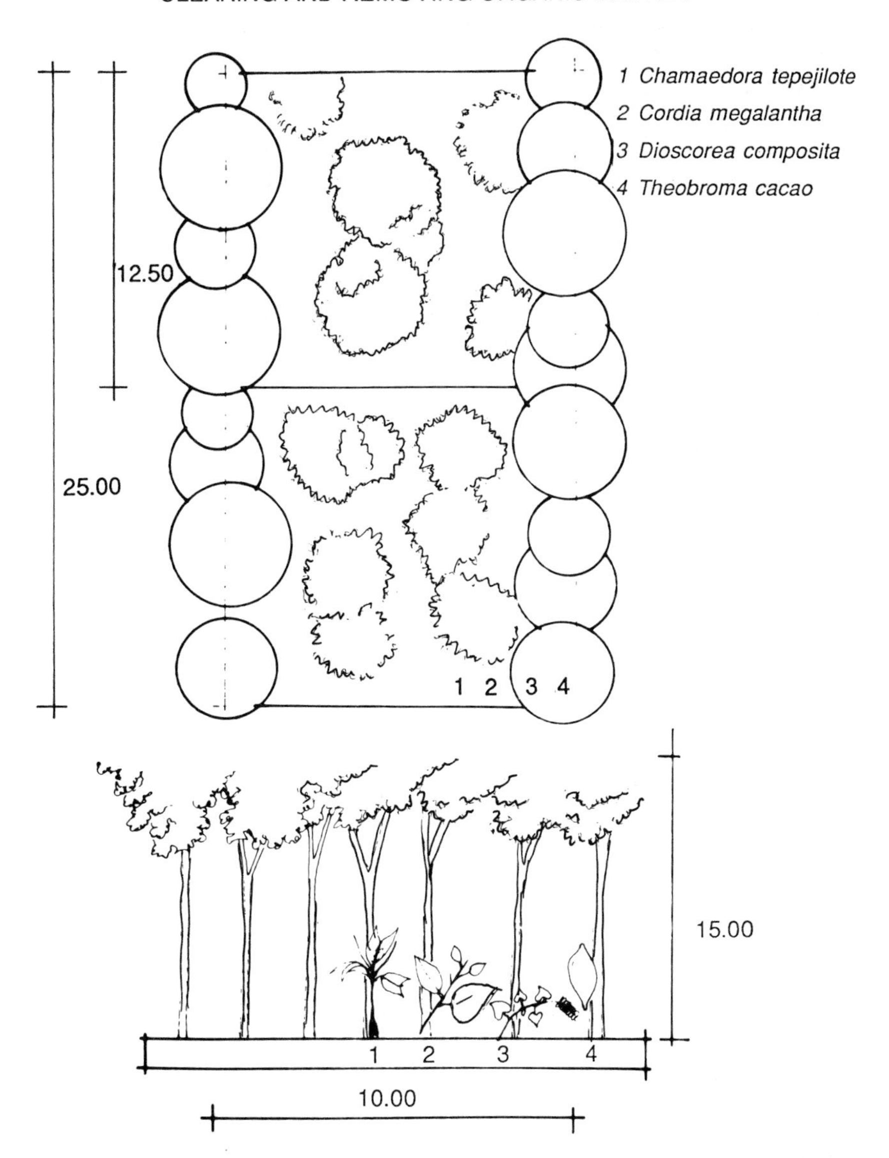

Fig. 22.1 Graphic representation of three treatments at the 9-year-old orchard stage: I – clearing and removing organic matter; II – clearing and leaving organic matter; III – clearing understorey and leaving organic matter. For each treatment, an area lacking introduced species is compared with one enriched by introduced species of different life cycles and habits. Among the measurements taken are rate of growth, biomass, floristic potential and differences in floristic composition

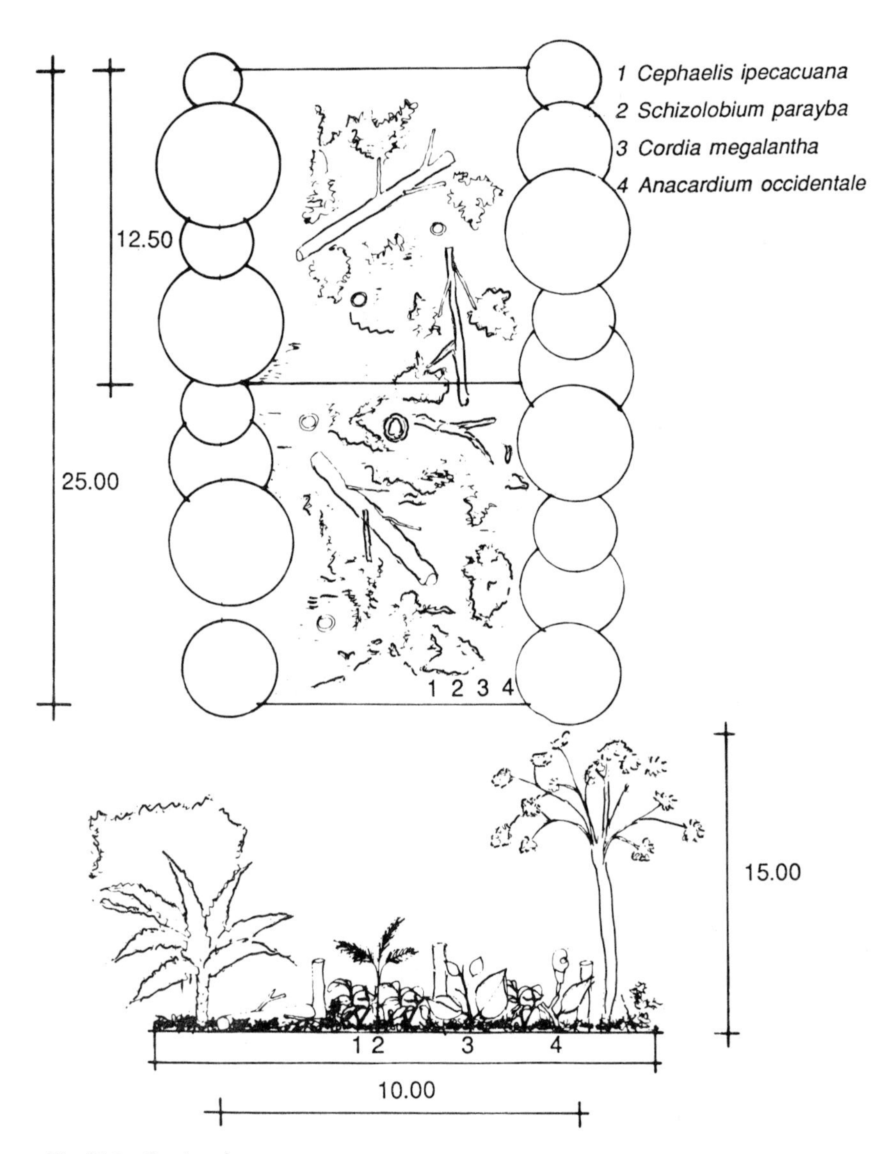

Fig. 22.1 *Continued*

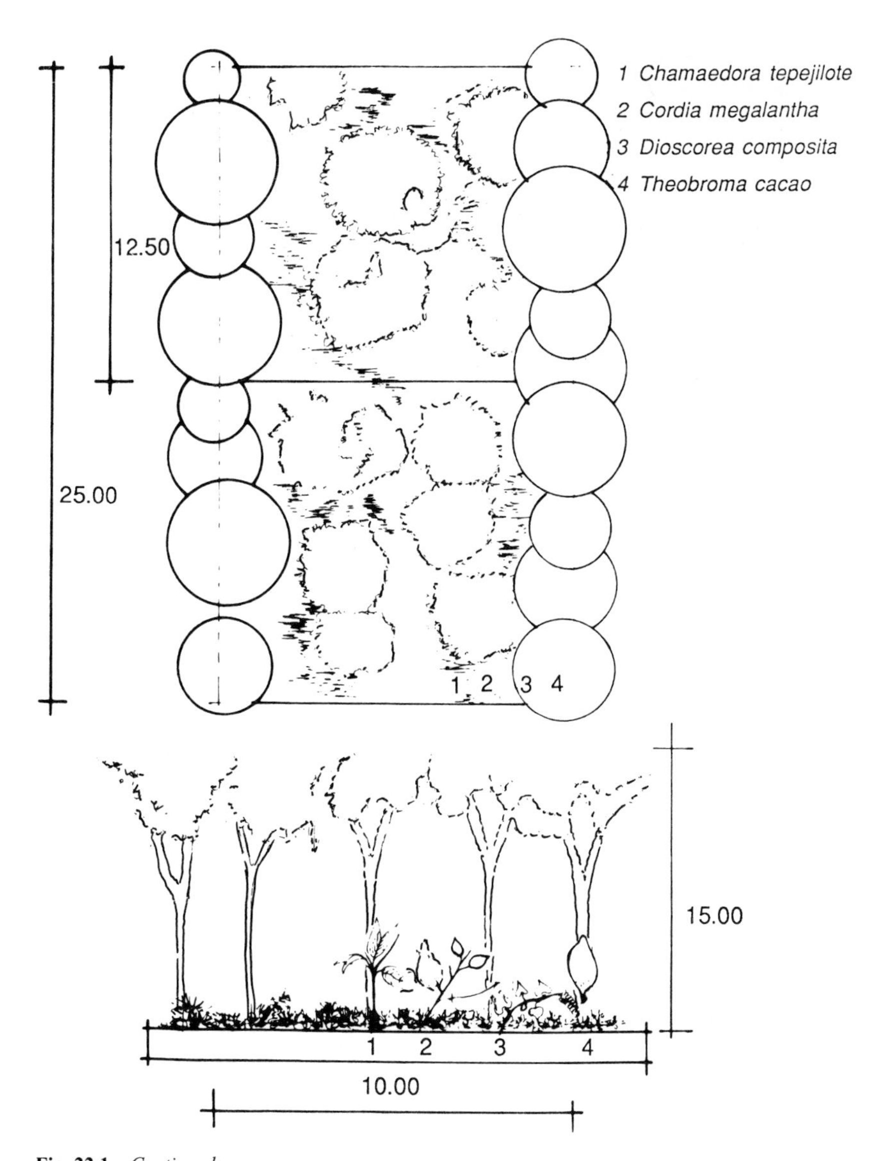

Fig. 22.1 *Continued*

with aqueous extract. The fungal growth was strongly inhibited by the aqueous extract, and this inhibitory activity was maintained throughout the course of the experiment. In the bioassays with methanolic and ethanolic extracts, a tendency towards recuperation could be observed in some of the fungi tested. The diluted aqueous fraction (1:10) of the methanolic extract produced a stimulating effect in the growth of all the fungi tested. The aqueous extract inhibited only the growth of the sole gram-positive bacteria among the species tested (Nava *et al.* 1987). The results obtained in this study suggest that *P. aquilinum* could offer a wide range of possible uses as a natural herbicide and fungicide.

REFERENCES

Gómez-Pompa, A. (1979). Antecedents de las investigaciones botánico-ecológicas en la región del Río Uxpanapa, Ver. *Biotica,* **4**(**8**), 127–33

Nava, V.R.E., Fernández, L. and del Amo R., S. (1987). Allelopathic effects of green fronds of *Pteridium aquilinum* on cultivated plants, weeds, phytopathogenic fungi and bacteria. *Agriculture, Ecosystems and Environment,* **18**, 357–79

CHAPTER 23

FOREST MANAGEMENT STRATEGIES BY RURAL INHABITANTS IN THE AMAZON ESTUARY

A. Anderson

ABSTRACT

*This paper provides a case study of forest management by rural inhabitants (*caboclos*) in the floodplain of the Amazon estuary. The relatively low biological diversity and high concentration of economic tree species make these forests amenable to so-called 'tolerant' forms of management, in which the native vegetation is largely conserved or reconstituted through successional processes. Rural inhabitants implement tolerant forms of forest management by selective weeding or thinning of less desirable competitors, and by promoting the regeneration or productivity of desirable species. Many of the latter are rarely planted due to their natural abundance in floodplain forests (e.g.* Euterpe oleracea, Hevea brasiliensis, Carapa guianensis, Spondias mombin*). As a result, forest stands in which virtually all species are useful can be generated or maintained with minimal effort. This form of land-use illustrates how extraction and forest management can be reconciled in ways that minimize risk and maximize sustainability.*

INTRODUCTION

Extraction can be defined as the removal of natural resources, with no provision for their replacement. In the Amazon Basin, the term has become historically associated with resource depletion, environmental degradation, socio-economic disruption, and cultural decimation. Since the arrival of the first European traders in the late 16th century, extraction has provided the foundation for virtually all major economic activities in the region and continues to do so today.

Despite its sordid reputation in Amazonia, and in other tropical regions, extraction has occasionally been coupled with sustained management of natural resources (e.g. Fox 1977, Posey *et al.* 1984). Such coupling not only promotes resource conservation, but may eliminate many of the socio-economic ills associated with purely extractive forms of resource exploitation.

This paper examines one such case of extraction and forest management by rural inhabitants (*caboclos*) in the floodplain of the Amazon estuary. This case study is of special interest for several reasons. First, it occurs in the Amazon floodplain (*várzea*), a zone known for its high agricultural potential (e.g. Meggers 1971, Roosevelt 1980) but not generally recognized as favourable for long-term forest management. Second, it involves *caboclos*, a group generally maligned for their purported lack of industry, 'collecting mentality', Indian blood, and 'primitive' land-use practices (cf. Ross 1978). Although *caboclos* are acknowledged to be the cultural representatives of the Amazon region since the early 19th century (Parker 1986), most documented case studies of sustained land-use management in Amazonia have involved Indians (e.g. Denevan *et al.* 1984, Posey 1983). Finally, this study shows that extraction, despite its historical defects, has the potential to be incorporated into economically and ecologically rational forms of land-use.

THE ECOLOGICAL SETTING

The term *várzea* refers to the floodplain of the Amazon River and its major white-water tributaries. Although the extent of the *várzea* is questionable, it covers a relatively small proportion of the Amazon Basin as a whole: the most frequently cited figures are 1–2% (Sombroek 1966, Sternberg 1975). Despite its relatively limited extent, the *várzea* has traditionally contained the region's highest population densities and offers the greatest opportunity for expanding intensive agriculture. Its high agricultural potential results from the annual deposition of nutrient rich sediments that originate from the geologically young (Quaternary) Andes.

Although its agricultural potential is high, the *várzea* is a problematic site for agriculture, precisely because it is subject to flooding. Excessive soil moisture is the key limiting factor to plant growth in this ecosystem. Depending on topography and hydrological conditions, floodplain soils can be water-logged from a few days per year to months at a time or even constantly throughout the year. Insufficient drainage limits plant growth primarily by restricting the availability of oxygen to roots. Many tree species in the *várzea* exhibit an exceptional abundance of structures (e.g. aerial roots, pneumatophores, lenticels, etc.) that are probably adaptations to oxygen-poor soils. Few species appear to thrive under these conditions, and, as a result, the floodplain forest is characterized by relatively low biological diversity and pronounced dominance of only a few tree species, many of which are of economic importance.

Forest resources in the *várzea* seem to be highly productive. Many of the dominants – such as açaí palm (*Euterpe oleracea* Mart.), rubber (*Hevea brasiliensis* (Willd. ex A.Juss. M.Arg.), and numerous timber species – appear to be fast growing, possibly due to relatively high soil fertility. Floodplain forests as a whole are extremely dynamic (cf. Frangi and Lugo 1985). Shallow root systems due to impeded drainage, combined with constant soil movement associated with flooding, apparently result in a high frequency of tree-falls, and abundant light gaps provide ample opportunity for forest regeneration. Hence, the floodplain forest seems to be able to support sustained short-cycle timber extraction.

'TOLERANT' FOREST MANAGEMENT

Around the dwellings of rural inhabitants in the floodplain of the Amazon estuary, one can observe a recurring pattern of managed zones, described in detail by Anderson *et al.* (1985). A house-garden (*terreiro*) typically surrounds the dwelling. This intensively managed, relatively open zone is used for raising domesticated animals and cultivating a wide variety of exotic and native plants. The house-garden, which is usually less than 1 ha in area, contrasts sharply with the far denser and more extensive floodplain forest (*mata de várzea*), which is covered by a closed canopy of trees. Inhabitants practice no discernible management in this latter zone, which is used primarily for gathering forest products and for hunting. At scattered locations within the dense floodplain forest, swidden plots (*rocas*), are cut for cultivation of annual subsistence crops such as rice, corn, and beans. Cash crops, such as sugarcane and rice, are sometimes grown on a more extensive scale.

The three major zones (house-garden, dense floodplain forest and swidden plot) are easily distinguishable entities that are clearly designated with local terms. Other, less clearly defined zones occur. These occupy a variety of locations in relation to the household, originate by different means, and are designated by a variety of terms. Despite the variable nature of these zones, they have an important element in common: they are all subject to long-term forest management. For purposes of simplicity, I refer to these various zones as 'managed forests' and distinguish them from dense floodplain forest that is not subject to discernible management, which I call 'unmanaged forest'.

Managed forests exhibit a number of distinctive features. For example, they typically contain a more or less continuous tree cover, in contrast to the relatively open house garden or swidden plot; conversely, the vegetation is considerably less dense than in unmanaged forest. Managed forests are also subjected to less intensive management than either the house garden or the swidden plot, where weeding is a more or less constant activity. Weeding does take place in managed forests, but to a lesser degree, as the relatively deep shade in this zone effectively reduces weed establishment and growth. By contrast, unmanaged forests are, by definition, not subjected to weeding or other discernible forms of management.

Managed forests exhibit considerable variability in structure and composition. At one extreme, they may appear as orderly plantations of exotic trees on sites where the original floodplain forest has been completely removed. Alternatively, they may be almost indistinguishable from the native floodplain forest, and comprised entirely of native forest species that are only subject to occasional thinning. These two extremes reflect contrasting management strategies. I refer to the first extreme as *intrusive* forest management, in which the native vegetation is replaced by tree plantations that are maintained by long-term care. The second extreme exemplifies *tolerant* forest management, in which the native vegetation is largely conserved or reconstituted through successional processes. The degree to which the intact native forest is preserved in these systems is inversely related to the intensity of human interference.

A managed forest studied intensively on Ilha das Oncas adjacent to Belém (Anderson *et al.* 1985) illustrates the principal features of tolerant forms of forest management. This zone consists of a 1.1 ha area located adjacent to the house-garden; according to informants, this area has been subjected to intermittent management since at least the beginning of this century.

Local residents implement two basic management strategies in this zone. The first strategy is to favour desirable species indirectly by weeding or thinning less desirable competitors, such as vines and tree species used exclusively for timber and firewood. This weeding not only favours desirable species that are already established by reducing competition, but also promotes their regeneration selectively by sparing their seedlings. Many species are neither eliminated by weeding nor subsequently favoured by discernible management practices. These so-called tolerated species are important sources of extracted products: examples include *taperebá* (*Spondias mombin* Urb.), *ingá* (*Inga* spp.), and *miriti* (*Mauritia flexuosa* L.), which bear edible fruits and also attract game; and '*seringueira*' (*Hevea brasiliensis*), which provides latex used to make rubber. Tolerated species are a key feature in the managed forest under consideration, whereas they are virtually absent in systems characterized by intrusive management.

The second management strategy in the managed forest under consideration is to directly favour desirable species by promoting their regeneration or productivity. This strategy is implemented in a number of ways. Local residents introduce seeds of desirable species by conscious planting or unconscious dispersal. They also introduce species by planting or transplanting seedlings and cuttings. For example, consciously planted species include *cacao* (*Theobroma cacau* L.), *cupuaçu* (*Theobroma grandiflorum* K. Sch.), coconut (*Cocos nucifera* L.), *genipapo* (*Genipa americana* L.), and mango (*Mangifera indica* L.); several varieties of banana (*Musa* spp.) have also been introduced. Species that are propagated by unconscious human dispersal probably include *açaí* (*Euterpe oleracea*), mango, *ingá*, *taperebá* and *urucú* (*Bixa orellana* L.). Seedlings of desirable species (e.g. *cacao*, *cupuaçu*, and coconut) are actively protected by more intensive weeding and construction of improvised fences (frequently made of palm leaves or stems) around them. Organic material, usually comprised of decaying inflorescences and leaves of the abundant *açaí* palm, are concentrated at the base of favoured plants. The stems of fruit trees such as mango and *genipapo* are often scored with a number of shallow cuts, which is claimed to promote fruit production. Local residents regularly prune the multistemmed *açaí* to harvest palm heart for sale to local industries. This practice, when carried out selectively, is claimed to enhance fruit production in the remaining stems. *Açaí* fruits are used to make a thick beverage that is widely consumed throughout the Amazon estuary.

The effects of these practices can be seen by comparing areas of managed and unmanaged floodplain forests on the Ilha das Oncas. The most notable difference between these zones is the much higher biomass of the unmanaged forest, due to selective thinning in managed forest, especially in the understorey. Species diversity is also reduced under management. In 0.25 ha samples, for example, a total of 52 species were recorded in the unmanaged forest, compared with 28 in the managed forest (Tables 23.1–23.2). Comparing resources in the forest types, sources of food and drink had a considerably higher (> 10%) relative importance on the managed site, whereas sources of wood (i.e. timber) and energy (i.e. firewood and charcoal) had a considerably higher importance on the unmanaged site (Table 23.3). As mentioned previously, trees that provide fuel are selectively eliminated under management.

According to informants, one of the principal motives for managing floodplain forests is to make resources more available. As shown above, forest thinning results in a reduction of biomass and vastly improved access. In addition, management provides improved conditions for consciously favoured species. Crops such as mango, *cacao*, and coconut, which are introduced as seeds or seedlings in the understorey of managed forests, subsequently provide both market and subsistence products. One of the most important of the favoured species is the *açaí* palm. Tolerant forms of forest management preserve a wide range of tolerated species. As mentioned previously, many of these species are economically useful as sources of extractive products such as fruit trees and rubber. Other tolerated

Table 23.1 Species with diameter at breast height (dbh) > 5 cm, collected in a 0.25 ha area of managed floodplain forest on Uha das Ongas with scientific and vernacular names, and data on ecology and uses. Numbers (N) refer to collections of Anthony B. Anderson *et al.* Asterisks indicate species also present in the inventory of unmanaged floodplain forest (Table 23.2). F = food, D = drink, M = medicine, W = wood for construction of furniture, f = fibres, A = game attractant, E = energy in the form of firewood or charcoal, O = organic material used as fertilizer, U = utensils, X = other uses

		Vernacular	*Abundance*		*Frequency*		*Dominance*		*Importance*	
N	*Scientific name*	*name*	n	%	n	%	cm^2	%	%	*Uses*
1066	**Euterpe oleracea* Mart.	açaí	163	50.5	24	19.0	11 919	15.2	28.4	F.D.M.W.f.O.U.
1083	**Hevea brasiliensis* (Willd.ex A.Juss) M.Arg.	serinqueira	27	8.4	15	11.9	23 746	30.2	16.9	M.A.X.
1129	*Theobroma cacao* L.	cacau	48	14.9	18	14.3	2359	3.0	10.8	F.D.
1095	*Inga edulis* Mart.	ingá cipó	27	8.4	13	10.3	8009	10.2	9.7	F.A.E.
1084	**Spondias mombin* Urb.	taperebá	5	1.6	5	4.0	9526	12.1	5.9	F.D.M.W.A.E.
1201	*Ficus* cf. *paraensis* (Miq.) Miq.	apuí	1	0.3	1	0.8	8511	10.8	4.0	A.
1152	**Inga* cf. *alba* Willd.	ingá xichica	8	2.5	6	4.8	2724	3.5	3.6	F.A.E.
1137	*Cordia* cf. *bicolor* A. DC.	–	5	1.6	5	4.0	1064	1.4	2.3	E.
1171	**Pentaclethra macroloba* (Willd.) Kuntze	pracaxi	5	1.6	5	4.0	1035	1.3	2.3	M.E.X.
1056	*Cecropia* cf. *obtusa* Trec.	imbauba	3	0.9	3	2.4	2229	2.8	2.0	M.A.E.
1128	*Theobroma grandiflorum* K.Sch.	cupuaçu	3	0.9	3	2.4	1215	1.6	1.6	F.D.
1085	**Astrocaryum murumuru* Mart.	murumuru	3	0.9	3	2.4	721	0.9	1.4	F.A.E.X.
1200	*Guarea* cf. *guidona* (L.) Sleumer	bototeiro	2	0.6	2	1.6	1342	1.7	1.3	E.
1146	*Pithecellobium glomeratum* (DC.) Benth.	jarandeua	3	0.9	3	2.4	102	0.1	1.1	M.
1116	*Allophyllus mollis* Radlk.	–	3	0.9	3	1.6	236	0.3	0.9	–
1074	*Genipa americana* L.	genipapo	1	0.3	1	0.8	1176	1.5	0.9	F.D.M.W.O.U.
–	*Mangifera indica* L.	manga	2	0.6	2	1.6	487	0.6	0.9	F.M.A.
1131	*Aegiphila* cf. *arborescens* Vahl.	–	2	0.6	2	1.6	146	0.2	0.8	–
1140	*Mauritia flexuosa* L.	miriti	1	0.3	1	0.8	951	1.2	0.8	F.D.W.f.A.U.
1150	**Virola surinamensis* (Rol.) Warb.	ucuuba branca	2	0.6	2	1.6	129	0.2	0.8	M.W.A.E.X.
	Other species (8)		9	2.8	9	7.1	943	1.2	3.6	
	Total		323	100.1	126	99.4	78 570	100.0	100.0	

Table 23.2 Species with diameter at breast height (dbh) > 5 cm, collected in a 0.25 ha of unmanaged floodplain forest on Ilha das Ongas with scientific and vernacular names, and data on ecology and uses. Numbers (N) refer to collections of Anthony B. Anderson *et al.* Asterisks indicate species also present in the inventory of managed floodplain forest (Table 23.1). F = food; D = drink; M = medicine; W = wood for construction or furniture; f = fibres; A = game attractant; E = energy in the form of firewood or charcoal; O = organic material used as fertilizer; U = utensils; X = other uses

N	*Scientific name*	*Vernacular name*	*Abundance* n	*Abundance* %	*Frequency* n	*Frequency* %	*Dominance* cm^2	*Dominance* %	*Importance* %	*Uses*
1066	**Euterpe oleracea* Mart.	açaí	236	59.9	10	8.1	13 814	13.8	27.1	F.D.M.W.f.O.U.
1405	*Pterocarpus officinalis* Jacq.	mututí	6	1.5	5	4.0	21 408	21.4	9.0	A.E.U.
1437	**Spondias mombin* Urb.	taperebá	4	1.0	3	2.4	16 770	16.8	6.7	F.D.M.W.A.E.
1402	**Pithecellobium glomeratum* (DC.) Benth.	jarandeua	23	5.8	9	7.3	1587	1.6	4.9	E.
1407	*Carapa guianensis* Aubl.	andiroba	8	2.0	6	4.8	6357	6.4	4.4	M.W.A.
–	**Astrocaryum murumuru*	murumuru	12	3.0	7	5.6	2223	2.2	3.6	F.A.E.X.
1417	**Hevea brasiliensis* (Willd.ex A.Juss.)M.Arg.	seringueira	5	1.3	4	3.2	4844	4.8	3.1	M.A.X.
1424	*Cynometra marginata* Benth.	jutairana	4	1.0	3	2.4	5430	5.4	2.9	W.A.E.
1413	*Macrolobium angustifolium* (Benth.) Cowan	ipê da várze	4	1.0	4	3.2	4343	4.3	2.9	M.W.E.
1416	**Inga* cf. *alba* Willd.	ingá xichica	8	2.0	5	4.0	2143	2.1	2.7	F.A.E.
1408	**Penthaclethra macroloba* (Willd.) Kuntze	pracaxi	6	1.5	4	3.2	3259	3.3	2.7	M.E.X.
1404	*Matisia paraense* Huber	cupuaçurana	6	1.5	5	4.0	2333	2.3	2.6	E.
1412	*Quararibea guianensis* Aubl.	inajarana	6	1.5	4	3.2	1155	1.2	2.0	M.W.E.
1414	*Terminalia dichotomi* Aubl.	cuiarana	2	0.5	2	1.6	3740	3.7	2.0	W.E.U.
1447	*Crudia* sp.	rim de paca	3	0.8	3	2.4	1977	2.0	1.7	M.W.E.
1420	*Dalbargia monetaria* L.f.	pó verônica	7	1.8	4	3.2	235	0.2	1.7	M.
1411	*Protium* cf. *polybotrium* (Turcz.) Engl.	breu branco	7	1.8	3	2.4	989	1.0	1.7	A.E.
1410	*Mora paraensis* Ducke	pracuuba	1	0.2	1	0.8	2463	2.5	1.2	M.W.A.
1428	**Virola surinamensis* (Rol.) Warb.	ucuuba branca	2	0.5	2	1.6	612	0.6	0.9	M.W.A.E.X.
1422	*Crudia oblonga* Benth.	rim de paca	1	0.2	1	0.8	1176	1.2	0.7	M.W.E.
	Other species (32)		43	10.9	39	31.4	3090	3.1	15.1	
Total			394	99.7	124	99.6	99 948	99.9	99.6	

Table 23.3 Relative importance of species and resources in areas of managed and unmanaged floodplain forests on the Ilha das Onças. Data summarized from Tables 23.1 and 23.2

	Item Importance (%)	
	Managed forest	*Unmanaged forest*
Species (20)	96.4	84.9
Food	64.0	40.1
Drink	48.4	33.8
Medicine	59.2	55.1
Wood	36.8	52.5
Game attractant	46.0	36.2
Energy	29.3	47.0
Fertilizer	29.3	27.1
Utensils	30.1	38.1
Fibres	29.2	27.1
Other uses	21.4	10.3

species that were neither discernibly managed nor eliminated from the managed forest included fruit trees such as *taperebá*, *ingá*, and *miriti*. Although important sources of timber such as *andiroba* (*Carapa guianensis* Aubl.) and *ucuuba* (*Virola* spp.) were not abundant in the managed forest under consideration, seedlings of these species were also reportedly tolerated by local inhabitants.

Most of the species utilized in the managed forest under consideration were not even planted, much less tended, in ways commonly associated with more intrusive forms of land-use. In fact, the very nature of management in this forest is frequently elusive. For example, a merely tolerated species such as rubber is hardly managed in a conventional sense. Yet its tolerance in a managed zone appears to lead to higher representation than in an unmanaged zone. Is this species undergoing management or mere extraction?

The elusive nature of forest management in its tolerant forms extends beyond individual plants to entire plant communities. Such management takes place along a gradient between other zones, and natural succession is the driving force along this gradient. A managed forest may originate from a house-garden or a swidden plot through selective regeneration; when regeneration ceases to be selective, it reverts to unmanaged forest.

The central role that forest succession is allowed to play under tolerant forms of management has important implications. First, it provides a greater diversity of resources and economic opportunities. In the case study under consideration, the co-existence of various transitional zones in close proximity to the household seems to be part of a general strategy of increasing opportunities for resource exploitation. These opportunities include domestic and recreational activities; fishing, hunting, and livestock rearing; and use of fruits, palm heart, wood, fertilizer, ornamental plants; fibres, latex, honey, oilseeds, medicinals, utensils, and so on. No single zone offers all these opportunities, but their combination does.

Second, the incorporation of succession in tolerant forms of management seems to have important functional implications. These systems require relatively minimal labour for their establishment and maintenance. In floodplain forests, relatively low biological diversity and high concentration of economic species permit minimal thinning and, as a result, high tolerance of regenerating vegetation. Such tolerance probably contributes to the long-term sustainability of local plant resources.

CONCLUSIONS

Extraction and management seem to be irreconcilable land-use strategies. Extraction can be defined as the removal of natural resources, with no provision for their replacement. Management, in contrast, refers to human manipulations that promote the maintenance and/or sustainability of resources at a given locale. Yet in describing a so-called tolerant strategy of forest management, I have entered domains that many would consider closer to extraction. Although the precise limits between management and extraction are not entirely clear, both processes clearly co-exist within the floodplain forest site described in this paper. I would suggest that such systems occupy a middle ground between conventional, intrusive land-uses on the one hand, and exclusively extractive land-uses on the other.

Reconciliation of extraction and forest management can probably occur only under certain circumstances. Such integration is probably facilitated when forests contain a relatively low biological diversity and high concentration of economic species. The essential ingredient for reconciling extraction and forest management, however, is a rural population with both a knowledge of and respect for forest resources. The Amazon *caboclo* is often considered to be culturally marginalized by both academics and policymakers. Yet, in his ignorance, he has developed sophisticated systems of forest management that appear to be both low in risk and ecologically sustainable.

ACKNOWLEDGEMENTS

The research for this paper was supported by the Ford Foundation, the World Wildlife Fund, and the Conselho Nacional de Desenvolvimento Científico e Tecnológico. I express my gratitude to Anne Gély, Bill Balée, and Darrell Posey for their helpful comments.

REFERENCES

Anderson, A.B., Gély, A., Strudwick, J., Sobel, G.L. and Pinto, M.G.C. (1985). Um sistema agroflorestal na várzea do estuário amazônico (Ilha das Onças, Município de Barcarena, Estado do Pará. *Acta Amazonica,* **15**, 195–224

Denevan, W.M., Treacy, J.M., Alcorn, J.B., Padoch, C., Denslow J. and Paitan, S.F. (1984). Indigenous agroforestry in the Peruvian Amazon: Bora Indian management of swidden fields. *Interciencia,* **9(6)**, 346–57

Fox, J.J. (1977). *Harvest of the Palm.* (Harvard University Press: Cambridge, Mass)

Frangi, J.L. and Lugo, A.E. (1985). Ecosystem dynamics of a subtropical floodplain forest. *Ecological Monographs,* **55(3)**, 351–69

Hallé, F., Oldeman, R.A.A. and Tomlinson, P.B. (1978). *Tropical Trees and Forests; an Architectural Analysis.* (Springer Verlag: New York)

Meggers, B.J. (1971). *Amazonia: Man and Culture in a Counterfeit Paradise.* (Aldine-Atherton: Chicago)

Parker, E.P. (1986). Caboclоization: the transformation of the Amerindian in Amazonia 1615–1800. In Parker, E.P. (ed.) *Peasantry of the Brazilian Amazon: Historical and Theoretical Perspectives.* Studies in Third World Societies Publication Series, vol. **29**. (William and Mary Press: Williamsburg)

Posey, D.A. (1983). Indigenous knowledge and development: an ideological bridge to the future. *Ciência e Cultura,* **35(7)**, 877–94

Posey, A.D., Frechione, J., Eddins, J., Francelino da Silva, L., with Myers, D., Case, D., and Macbeath, P. (1984). Ethnoecology as applied anthropology in Amazonian development. *Human Organization,* **43(2)**, 95–107

Roosevelt, A.C. (1980). *Parmana: Prehistoric Maize and Manioc Subsistence along the Amazon and Orinoco.* (Academic Press: New York)

Ross, F.B. (1978). The evolution of the Amazon peasantry. *Journal of Latin American Studies,* **10(2)**, 193–218

Sombroek, W.G. (1966). *Amazon Soils.* (Centre for Agricultural Publications and Documentation: Wageningen)

Sternberg, H.O. (1975). The Amazon River of Brazil. *Geographische Zeitschrift,* **40**, Weisbaden.

CHAPTER 24

NATURAL REGENERATION AND ITS IMPLICATIONS FOR FOREST MANAGEMENT IN THE DIPTEROCARP FORESTS OF PENINSULAR MALAYSIA

S. Appanah and Salleh Mohd. Nor

ABSTRACT

Knowledge on the regeneration of dipterocarp forests in Peninsular Malaysia, contrary to many other parts of the tropics, is adequate for the formulation of appropriate management systems for these forests. Many of the dipterocarps, which form the bulk of the timber trees of these forests, possess a coterie of characteristics (e.g. gregarious fruiting, dense seedling populations, quick response by seedlings to light openings, relatively fast growth, excellent form, gregarious stands, etc.) that allow foresters to manipulate these forests at little cost in order to produce a sustained yield of timber and other forest products. However, most of the productive forests left in Peninsular Malaysia today are restricted to hilly terrain where the working conditions are difficult and stocking is poorer. Nevertheless, it is possible to apply the knowledge inherited from lowland dipterocarp forests to the proper management of these hill dipterocarp forests. Furthermore, with the dwindling of timber and forested areas, intensive management for sustained yield may be the only option left for preserving these rich and productive forests.

INTRODUCTION

The land area of Peninsular Malaysia (PM) totals 13.2 million ha. Of this, the area still under forest cover amounts to 5.18 million ha; within this area, 4.70 million ha have been set up as the National Permanent Forest Estate (Fig. 24.1). This forest estate includes 3.28 million ha of productive forests, the majority of which are the hill dipterocarp forests.

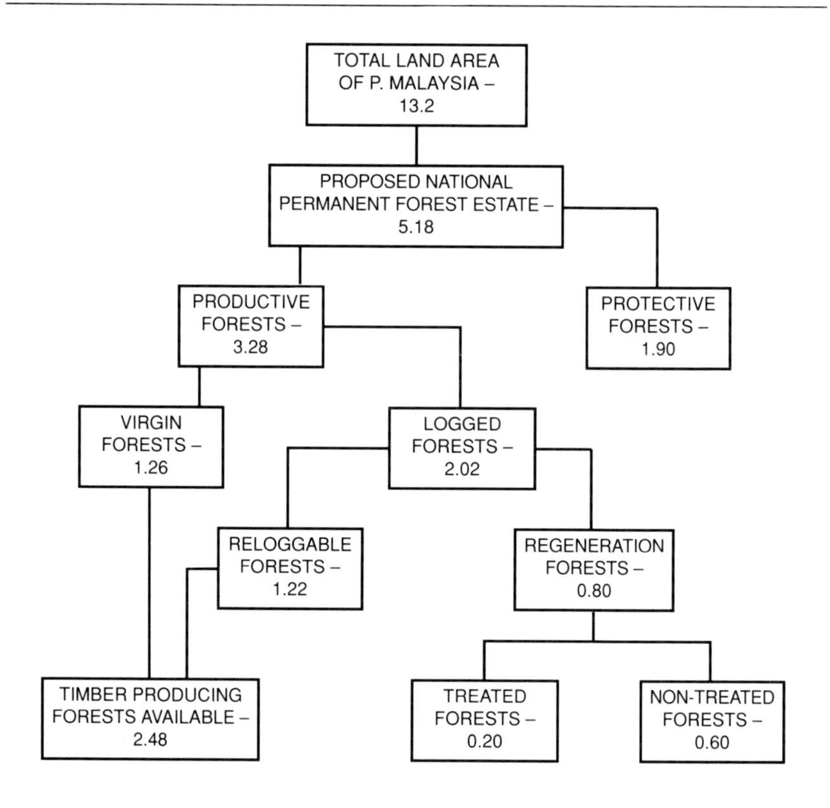

Fig. 24.1 Source of timber supply for Peninsular Malaysia after 1987 (Figures in million ha; estimated only)

Natural forests provide one of the best options for solving the problems of tropical resource management in PM. While the apparent high internal rates of return or high cost/benefit ratios of plantation forestry appear fascinating, in reality, it should be recognized that the annual increase in value of the natural forests is just as competitive in socio-economic terms if the other services such as sources of fresh water, animals, and other 'minor products', are given a fiscal value. The natural forests still remain the nub of forestry in the peninsula, and every effort should be undertaken to raise their yields and increase their utility. To transform them into managed forest estates, a deep understanding of their ecosystems is required so as skilfully to translate ecological knowledge into silvicultural practices.

NATURAL REGENERATION OF DIPTEROCARP FORESTS OF PENINSULAR MALAYSIA

Flowering and fruiting patterns

The rule among canopy species, including the majority of dipterocarps in PM, is to flower and fruit gregariously, massively, at about 2 to 10 year intervals (see review by Appanah 1985). Isolated or sporadic flowerings of dipterocarps do occur, but at low intensities, and the fruiting is poor. During heavy flowering years, 40–50% of the mature trees flower, and the floor of the forest is later littered with fruits. Most of the fruits are poorly dispersed by wind, and the majority remain within 60 m of the mother tree (Burgess 1970). While relatively young trees are known to flower in the arboreta (Ng 1966), often only the mature ones flower in nature (Fox 1972). The seeds of the gregarious fruiters are heavily parasitized by predators (Burgess 1970, 1975, Daljeet-Singh 1974), this parasitism being much higher during isolated fruiting years.

Some dipterocarps have been shown to be pollinated by tiny, fecund, common, flower-feeding insects such as thrips (Appanah and Chan 1981), and probably other similar groups of insects (Appanah 1987). The dipterocarps pollinated by these groups of insects which have been investigated are mostly outbreeders (Chan 1981), and breeding clumps are needed for high fruit set (Ashton 1969, Chan 1980). Many of the canopy species that flower more regularly (e.g. *Dryobalanops* species, *Neobalanocarpus*) are pollinated by higher-energetic insects, and may not have such a dominance.

Seed dormancy and germination

The majority of seeds of canopy species in the dipterocarp forests lack dormancy, are recalcitrant (Jensen 1971), and will germinate within 2 weeks (Ng 1980). However, *Shorea* species of the seasonally dry, north-western part of PM have seeds with slightly longer viability (Wyatt-Smith 1963). Interestingly, many of the legumes like *Koompassia* have staggered germination (Wyatt-Smith 1963). In contrast, the seeds of pioneer species possess dormancy, and are usually incorporated into the top soils of the forest (Symington 1933). These include grasses, herbs, shrubs, climbers, palms (including rattans) and secondary tree species. These seeds only germinate in a gap on the stimulation of bright light and/or the increase in temperature.

Many primary forest species can germinate in the shade of the canopy; dipterocarps that have been investigated can establish under 50% shade but need progressively more illuminance for maximum growth (Mori

1980). The forests are usually carpeted with seedlings following a flowering; they progressively decline in numbers, and may even reach zero in some forests (Wyatt-Smith 1963) until the next mass flowering boosts their populations again.

Growth of seedlings into trees

It now appears that almost every forest species has to go through some growth in the gap phase of the forest before it attains maturity. Three or perhaps four categories of trees can be recognized depending on their response to the creation of a gap, while the size of the gap strongly influences the composition of the species and spatial patterns in these forests. Many seedlings of the light hardwoods grow very little under the shade of the forest cover (Nicholson 1965), occur as a seedling-bank (Whitmore 1984), and progressively die off until a gap opens up and boosts their growth. In contrast, some of the heavy hardwood species are extremely shade tolerant, grow very slowly and perish if exposed, as in a large tree-fall gap. At the other extreme are the pioneer species which generally regenerate in large gaps; they mostly originate from the buried soil seed-bank, with a small input from the seed rain coming into the gaps after their formation (Putz and Appanah 1987). These pioneer tree species, strong light demanders, grow rapidly and pre-empt the gap for themselves. Large gaps are also the sites for the growth of a thick tangle of woody climbers. The pioneer vegetation persists for about 2 decades before the primary vegetation overtops it, after which the pioneers perish. Another group of pioneers can be differentiated from the above: these are the long-lived ones such as *Endospermum*, *Anthocephalus* and *Octomeles*.

Growth rates

Growth of the canopy species takes place initially mostly in terms of height. Upon reaching the canopy, the crown expands out and girth growth then advances rapidly. One enigmatic phenomenon observed in the tropics is that girth increment in the closed forests is believed to increase with size (Nicholson 1965). Annual girth measurements maintained from 1937 to 1973 on a stand of dominants and co-dominants of *Shorea leprosula* and *S. parvifolia* in the Merlimau Forest Reserve, Malacca, illustrate this phenomenon clearly. There is great variation in the annual growth rates between the individuals. Small trees grew very slowly. Conversely, the bigger individuals grew relatively faster. This evidence leads us to speculate that such diverse growth rates probably stem from genotype differences, since the stand was selected to have minimum microsite differences.

The growth rates of the different species also vary tremendously. The light demanding, light hardwood dipterocarps reach mature habit in about 60 years and have a life span of about 250 years. The shade tolerant species grow very slowly, reach maturity in about 250 years and may have a life span in excess of about 1000 years.

SILVICULTURAL IMPLICATIONS

From knowledge of the flowering and fruiting patterns in the dipterocarp forests in PM, it is clear that the effective fruiting years are the massive long-intervalled ones, and that the seedling populations between fruitings are minimal in some forests. Logging all the mature fruiting individuals can be disastrous after a couple of cutting cycles. Thus, it is preferable to log a forest following a mass fruiting year. However, foresters may baulk at such a limitation as they may have to concentrate logging activities at long intervals, sometimes 10 years apart. Girdling operations to kill unwanted or weed species and to liberate juveniles of desired species can also be increased during the year following fruiting, in order to maximize the benefits of girdling.

Since logging generally removes most of the mature fruiting individuals in the forests, it becomes imperative to leave behind mother trees to maintain the seed production. However, the drawback is that many of the desired species seem to be outbreeders and reproduction is probably mostly within the clump. So, in order to regenerate these species, islands of mature individuals need to be retained during logging. Failure to meet such requirements will probably result in gradual change in the composition of the dominant species after several loggings. Such logging will tend to favour species that flower more regularly, those capable of selfing, those that produce multiple seedlings, and those with higher-energetic pollinators. Examples of dipterocarps with some of these characters are *Shorea ovalis*, *S. macroptera*, *Neobalanocarpus* and *Dryobalanops*.

If the flowering trigger is a cold shock, as postulated by Ashton *et al.* (1988), then it should be possible to induce more regular fruitings if dipterocarps from the aseasonal zone were planted in the slightly more seasonal north-western zone of the peninsula, where annual flowering of dipterocarps is usual. Under plantation conditions, these trees would also be more likely to fruit at an earlier age. Such plantations could provide a regular source of seeds for other planting programmes and for research.

To date, attempts to extend the seed dormancy of canopy species using low temperatures, controlled humidity, and chemicals have failed or proved to be expensive (Tamari 1976). Thus, foresters might usefully exploit the phenomenon of mast fruiting. Viable fruits from virgin forests could be collected and broadcast in partially open areas of logged forests. It might also be worthwhile to take the trouble to redistribute the large number of fruits that fall under the mother tree to the more open areas nearby, where their chances of survival would be heightened. Another option would be to introduce species with longer dormancy, such as the *Shorea* species, from the seasonally dry north-western parts of PM into these everwet forests.

The practice of girdling unwanted, deformed or weed trees to create gaps for the liberation of young saplings and seedlings of desired species was started as early as 1900 in PM (Hill 1900). Girdling is still a much cherished management tool, but its uncontrolled usage is currently under question (Appanah 1986). With today's higher levels of exploitation, the greater opening of the canopy, the wider range of marketable species, and improved milling techniques to handle poor quality logs, fewer trees need to be girdled. Nevertheless, a judicious application of girdling, which would result in the release of many desirable juveniles, is still tenable. In this respect, the careful manipulation of the canopy to promote the growth of selected species is one area that needs extensive investigation. Of the commercial species, the light hardwoods grow faster upon exposure to gaps and reach maturity within 60–70 years (cf. 250 years for the slower-growing heavy hardwoods). Thus, the present logging systems create light regimes which favour the light hardwoods, and forest management should exploit this favourable situation.

Another serious, and some may say dubious, problem is that of woody climbers: is there a need to eradicate them? In large gaps, climbers establish and grow rapidly, pressing down the early regrowth of primary species. In the mature vegetation, the climbers bend or snap saplings and poles, deform, weigh down or compete for leaf space with older trees. Wyatt-Smith (1963) was of the opinion that the regenerating primary species overcome the climber problem rapidly, and he recommended only very light post-felling climber-cutting. Even such light climber-cutting may have dubious value if the timing of the prescription is inappropriate: Fox (1968) and Appanah and Putz (1985) have shown that felling trees laden with woody climbers that are entwining neighbouring trees causes much logging damage, and that selective climber-cutting prior to felling significantly reduces logging damage.

The discovery that trees are small because they have slower growth rates, and the tendency to log the larger and faster growing species, will result in progressively poorer genotypes being selected after a few cutting cycles. One way to reverse the trend is to thin the stand of the slow growing individuals. Such an exercise will lead to selection of the better genotypes, but how do you age a tree to be able to differentiate between a small old tree and a small young one? Attempts to age trees in the tropics have not been successful.

Thinning trials have been performed in the past in order to encourage a more uniform stand and a better stocking of desirable species. Such practices can be calamitious if they are not done with cognizance of the soil–water relationships and water losses caused by evapotranspiration of the vegetation (P.S. Ashton, *pers. comm.*). For example, the Senaling Inas Forest Reserve in Negri Sembilan was selectively logged and subsequently treated to develop a rich and somewhat uniform stand of *Shorea curtissii* and *S. platyclados*. During the severe drought of 1975–76, the whole hill of the rich stand of regenerated dipterocarps died. It is speculated that the enriched uniform stand of trees resulted in excessive loss of ground water from evapotranspiration, and they succumbed during the drought.

The large-scale, mechanized logging that is currently practised tends to produce huge gaps in the forest canopy, and the resulting succession of pioneers can retard the regeneration of primary species by as much as 2 decades. The heavily compacted or scraped soils remain barren and are vulnerable to erosion and loss of nutrients, or *Gleichenia* fern develops into a dense mat and prevents colonization of primary species. In attempting to prevent such problems arising, one practice currently adopted by some forest concessions in PM is the planting of long-lived pioneers (indigenous species belonging to *Endospermum* and *Anthocephalus*, and exotics such as *Acacia mangium* and *Eucalyptus* species) soon after logging, while the barren sites are still accessible. One provocative question is whether trees like *A. mangium* will be able to outcompete the other pioneers, incorporate into the soil seed-bank, and become feral. This practice may prove a boon or bane, depending on how the species affect the functional ecosystem of the forest. It might be worthwhile to try out some indigenous fast growing dipterocarps like *Shorea leprosula* and *Dryobalanops* species, and the late-succession dipterocarps from Papua New Guinea such as *Anisoptera* and *Hopea* (Johns 1987).

Present-day logging techniques cause extensive damage to residuals, and can render the forests depauperate of saplings and seedlings of commercial timber species (Fox 1968, Borhan 1985). Where natural regeneration can no longer be relied upon for the renewal of the timber crop, artificial regeneration has to be sought. Reforestation of whole compartments with *Shorea leprosula* and *S. parvifolia* was performed in the states of Selangor and Perak (Tang and Wadley 1976). Since seed survivability for most dipterocarps is low, and most do not coppice in the wet tropics, planting must necessarily rely upon potted seedlings. These seedlings are still raised from seeds, as vegetative propagation is still intractable, though experimental work on bare root planting and vegetative propagation is beginning to show promise (Sasaki 1980). Planting such poor forests with fast growing dipterocarps from other regions is worth examining too. For example, *Shorea trapezifolia* from Sri Lanka, which can maintain diameter growth even during flushing, flowering and fruiting (N. Gunatilleke, *pers. comm.*) would be an excellent candidate for such planting in the peninsula.

That the future of the dipterocarp forests in PM lies in managing them for a sustained yield of timber and other forest products holds much validity. The important question to ask now is, "How much interference can these forests take?" Can the faster growing species supplant the slow growing ones? Is there a need to maintain the present diversity, or can many of the understorey species be removed, leaving behind only the dominants for a more productive forest? Or is it possible to replace the numerous species of dipterocarps in the overstorey with fewer species for easier management of these forests? The solutions to these questions lie in the answer to a more primeval issue, 'Is this assemblage of tree species in the dipterocarp forests, with their complexity and diversity, the result of purely chance factors, or solely selection, or an admixture of both?' If it is prevalently a chance factor, then it is possible to alter the forest to semi-forest plantations without undue concern. If it is otherwise, excessive manipulation may lead to catastrophes. P.S.Ashton (*pers. comm.*) is of the opinion that the rare or uncommon species are a product of chance, while the majority of the dominants in the overstorey, the ones that concern us here, are a product of high biotic selection. If this is true, then it should be possible to encourage a more productive forest of altered structure and composition, by reducing the understorey vegetation and increasing overstorey elements. It is hoped that some answers to this fundamental question will be provided by a long-term, large-scale demographic study of a 50 ha plot in Pasoh Nature Reserve, launched in 1986.

ACKNOWLEDGEMENTS

We would like to thank J. Racz, L. Chew and M. Borhan for the valuable discussions on the subject of dipterocarp forest management.

REFERENCES

Appanah, S. (1985). General flowering in the climax rain forests of South-east Asia. *Journal of Tropical Ecology,* **1**, 225–40

Appanah, S. (1986). Is there a need to do GCL in Malaysian forests now? In Hadi, J., Kamis Awang, Nik Muhamad Majid and Shukri Mohamed (eds.) *Impact of Man's Activities on Tropical Upland Forest Ecosystems*, pp. 33–9. (Universiti Pertanian Malaysia: Serdang)

Appanah, S. (1987). Insect pollination and the diversity of the dipterocarps. In Kostermans, A.J.G.H. (ed.) *Proceedings of the Third Round Table Conference on Dipterocarps.* Samarinda, Indonesia. 16–20 April 1985, pp. 227–39. (Unesco: Jakarta)

Appanah, S. and Chan, H.T. (1981). Thrips: the pollinators of some dipterocarps. *Malaysian Forester,* **44**, 234–52

Appanah, S. and Putz, F.E.. (1985).Abundance of woody climbers in a lowland dipterocarp forest and the effect of pre-felling climber cutting on logging damage. *Malaysian Forester,* **47**, 335–42

Ashton P.S. (1964). *Ecological Studies in the Mixed Dipterocarp Forests in the Brunei State.* Oxford Forestry Memoirs 25.

Ashton, P.S. (1969). Speciation among tropical forest trees: some deductions in the light of new evidence. *Biological Journal of Linnean Society*, London, **1**, 155–96

Ashton, P.S., Givnish, T.J. and Appanah, S. (1988). Staggered flowering in the Dipterocarpaceae: New insights into floral induction and the evolution of mast fruiting in the aseasonal tropics. *American Naturalist*, **132**(1), 44–66

Borhan, M. (1985). *Some Aspects of the Silviculture and Management of the Hill Dipterocarp Forests of Peninsular Malaysia.* M. Sc. Thesis, University of Wales.

Burgess, P.F. (1970). An approach towards a silvicultural system for the hill forests of the Malay Peninsula. *Malayan Forester,* **33**, 126–34

Burgess, P.F. (1975). *Silviculture in the Hill Forests of the Malay Peninsula.* Malaysian Forestry Department Research Pamphlet 66.

Chan, H.T. (1980). Reproductive biology of some Malaysian dipterocarps. II. Fruiting biology and seedling studies. *Malaysian Forester,* **43**, 438–51

Chan, H.T. (1981). Reproductive biology of some Malaysian dipterocarps. III. Breeding systems. *Malaysian Forester,* **44**, 28–36

Daljeet-Singh, K. (1974). Seed pests of some dipterocarps. *Malaysian Forester,* **37**, 24–36

Fox, J.E.D. (1968). Logging damage and the influence of climber cutting prior to logging in the lowland dipterocarp forest of Sabah. *Malayan Forester,* **31**, 326–47

Fox, J.E.D. (1972). *The Natural Vegetation of Sabah and Natural Regeneration of the Dipterocarp Forests*. Ph.D. Thesis, University of Wales.

Hill, H.C. (1900). *The Present System of Forest Administration in the Federated Malay States with Suggestion for Future Management.* (Government Printer: Selangor)

Jensen, L.A. (1971). Observations on the viability of Borneo camphorwood *Dryobalanops aromatica* Gaertn. f. *Proceedings of the International Seed Testing Association,* **36**, 141–6

Johns, R.J. (1987). The natural regeneration of *Anisoptera* and *Hopea* in Papua New Guinea. In Kostermans, A.J.G.H. (ed.) *Proceedings of the Third Round Table Conference on Dipterocarps.* Samarinda, Indonesia, 16–20 April 1985, pp. 172–90. (Unesco: Jakarta)

Mori, T. (1980). *Physiological Studies of some Dipterocarp Species of Peninsular Malaysia as a Basis for Artificial Regeneration.* Malaysian Forestry Department Research Pamphlet **78**.

Ng, F.S.P. (1966). Age at first flowering in dipterocarps. *Malayan Forester,* **29**, 290–5

Ng, F.S.P. (1980). Germination ecology of Malaysian woody plants. *Malaysian Forester,* **43**, 406–37

Nicholson, D.I. (1965). A study of virgin forest near Sandakan, North Borneo. In *Proceedings of Symposium on Ecological Research in Humid Tropics Vegetation.* Kuching, Sarawak. 2–10 July 1963, pp. 67–87. (Unesco: Jakarta)

Putz, F.E. and Appanah, S. (1987). Buried seeds, dispersed seeds, and the dynamics of a lowland dipterocarp forest in Malaysia. *Biotropica,* **19**, 326–33

Sasaki, S. (1980). Storage and germination of dipterocarp seeds. *Malaysian Forester,* **43**, 290–308

Symington, C.F. (1933). The study of secondary growth on rain forest sites in Malaya. *Malayan Forester,* **2**, 107–17

Tamari, C. (1976). *Phenology and Seed Storage Trials of Dipterocarps.* Malaysian Forestry Department Research Pamphlet **69**.

Tang, H.T. and Wadley, H.E. (1976). *A Guide to Artificial Regeneration with Particular Reference to Line-Planting in Peninsular Malaysia.* Malaysian Forestry Department Research Pamphlet **68**.

Whitmore, T.C. (1984). *Tropical Rain Forests of the Far East.* Second Edition. (Clarendon Press: Oxford)

Wyatt-Smith, J. (1963). *Manual of Malayan Silviculture for Inland Forests.* (2 vols.). Malayan Forest Record **23**.

CHAPTER 25

A COMPARATIVE ACCOUNT OF SILVICULTURE IN THE TROPICAL WET EVERGREEN FORESTS OF KERALA, ANDAMAN ISLANDS AND ASSAM

C.T.S. Nair

ABSTRACT

Tropical evergreen forests in three widely separated regions of India (Western Ghats, Andaman and Nicobar Islands, and the North Eastern region) have a long history of management. Over time, wood production, by both polycyclic and monocyclic systems, has tended to receive priority over multiple use management. Changing patterns of forest management in the three regions are described in terms of objectives, regulation of yield, and regeneration, and a summary account provided of management for other objectives (protection of catchments and slopes, non-timber products, etc.). One conclusion of Indian experience in rain forest management is that prescriptions are most often not based on the growth rates and regeneration requirements of the tree species, but on exigencies to increase wood supply in the short-term.

INTRODUCTION

Tropical wet evergreen forests in India form the western outlay of the Indo-Malayan rain forest formation. Extending over an area of about 4.5 million ha, they occur in three widely separated regions, namely, the Western Ghats (especially the states of Kerala and Karnataka), Andaman and Nicobar Islands and the North-eastern region (particularly the states of Assam, Arunachal Pradesh, Nagaland and Manipur). These forests are found in the high rainfall zone, where annual precipitation exceeds 2500 mm. In the Western Ghats and the North Eastern region, there is a marked seasonality and monsoon forest (tropical moist deciduous forest) is the predominant vegetation type. Evergreen forests are found where physiographical and edaphic factors reduce the moisture stress during dry months. Seasonality is much less pronounced in the Andaman Islands.

While structural and physiognomic similarities exist between forests in the three regions, there are significant differences in the floristics (Champion and Seth 1968, Puri *et al.* 1983). Both in the upper Assam forests and the wet evergreen forests of Andamans, dipterocarps form an important component in the top canopy. *Dipterocarpus macrocarpus* and *Shorea assamica* sometimes account for 50% of the total trees above 90 cm girth in the former. Evergreen forests in the Western Ghats also have dipterocarps, but the three main genera, *Dipterocarpus*, *Vateria* and *Hopea*, seldom account for more than 6–10% of the total number of trees (>30 cm girth) in any given area.

Forests in all these three regions have a long history of management. Traditionally, they were utilized to realize a variety of benefits, but, over time, the wood production objective has gained primacy. While subtle differences exist in the silvicultural systems, they fit into a broad pattern, largely determined by the socio-economic environment. This contribution briefly discusses the present systems of utilization and analyses their implications with regard to their conservation and management.

MANAGEMENT FOR WOOD PRODUCTION

Both polycyclic and monocyclic systems of management are followed in the evergreen forests. Historically, in all these forests, timber extraction commenced with a polycyclic selective felling system (FRI and Colleges 1961). Low demand, limited number of marketable species, and poor accessibility favoured the adoption of such a system. With the increasing demand for timber, systems of concentrated felling have been followed, and even clear-felling has been resorted to in extreme cases (Nair and Chundamannil 1985).

Selective felling in the evergreen forests of Kerala

Selective felling was in vogue in the evergreen forests of Western Ghats from the beginning of the century. Extraction was limited to hardwood species suitable for railway sleepers (e.g. *Hopea parviflora*, *Mesua nagassarium*) and ship masts (e.g. *Calophyllum* spp.). Growth in demand for timber during the First World War, and consequent improvement in accessibility, facilitated introduction of systems of concentrated felling in certain areas (Champion and Osmaston 1962). Failure of regeneration, and more particularly the slump in demand during the Inter-war recession, led to abandonment of such trials. Growth of the plywood industry during the Second World War period and the increased demand for railway sleepers led to large-scale adoption of selective felling. Presently, it is the most widely adopted system of working the wet evergreen forests in Kerala.

Objectives of management

Under the system, forests earmarked for felling and regeneration are included under the selection or selection-cum-improvement working circle. Maintenance of tree cover to protect the soil and to regulate water yield in catchments, increasing the supply of wood and other products, and realization of maximum revenue subject to the above constraints are the most commonly stated objectives of management (Achuthan 1982, Pillai 1984). Accessible areas alone are taken up for working, while inaccessible areas are included under the protection working circle. Selective removal is primarily designed for a situation where the proportion of commercially important species is low and accessibility is poor.

Yield regulation

In theory, under a polycyclic system, the same area is visited at periodic intervals and trees reaching specified exploitable girth are removed. The length of the felling cycle is determined by the time taken for trees in the pre-exploitable class to reach the exploitable class, the proportion of mature and overmature trees, and the infrastructural facilities. In Kerala, the felling cycle varies from 10 to 30 years. Removal is further subjected to a girth limit, which is determined taking into account the nature of the demand and the size class distribution of trees. For unworked stands with a preponderance of large sized trees, girth limit is fixed high. Depending upon the species removed, it varies from 120 cm to 210 cm.

A further restriction is imposed by fixing the number of trees that can be removed from the exploitable class. Often this number is prescribed arbitrarily, based on past practices. A more refined approach is to fix the harvestable number as a percentage of the total number of trees in the exploitable class. Including those that are likely to become exploitable during the felling cycle, using the Smythie's Safeguarding formula (see FAO 1984, page 66), the number of trees prescribed for removal varies from 5–12 ha^{-1}.

Timber extraction is carried out either by industries (especially plywood and match units) to whom an annual quota of timber is allotted, or by the forest department which engages logging contractors. The system of giving long-term lease is not in vogue. Partly due to the rugged terrain, and partly due to the system of short-term contracts, capital intensive techniques are not adopted for logging. Traditional tools are used and short-distance haulage is done by elephants.

Some of the important felling rules prescribed in management plans include: a minimum distance of 20 m should be kept between two marked trees to ensure that felling is not concentrated in pockets; climber cutting should be done at the time of marking to minimize felling damage; no felling should be carried out for a width of 20 m on either side of water courses to prevent soil erosion; marking should be carried out in such a way as not to cause any permanent gaps in the canopy; and only dead and dying trees should be marked on steep slopes.

Although about 30 species are listed for felling, in practice, a disproportionately large number of trees belonging to three or four species required for specific end uses are removed. When railway sleepers are to be supplied, the most preferred species are *Dichopsis ellipticum*, *Cullenia exarillata* and *Mesua nagassarium*. Sometimes, these three species account for about 90% of the trees felled. When the demand is from the plywood industry, the focus shifts to prime veneer species like *Vateria indica* and *Dipterocarpus* spp. (FAO 1984).

Regeneration

Improvement in light conditions following selective cutting is assumed to help the establishment of regeneration and the growth of trees from the pre-exploitable class to the exploitable class during the felling cycle. Prescriptions aimed at promoting regeneration include the cutting back of all broken and completely damaged trees. A regeneration map should be prepared for each annual coupe and treatments appropriate to the status of regeneration carried out. Most working plans prescribe gap planting in areas deficient in natural regeneration. The preferred species are *Vateria indica*, *Dichopsis ellipticum*. *Toona ciliata*, *Dysoxylon malabaricum*, *Lophopetalum wightianum*, *Canarium strictum* and *Persea macrantha*. In congested patches of pole crop, thinning is prescribed.

However, institutional, technical and financial constraints hamper implementation of these prescriptions, and invariably the felled areas are neglected. Augmentation of regeneration undertaken by strip and gap planting covers only a small fraction of the area felled annually. The main factors that inhibit natural restocking are: absence of regeneration in the form of seedlings and saplings; felling damage to saplings, poles, and even mature trees; competition from colonizers, which come up in the openings.

Regeneration is generally poor in most of the evergreen forests. Even when seedlings are present, drastic changes in light and moisture conditions consequent to felling adversely affect their establishment and growth. Openings are invariably occupied by heliophilous colonizers, like *Macaranga peltata*, *Leea sambucina* and *Trema orientalis* (Rai 1979). Left undisturbed for a sufficiently long time, the primary colonizers would doubtless be replaced by other species, including those which are commercially important. But, with short felling cycles, such a possibility can be ruled out.

Evaluation
Long-run sustainability of a polycyclic system depends on the intensity of timber harvesting and the success of regeneration. Examples where the system has been applied over a number of cycles continuously are not available. At every revision of the management plan, new areas are included for timber extraction and the felling cycle and girth limits are reduced to enhance the availability of timber. Removal is focused on species that were not harvested earlier. In the absence of effective steps to augment regeneration, selective felling at short intervals leads to degradation, seriously affecting wood production in the long run. From the experience in Kerala, it would appear that selective felling is a passing phase and is soon replaced by alternative systems (especially clear-felling) or alternative land-uses. Improvement in accessibility, increase in the number of marketable species and acceptability of low girth logs for industrial uses have contributed to intensive exploitation.

The Andaman canopy lifting system

Until the 1950s, when demand was limited to a few species such as padauk (*Pterocarpus dalbergioides*), gurjan (*Dipterocarpus* spp.) and white chuglam (*Terminalia manii*), a selective felling was being followed in the Andaman Islands. Growth of the plywood industry enhanced marketability of species and improved accessibility favoured the adoption of a monocyclic system, the Andaman canopy lifting system, a variant of the Indian irregular shelterwood system (Anon 1983). Prescriptions dealing with yield regulation and regeneration are summarized below.

Yield regulation
Conversion of the irregular forest into a normal forest and realization of the maximum yield of timber are the stated objectives of management (Sharma 1979). There are about 200 tree species, but only 40 are commercially valuable (Bathew 1983). A rotation of 100 years is prescribed; but the annual area for felling is worked out adopting a shorter conversion period of 75 years to enable the harvesting of mature and overmature trees to be achieved quickly. A system of floating periodic blocks is adopted. An area identified for harvesting and regeneration during the tenure of a working plan is allotted to PB I. All other areas are included under PB unallotted.

A combination of area, volume, girth, and number checks is used for yield regulation. Total volume of trees in the exploitable girth (which varies from

120 to 180 cm) in PB I area is estimated by a 10% partial enumeration. From this enumeration, the annual availability is worked out and a further check is exercised by fixing an upper limit of 15 trees ha^{-1}. Important marking rules adopted in PB I areas are as follows: trees marked for felling should as far as possible be evenly spaced; felling should not be carried out on steep slopes and blanks where regeneration is deficient; where regeneration is insufficient, at least 10 sound healthy trees of commercial species should be retained per hectare as seed trees; no felling should be carried out for a width of 40 metres on either side of streams. This is the situation in principle. In practice, implementation of these rules is tardy, and felling damage is severe.

Regeneration

All sound trees of commercial species below the prescribed exploitable girth are retained as advance growth and will form part of the future crop. Several operations spread over the first 3 years are prescribed to promote natural regeneration (Table 25.1).

Natural regeneration of *Dipterocarpus* spp. is satisfactory. Timber extraction leaves the canopy more or less completely open, but weed growth is not very dense as in other evergreen forest areas in the country. If weed growth is kept under check during the first few years, the light demanding species, especially *Dipterocarpus* spp., are able to establish quickly. Light crown thinnings have been prescribed at years 6, 15, 30 and 50. In an uneven aged crop comprising large trees and poles, thinning is difficult and hence tends to be neglected. Absence of adequate follow-up operations affect the establishment of regeneration, undermining long run viability of wood production.

Table 25.1 Regeneration operations under Andaman canopy lifting system

Year	*Month*		*Operations*
1	March–April	(i)	Completion of timber extraction
	October	(ii)	Brushwood cutting
		(iii)	Felling of undergrowth and poles up to a height of 10 m
		(iv)	Girdling of trees between 10–20 m
2	April–May	(i)	Broadcast sowing of seeds of commercial species if natural seeding is considered inadequate
		(ii)	Weeding
	September	(iii)	Climber cutting and weeding
		(iv)	Girdling of understorey trees to permit more light
3	March–April	(i)	Weeding
	September	(ii)	Final felling and girdling of unwanted trees
		(iii)	Weeding

Evaluation
Industrial wood demand is the main factor influencing forest management in the Andaman Islands. As no area has yet been worked under the second rotation, it is difficult to assess the long run sustainability of the system. The yield is likely to decline during the second cycle and could be far less than that obtained during the initial phase of conversion.

A critical evaluation of the environmental effects of the system has yet to be made. The soil is extremely fragile and erodable. Microclimatic and edaphic changes are said to result in the colonization of deciduous species, particularly, *Pterocarpus dalbergiodes*. The intensity and scale of fellings have increased in the recent past, but regeneration operations have not changed very much. The suitability of the system to fulfil the multifarious objectives is yet to be evaluated.

Indian irregular shelterwood system in Assam

The Upper Assam evergreen forests found in Assam, Arunachal Pradesh and portions of Nagaland are characterized by the predominance of two species, namely Hollong (*Dipterocarpus macrocarpus*) and Makai (*Shorea assamica*). Initially, accessible areas of these forests were worked for railway sleepers under a selective felling system. As elsewhere, growth of industrial demand reoriented management towards the production of veneer logs (KFRI 1978). The North Eastern region, in particular the Upper Assam tract, accounts for about 55% of the installed capacity of the plywood industry in the country. Naturally, the main objective of management, as stated in the management plans, is to meet the increasing demand for veneer logs. Areas earmarked for felling and regeneration are constituted as Hollong-Makai regeneration working circle. Important prescriptions relating to yield regulation and regeneration are described below.

Yield regulation
In a situation of low demand, the rotation and regeneration period adopted were 150 years and 30 years respectively. Subsequently, they were reduced to 120 and 20 and in the 1974 revision (Das 1974) they were further reduced to 84 years and 12 years. Each felling series is divided into PB allotted (consisting of the area taken up for felling and regeneration) and PB unallotted. PB allotted consists of areas with advance growth and mature and overmature trees.

Yield is prescribed for the whole felling series and is obtained from PB I areas by way of regeneration fellings and also from PB unallotted through selective fellings on a cycle of 12 years. In both cases, yield is regulated firstly by area and secondly by a girth limit check. In PB I areas, the girth limit adopted is 150 cm and all trees below this are retained as advance growth to form part of the future crop. In PB unallotted, a girth limit of 300 cm is prescribed. If trees above 300 cm are not available, the girth limit can be lowered to 270 cm. Interestingly, such selective removal from the PB unallotted provides the bulk of the timber yield, sometimes accounting for about 65% of the annual yield (KFRI 1978).

To facilitate regeneration, fellings are staggered in three stages. In the initial stage, all trees above the exploitable diameter are removed. If advance growth is absent, stems above 180 cm girth are retained as mother trees. Underwood is removed leaving a sufficient number of trees to keep down weed growth. During the second intermediate stage, there is periodic removal of underwood and overwood as regeneration becomes established. In the final stage, underwood and overwood are removed, except for that retained as part of the future crop.

Based on the stocking of mother trees, PB I coupes are divided into three categories and felling is prescribed taking into account the status of regeneration. Profuse weed growth, particularly of *Mikania* sp. is a serious problem. Drastic openings promote weed growth, totally smothering regeneration.

Regeneration

In PB I areas, there is complete reliance on natural regeneration. All Hollong and Makai trees below 150 cm girth are retained as advance growth. Prescriptions aimed to promote regeneration are as follows: all marked trees not removed during main fellings are felled or girdled; damaged seedlings are coppiced; thinning is prescribed in congested groups of poles and advance growth thus freed; weeding and climber cuttings are to be carried out for 3 consecutive years after main felling; after the 3rd year, weeding is undertaken every alternate year and climber cutting once every 3 years until the 9th year; where regeneration is deficient, artificial planting should be done in strips by transplanting seedlings of 60 cm and up at a spacing of 50 x 50 cm.

Failure of regeneration in areas worked previously has necessitated artificial regeneration. Weed growth, especially *Mikania*, is a major problem. Degraded forests and blanks, which were earlier worked under the shelterwood system, have been constituted into the Hollong plantation working circle with the objective of converting them into Hollong and Makai plantations.

Evaluation

Management of Upper Assam evergreen forests is entirely influenced by the demand for veneer logs. Rotation, regeneration period, exploitable girth, etc. have been revised periodically to enhance immediate wood supply. Even with such drastic changes, the forests are unable to meet the growing wood demand. On account of the significant reliance on selective removal from PB unallotted to augment the annual yield, the system cannot be strictly called a monocyclic shelterwood system. Retention of advance growth up to 150 cm girth in PB I areas defeats the avowed purpose of introducing a monocyclic system.

Implementation of regeneration prescriptions has been far from satisfactory. Consequently, most of the forests in the Assam valley have been severely degraded, and increasingly the plywood industry has to rely upon resources available from the adjoining forests in Arunachal Pradesh and Nagaland. Management practices adopted in these states continue to be the same and it is only a matter of time before these forests are also depleted. Failure of natural regeneration has necessitated the adoption of artificial regeneration.

Constraints in sustainable management

To summarize, the different management systems adopted in the wet evergreen forests in India are primarily geared to enhance wood production in the short run. Consequently, some of the theoretical differences between silvicultural systems disappear at the time of implementation. Reliance on selective removal under the irregular shelterwood system to increase the yield and retention of all trees below the exploitable girth as advance growth are not consistent with the principles of monocyclic systems. Rotation, regeneration period, felling cycle, harvestable girth limits, etc., are not based on the growth rates and regeneration requirements of the species, but on compulsions to augment immediate wood supply.

Regeneration, if at all successful, is due to the accidental coincidence of favourable conditions and not something planned deliberately. Success of regeneration is hence an exception rather than the rule. Seed and seedling biology, the effect of changes in light and moisture conditions due to felling and the response of various species to the changed conditions are not understood adequately. In view of the failure of regeneration in most areas, one cannot be too optimistic about the long run sustainability of the present approaches. As already discussed, the trend is towards abandonment of natural regeneration in favour of artificial regeneration, with species whose silviculture and management are well understood. Extensive forests have been clear-felled and planted with species like teak and eucalypts.

MANAGEMENT FOR OTHER OBJECTIVES

Protection of catchments and steep slopes and the production of non-timber products are other important objectives of management of the evergreen forests. Management for fulfilling these functions can be illustrated by the practices adopted in Kerala.

Protective functions

Areas which are difficult to work are constituted into protection working circles. No timber extraction is carried out in these areas, thus incidentally fulfilling functions which are incompatible with intensive wood production. Interestingly, no consistent criteria are adopted for the purpose of identifying the protection working circle. Strictly, factors like terrain, soil characteristics, rainfall intensity and watershed values should be taken into account. In practice, with every revision of the management plan, areas which were previously under protection working circle tend to be reclassified to enable timber extraction, depending upon the demand for timber and improvement in accessibility. In short, the area classified as a protection working circle provides a reserve to be tapped as and when the demand increases (FAO 1984).

Production of non-timber forest products

Collection of non-timber products is an important use of evergreen forests. Availability of a variety of products – medicinal plants, tanning and dyeing materials, gums, resins, fruits, roots, etc. – is directly linked to the high species diversity. Traditionally, forest dwelling communities collect these items and use a major part for subsistence consumption. Low intensity collection of minor forest products is compatible with other uses and management prescriptions are primarily regulatory in nature.

Exploitation of certain products is intensified when their commercial value increases. They are separated from the general list of minor forest products, and rules and regulations are formulated for their management, as in the case of rattan, bamboo, reeds, etc. Like those prescribed for timber working, they are more often rules of thumb, not based on technical information, and their implementation is tardy. The emphasis is on harvesting what is available naturally, and short-term profit motives dominate decision-making to the extent of drastically undermining the basis for sustainable production (Nair 1985).

Other uses

In the extreme case of commercial orientation, organized cultivation is taken up if the produce is of high market value. A typical example is that of cardamom cultivation in the high level evergreen forests in Kerala. Incompatibility between intensive cardamom cultivation and realization of other values from tropical evergreen forests is discussed elsewhere (FAO

1984). Low returns from cardamom – either due to decline in productivity, or due to periodic decrease in world prices – has compelled owners to alter the land-use pattern (Nair 1984). Clear-felling of evergreen forests and their conversion to plantations of coffee, tea, rubber, oil palm, etc. is one step further. This is an important factor contributing to the deforestation in all the three tropical humid regions in India.

The home-garden system found in Kerala and other tropical humid areas attempts to simulate the evergreen forest, both structurally and functionally, and this seems to be an ecologically acceptable option in the context of population pressure. The traditional home-garden system has several interlinked components and caters to a variety of household consumption needs. Interestingly, this system is also undergoing changes in response to economic pressures (Nair and Krishnankutty 1984) resulting in a simplification of the system, a reduction in species diversity and the adoption of intensive cash crop monoculture.

CONCLUSION

Management of the tropical wet evergreen forests in India has been dictated by the rapidly changing socio-economic environment. Growing industrial demand has accelerated the rate of exploitation and very little effort has been made to restock felled areas. Management prescriptions have been changed frequently to enhance immediate wood production. Further, the pressure for conversion to other forms of use, in particular cash crop plantations, has increased.

The non-sustainability of such approaches has already become evident. In the recent past this has led to a near complete reversal of the policy governing forest management in the country. The forest policy resolution of 1988 has given a high priority to maintenance of environmental stability and proposes to protect completely the tropical evergreen forests, especially those in Kerala, Andaman and Nicobar Islands and Arunchal Pradesh in the North Eastern Region. A complete ban on felling in natural forests has been agreed to, but the implementation details have yet to be worked out. Import of logs has been liberalized and increasingly wood based industries are depending on wood from distant sources, in particular South East Asia. In a way this only transfers the problem to another area and is not a viable option in the long run. An ecologically sound system to resource management focusing on the balanced use of the large number of products, including timber, is yet to emerge.

Avoiding the blind alleys of extreme approaches requires concerted efforts to develop intensive multiple use management practices. No doubt, it will take considerable time to develop and adopt them. Establishment of a network of biosphere reserves covering all representative ecosystems is an important approach to meet this challenge. While the core areas will help to protect intact ecosystems, they would also provide tremendous scope for research on natural processes designed to enhance our understanding of the system as a whole. Buffer zones could be utilized to develop and test sustainable management systems with the active participation of local communities. Survival of the few remaining pockets of tropical evergreen forests is crucially dependent on immediate action in this direction.

REFERENCES

Achuthan, K. (1982). *Working Plan for Thenmala Division, 1981–82 to 1990–91*. (Kerala Forest Department: Trivandum)

Anon. (1983). *Hundred Years of Forestry in Andamans*. (Forest Department, Andaman and Nicobar Islands: Port Blair)

Bathew, W.B. (1983). Forward. In *Hundred Years of Forestry in Andamans*. (Forest Department, Andaman and Nicobar Islands: Port Blair)

Champion, H. G. and Osmaston, F.C. (1962). *The Forests of India*. Vol. **IV**. (Oxford University Press: Oxford)

Champion, H.G. and Seth, S.K. (1968). *A Revised Survey of Forest Types in India*. (Manager of Publications: New Delhi)

Das, A.C. (1974). *Working Plan for the Digbol Forest Division 1974–75 to 1985–86*. (Forest Department: Assam)

FAO. (1984). *Intensive Multiple-Use Forest Management in Kerala*. FAO Forestry Paper 53. FAO, Rome.

FRI and Colleges. (1961). *Hundred Years of Indian Forestry*. Vol. **II**. (Forest Research Institute and Colleges: Dehra Dun)

Nair, C.T.S. (1984). Land use conflicts in the catchment of Idukki reservoir: their implications on erosion and slope stability. In Loughlin, C.L.O. and Pearce, A.J. (ed.) *Proceedings of the Symposium on Effects of Forest Land Use on Erosion and Slope Stability*. Hawaii. 7–11 May 1984. (East-West Center: Hawaii)

Nair, C.T.S. (1986). Bamboo reed industry in Kerala State, India. In FAO (ed.) *Appropriate Forest Industries*, pp. 99–109. FAO Forestry Paper 68. (FAO: Rome)

Nair, C.T.S. and Chundamannil, M. (1985). *Forest Management Systems in the Tropical Mixed Forests of India*. Unpublished report. (Kerala Forest Research Institute: Peechi)

Nair, C.T.S. and Krishnankutty, C.N.. (1984). Socio-economic factors influencing farm forestry: a case study of tree cropping in the homesteads in Kerala, India, In Rao, Y.S. (ed.) *Community Forestry: Socio-Economic Aspects*. (FAO (RAPA): Bangkok)

Pillai, N.M. (1984). *Second Working Plan for Ranni Division 1974–1984*. (Kerala Forest Department: Trivandrum)

Puri, G.S., Mehr Homji, V.M., Gupta, R.K. and Puri, S. (1983). *Forest Ecology*. Vol. **I**. *Phytogeography and Forest Conservation*. (Oxford and IBH Publishing Co: New Delhi)

Rai, S.N. (1979). *Gap Regeneration in the Wet Evergreen Forests of Karnataka*. Research Paper KFD.2. (Karnataka Forest Department: Bangalore)

Sharma, S.K. (1979). *Working Plan for the North Andaman Forest Division 1979–80 to 1988–89*. (Forest Department, Andaman and Nicobar Islands: Port Blair)

CHAPTER 26

SILVICULTURAL INTERVENTIONS AND THEIR EFFECTS ON FOREST DYNAMICS AND PRODUCTION IN SOME RAIN FORESTS OF COTE D'IVOIRE

H.F. Maître

ABSTRACT

The reaction of the tropical rain forest to different silvicultural operations – logging and thinning – was investigated in three reserved forests of Côte d'Ivoire (one evergreen, one semi-deciduous, one transitional between the two). A similar 400 ha experimental area was laid out in each forest type, each divided into 25 plots of 16 ha. Three sets of measurements, made on more than 48 000 individual trees over a span of 4 years, provided data on diameter growth and annual increment of more than 50 merchantable species, thus yielding information formerly unavailable about the requirements and behaviour of these species. Taking into account growth, mortality, and regeneration of the entire stand, silvicultural operations frequently result in a doubling of annual production, and also favour the recruitment of the most desirable species. As a result of these findings, a larger scale pilot programme in forest management is being carried out in a 10 000 ha block of evergreen tropical forest at Yapo.

THE CHANGING FOREST RESOURCES IN COTE D'IVOIRE

The most realistic way of ensuring the survival of the forest of the humid tropics as sources of timber and other products and services is to demonstrate its economic usefulness. One way of doing this is rigorously to establish the production possibilities of these forests after traditional selective harvesting and simple silvicultural treatments. There are two key questions. What are the possibilities of rebuilding the productive base of forest stands affected by an initial exploitation regime? Are there simple, low-cost measures for encouraging the favourable development of valuable species and thus ensuring their potential production of timber?

This dual question has been posed in particularly vivid terms in Côte d'Ivoire, where considerable clearing of forested lands has occurred over the last 30 or 40 years. Thus, in 1983, the estimated total area of forest cover in Côte d'Ivoire was 3.5 million ha, compared with 14.5 million ha 40 years earlier. Between 1973 and 1982, some 3 million ha of forest were cleared, a rate of about 300 000 ha yr^{-1}. Conscious of this problem, the authorities of Côte d'Ivoire have decided to set up a permanent forest estate regrouping 2.5 million ha of relatively untouched gazetted forest.

In order to test and refine a continuing and practicable system of management based on a relatively simple set of practices, a triple series of silvicultural trials was initiated in 1976 by the national company responsible for the development of forest plantations (Société pour le Développement des Plantations Forestières, SODEFOR), with the technical support of the Technical Centre for Tropical Forestry (Centre Technique Forestier Tropical, CTFT). The overall design and initial results of these trials in natural dense forest form the basis of the present contribution. Further information on the work is provided in Maître (1986, 1987) and Maître and Hermeline (1985).

EXPERIMENTAL DESIGN

Research principles in natural forest

World-wide, the early research efforts of a range of different organizations in the heterogeneous environment of the mixed tropical forest were undertaken in a very dispersed manner. The means were practically always disproportionate to the stated objectives, and there was often confusion between research and application. In most of the tropical forest regions, large numbers of experimental and trial plots were set up. However, these were almost invariably too small in area, and there were no linkages between them due to the absence of a basic methodology and agreed research design. There was a lack of practical possibilities for interpretation and comparison of the data collected.

Over the years, the series of forest plots, often well founded in themselves, suffered from financial cut-backs and shifts in forest policy as a result of changing administrations. Nonetheless, these plots gave rise to a mass of documentation on the management of natural forests. However, these data were often incomplete and were not interpreted. The amount of useful information and action guidelines which flowed from them were not commensurate with the amount of data collected, and with the time and money invested in obtaining them.

These shortcomings were taken into account in the planning of a FAO-implemented project in the dense forest of Malaysia in the 1970s (Cailliez 1974). Among the design principles was, first, a concentration on a few plots of large size (several hectares), with the greatest possible number of replications in spatial terms; second, measurement of simple parameters (e.g. diameter and location of trees); third, provision of a certain capacity for storage and interpretation of data (entailing the use of computers). Such a conception of silvicultural research in dense tropical forest is certainly ambitious, above all in terms of the means required, but it is perhaps the only way at the present state of understanding to get round the bias and erroneous interpretations that result from the heterogeneous nature of the forest environment.

Research objectives in Côte d'Ivoire

It was within this general context that an experimental research operation was initiated in Côte d'Ivoire in 1976 by the national body responsible for forest development, SODEFOR. The objectives of the study were defined principally in terms of the considerations mentioned above, but also in the light of the successes and failures recorded in work on the improvement of natural forest stands undertaken in various parts of Africa (e.g. Côte d'Ivoire, Gabon, Ghana, Nigeria) between 1945 and 1965. In broad terms, this work had primarily aimed at making the forest more homogeneous, through intervening directly at the level of the natural regeneration by the more or less brutal opening of the forest canopy. Practically everywhere, such initiatives were faced with the problem of proliferation of light-demanding climbers, and the high cost of interventions, often numerous and phased in time, which were difficult to justify both on technical and economic grounds.

The present study in Côte d'Ivoire was primarily focused on stems and trees of diameter ≥ 10 cm diameter at breast height (dbh). The principal aim was to elucidate stand dynamics as affected by relatively simple, low-cost treatments that could be undertaken on a large scale. As regrowth at ground level was known to be difficult to manipulate and control, no particular treatment was envisaged for facilitating the early regeneration. It was considered that regeneration growth in a favourable or unfavourable direction was determined by the intensity and type of silvicultural treatment at a higher level in the vertical profile.

The principal themes of the study were thus as follows: to test and refine simple silvicultural techniques, particularly thinning and harvesting; to study the behaviour and growth of trees of different species as a function of these silvicultural treatments; to clarify the development of the forest stand as an ensemble (induced mortality, natural recruitment of young stems, effects on climbers and regrowth), always as a function of silvicultural treatments; to monitor the growth of young stems and seedlings within the regrowth; to quantify the effects of different interventions on production and to define the treatment(s) best adapted to the constraints of field site and production regime; and, finally, to determine the benefits gained through intervention compared to non-intervention.

Summary description of experimental areas

The same type of research procedure and experimental design was repeated at three sites, one for each major forest type found in Côte d'Ivoire. These were at Irobo (evergreen), at La Téné (semi-deciduous) and at Mopri (transitional). Though separated from each other geographically, each area had a similar basic research design, in terms of surface area and lay out. Each study area comprised a square of 2 km by 2 km, that is a total of 400 ha, subdivided into 25 plots each of 16 ha. The total research area was thus 1200 ha.

The principal species studied were those destined for merchantable production, either those actually in commercial use or little known species whose use was considered technically viable. The number of such species is in excess of 70 in Côte d'Ivoire, and some 50 were present within the various study areas. Other tree species were also studied, but without botanical identification and through grouping them as species of secondary interest (or secondary species). As a function of the richness of the forest and the type of exploitation previously carried out in the three forest blocks, the treatments performed in the 75 plots were as follows:

> Thirty plots (10 in each study area) were left untouched and served as controls;
>
> In 35 plots, thinning operations were carried out by the on-the-spot elimination of individuals of the so-called secondary species using standard poisoning techniques with arboricides (these refinements were carried out at two intensities, by the suppression either of 40% or 30% of the basal area, in a systematic fashion and by starting with the largest stems of secondary species until the desired percentage refinement was achieved);
>
> Ten plots (all located at La Téné) were exploited commercially by the removal and sale of trunks of desirable species with a diameter ≥ 80 cm (average extraction 53 m^3 ha^{-1}).

Application of these treatments coincided with the first series of general measures that marked the start of the experimentation process. Silvicultural treatments were undertaken over the whole 16 ha area of each plot unit, but observations and measurements were confined to the central 4 ha area, to avoid or minimize border effects. Only stems > 10 cm dbh were numbered, and all trees belonging to the principal species were localized precisely within each plot following the classical system of rectangular co-ordinates.

Periodic measurements were made of the circumference of all trees (both principal and secondary species) at 1.30 m or above the buttresses. These were made manually at 2 year intervals, beginning at the time of plot establishment and initiation of treatments (year 0). Such measurements were made on more than 48 000 individual trees. In addition, a subsample of 3750 individuals of the principal species was followed and precisely measured every 6 months, using metallic dendrometric ribbons permanently placed on the trees.

These measures provided information on individual and total growth in terms of diameter, basal area and volume (volumetric tables having been previously established). In addition, information was obtained on the recruitment of young trees having attained 10 cm diameter between two inventories, and the natural disappearance or mortality of trees (other than through harvesting or thinning) that occurred between two inventories. The analysis and deduction of trees of 2–10 cm diameter corresponding to the different initial stages of recent regrowth was carried out in 40 subplots of 100 m^2 systematically established within each central 4 ha study plot.

The mass of information thus obtained required a considerable effort in terms of computer storage and handling of data. The results summarized here are based on data obtained after the third general series of measurements (i.e. 4 years after the silvicultural treatments) for stands with trees >10 cm diameter at breast height (dbh), and 7 years after the treatments for the sub-plots in which studies were carried out on seedlings and young trees of diameter 2–10 cm.

MAIN RESULTS

Visual change in tree populations

Several months after silvicultural interventions (poisoning of secondary species or harvesting of principal species), there was a marked visual change in stand appearance – existence of chablis, dead or broken trees, and other gaps in the forest canopy induced by the harvesting of 50 m^3 ha^{-1} of commercial timber or through elimination of 30–40% of the basal area. The largest gaps were less evenly distributed within the forest stands where harvesting had been carried out than in stands thinned through poisoning. In effect, commercial harvesting isolated certain individual trees, while nearby there would be clumps of practically undisturbed individuals next to the control plot.

Nevertheless, within 4 years, the forest had almost regained its former mantle. The naked uprights of dead trees had for the most part disappeared, and the stand was again made up of healthy looking trees without climbers. The upper stratum had simply been transformed into a more open canopy, with a marked abundance of the more valuable species, particularly in those areas that had been thinned. In addition, the numerous stems of small diameter which had been given light by the partial opening of the dominant strata were emerging clearly from the understorey. An important amount of regrowth was also evident in the larger gaps. These comparisons were particularly evident as control and treated plots were adjacent to each other within the same experimental area. The fear at the beginning of the experiments that one might be inducing an irreversible destruction of the dense forest stands gave way, rapidly and surely, to the feeling that intervention had, in fact, favourably affected the dynamics of the forest.

Behaviour and response of principal species

The response of eight tree species to different treatments during a 4-year period is summarized in Table 26.1 (in which the two thinning intensities are grouped). A number of general conclusions can be drawn about the effects of the two main treatments on mean annual growth in diameter, percentage recruitment, and mortality for two size categories of trees.

First, principal species respond favourably to the thinning of secondary species. A marked increase (50–100%) in diameter growth is recorded, particularly for stems of modest size. In addition, thinning considerably accelerates the increase of young stems of value, particularly in the category > 10 cm dbh. The dynamics of the small stemmed individuals is markedly affected by the partial and homogeneous opening of the forest cover, which acts like a breath of fresh air for natural regeneration.

Second, simple commercial harvesting produced gains in growth and recruitment systematically inferior to those obtained through thinning, but often higher than those recorded in the untouched control plots. Harvesting leads to a reduction in basal area of the same order of magnitude as that in the thinning operation, but it occurs in a much more heterogeneous way, and without favouring the principal species over the secondary ones. This explains the relatively modest impact of harvesting on the growth of valuable species.

Third, in respect to mortality (both in terms of chablis and natural mortality of principal species), the interpretation is somewhat ambiguous.

Table 26.1 Responses of eight tree species to thinning and harvesting

	Plots	Number of trees measured	Mean annual increase in dbh for stems between		Percentage stems attaining 10 cm dbh (recruitment)	Percentage mortality during 4-year period of stems of diameter	
			10–25 cm	25–65 cm		10–25 cm	>25 cm
Khaya anthotheca	Control area	198	0.20	0.61	+6.0	-4.1	-2.3
	Thinned	308	0.37	0.97	+10.4	-5.4	-1.4
	Harvested	104	0.22	–	+2.3	-2.8	–
Gambeya delevoyi	Control area	741	0.29	0.32	+3.5	-4.4	-12.2
	Thinned	637	0.59	0.64	+10.4	-5.1	-8.3
	Harvested	253	0.21	0.38	+2.4	-4.8	-11.2
Scotellia sp. pl.	Control area	812	0.12	0.28	+3.1	-4.1	-4.4
	Thinned	910	0.29	0.42	+3.2	-4.1	-6.0
	Harvested	240	0.16	0.42	+3.5	-4.5	-3.7
Aningueria robusta	Control area	341	0.20	0.41	+4.5	-3.9	-3.1
	Thinned	371	0.39	0.58	+9.9	-4.2	-5.3
Guarea cedrata	Control area	325	0.25	0.31	+7.0	-5.5	-3.1
	Thinned	566	0.47	0.53	+16.5	-8.3	-6.7
Nesogordonia papaverifera	Control area	1082	0.27	0.23	+5.5	-2.0	-1.8
	Thinned	790	0.51	0.32	+12.5	-3.2	-2.2
	Harvested	644	0.25	0.19	+2.6	-3.7	-1.9
Tarrietia utilis	Control area	1280	0.28	0.57	+2.9	-2.8	-2.6
	Thinned	1942	0.58	0.91	+6.0	-3.9	-3.3
Triplochiton scleroxylon	Control area	640	0.60	0.79	+5.7	-2.8	-1.4
	Thinned	359	1.53	1.48	+15.6	-2.5	-1.4
	Harvested	511	0.87	1.13	+7.2	-5.7	-2.8

No clear and direct correlation could be established between the opening of the stand and an increase in mortality. Noteworthy is that tree mortality can induce important losses in volume that may result in a 'no increase' or 'negative balance' in production for entire stands.

Finally, precise and regular measurements of permanent dendrometric ribbons have demonstrated that the phenomenon of increased diameter growth is amplified from one year to another, and it is thus expected that this phenomenon would extend over a period of at least a decade.

Regeneration studies at ground level

Studies of different stages of the regrowth have been late starting, and it is still too early to draw conclusions. Nonetheless, several important ideas have already emerged. In comparison with the control plots, the silvicultural intervention has scarcely modified the floristic composition of the understory comprised of seedlings, shoots and small stems with diametersof 2–10 cm at height of 1.30 m. The dynamics of this stratum have even been slightly stimulated, particularly in respect to the frequency of young individuals. Climbers and regrowth have therefore in no way impeded the 'normal development' of the regeneration. The more valued species are poorly represented in this low stratum however; for example, in the evergreen forest at Irobo, only 16% of individuals belonged to species capable of growing to a diameter of 40 cm or more.

Production balance sheet

Estimates of changes in standing volume by individual species or groups of species (either principal or secondary) have demonstrated that untouched forest stands (i.e. those without silvicultural intervention) have a tendency to maintain a pretty constant volume of standing wood. Natural mortality is compensated in large part by overall stand growth, and the production balance sheet is a very modest one in the control plots. More particularly, each species has its own behaviour and well-defined pattern of dynamics. Certain species are very prone to producing chablis and of dying in an upright position, and this tendency is independent of silvicultural treatments. This phenomenon is particularly important in trees of large size, and losses in standing volume may have a devastating effect on the overall production of the surrounding stand. Thus, a controlled and directed commercial harvesting, which takes account of the behaviour of each species, is necessary in order to lessen potential losses of merchantable wood.

On the other hand, it appears that simple thinning operations (whose intensity is not decisive at a level of extraction higher than 30% of the basal area) creates a disequilibrium which is beneficial for the productivity of the remainder of the stand. This is particularly true for those commercial species whose increased productivity lies at the heart of the objectives of the thinning operation.

Treatment through simple exploitation leads to an increase in volumetric production of trees compared to that in the control plot. This increase is nevertheless lower than that induced by thinning, at a given level of wood removal. This apparent contradiction is explained by the fact that the impact of harvesting is very irregular. It leads to large holes, poorly distributed within the forest, and only well-situated trees can benefit from the intervention (and even then only if they have escaped residual logging damage). Moreover, the irregularity in harvesting treatment does not induce more vigorous growth of seedlings, young shoots and saplings in the most recent regrowth. In effect, the opening of the canopy leads, above all, to more active growth of small and medium size individuals, between which there is intense competition.

The woody production of secondary species is negligible, both in control and treated plots. However it is consistently important for the principal species, the annual production of which ranges between 0.5–2% of standing volume in the untouched plots, 1.5% in exploited plots and 2–3.5% in thinned areas. Thus, just for commercial species, annual increases in volume for individuals > 10 cm dbh are 0.7–1.8 m^3 ha^{-1} yr^{-1} for the untouched areas, about 2.5 m^3 ha^{-1} yr^{-1} in exploited areas, and 2.2–3.6 m^3 ha^{-1} yr^{-1} for thinned plots. There is thus a doubling in production, corresponding approximately to about 270 m^3 ha of standing volume, of which the principal species make up something in the order of 100–150 m^3 ha^{-1}.

The production of natural forest areas improved through silvicultural intervention bears comparison with that in man-made commercial plantations such as 4–5 m^3 ha^{-1} yr^{-1} obtained for teak (*Tectona grandis*) and 7–8 m^3 ha^{-1} yr^{-1} for *Terminalia ivorensis*. Paradoxically, the very heterogeneity of the forest can provide an additional advantage, since the multiplicity of hardwood species is capable of providing a large range of commercial products.

CONCLUDING REMARKS

This contribution provides a summary of the main results acquired 4 years after the setting-up of three experimental research plots for studying the dynamics of mixed forest stands in Côte d'Ivoire. Relatively precise data have enabled an initial balance sheet to be drawn-up for the mixed forest.

This system was previously little understood and interventions were often based on intuitive reasoning or deduced from observations restricted in both time and space. Though the data are considered to be reasonably precise, they remain preliminary and provisional. It is only with the benefit of more time and complementary measures that they will acquire the degree of reliability required for a deep understanding of stands of dense tropical forest.

The quantitative results obtained during the study give a clearer idea of the comparative development of stands as a function of simple silvicultural treatments, including the beneficial effects of such treatments on the growth of valuable species. The research objectives of the experimental design phase have largely been attained, in particular the primary objective, which was to highlight in quantitative terms the productivity of populations of trees, both for the various commercial species and for an entire stand.

The practical aspects of the silvicultural treatments such as thinning and harvesting, and the gains in production that result from such treatments, have been well established This has permitted the national body responsible for forestry development (SODEFOR) to launch a practical pilot programme in forest management in 10 000 ha of evergreen tropical forest at Yapo. Outstanding challenges include assessment of the duration of the effects of silvicultural treatments and of the possibilities for further silvicultural intervention. There is also a need to obtain a better understanding of the direct effects of treatments on regeneration. Assessment of management costs, economic profitability, and possibilities for large-scale application are among the questions to be answered through the pilot management scheme that is presently underway at Yapo.

The overall conclusion of the case study is that the dense tropical forest presents a continuing and vigorous dynamic system, which can be managed in a way that results in a higher and more useful production; and, in so doing, can justify the conservation of these valuable systems and turn them into a permanent source of useful products and services.

REFERENCES

Cailliez, F. (1974). *Experimental Design for Growth and Yield Experiments in Malaysia*. Report prepared for FAO/CTFT. (Centre Technique de Forestier Tropical: Nogent)

Maître, H.F. (1986). Recherches sur la dynamique et al production des peuplements naturels en forêt dense tropicale d'Afrique de l'Ouest. *Proceedings of 18th World Congress of IUFRO (Ljubljana, September 1986), Division* ***I****, Volume* ***2***, 438–50. (Yugoslav IUFRO World Congress Organizing Committee: Ljubljana)

Maître, H.F. (1987). Natural forest management in Côte d'Ivoire. *Unasylva,* **157/158**, 53–60

Maître, H.F. and Hermeline, M. (1985). *Dispositifs d'étude de l'évolution de la forêt dense ivoirienne suivant différentes modalités d'intervention sylvicole. Présentation des principaux résultats après quatre années d'expérimentation*. (SODEFOR: Abidjan, and CTFT: Nogent and Abidjan)

CHAPTER 27

MANAGING NATURAL REGENERATION FOR SUSTAINED TIMBER PRODUCTION IN SURINAME: THE CELOS SILVICULTURAL AND HARVESTING SYSTEM

N.R. de Graaf

ABSTRACT

The Celos Management System has been developed following four large experiments, carried out in Suriname for more than a decade, and holds promise as an economically and ecologically viable approach to managing highly mixed forest growing on chemically very poor soils, in areas of low population density. Controlled logging and repeated refinement are the corner-stones of the system, which is based on the harvesting of a restricted amount of about 20 m³ of quality timber from each hectare, once about every 20 years, in a well-controlled selection felling operation. Selection felling is followed by refinement using arboricides, three times during the cycle (in year 0, year 8 and year 16), to release commercial species, and provide economically sufficient increment. The main principles underpinning the system are minimum interference and maintenance of a high level of biomass to prevent leaching of nutrients from the ecosystem.

INTRODUCTION

The results of converting Amazonian rain forest into monocultural plantations have generally fallen short of expectations (for example, see contributions by Jordan and Palmer in this volume), An alternative approach to conversion is to seek to develop management systems adapted to the ecological conditions of the mixed natural vegetation and to the characteristics of each individual site. Such a system, called the Celos Management System, is the subject of this contribution. The central theme is that tropical lowland rain forest of the type studied in Suriname can be used for sustained production of quality sawlogs and peeler logs, as long as certain conditions are met.

Previously, it had been thought that this forest type was worthless after exploitation, because of low productivity. But 20 years of research and manipulation have shown the system to be experimentally manageable. That such a statement can be made has required an adjustment of management goals to ecological realities. Exaggerated demands from society, formerly taken seriously by the manager, had to be exposed as unrealistic, and destructive to the forest ecosystem.

Use of this forest for timber production calls for firm control of timber exploitation and of the harvesting of other products, and silvicultural freedom of action is restricted. The forest manager has to fend off large claims on the forest estate with one hand, and to guarantee and possibly increase sustained yield from the estate with the other. The experience in Suriname may provide insights useful for the management of comparable forests elsewhere in South America.

ROLE OF FORESTRY IN SURINAME

Suriname is situated in the humid tropics, on the north-eastern coast of South America, and contains 12 million hectares of extended forest estate out of a total land area of 16.3 million hectares. The very poor soils in the interior are not fit for permanent agricultural use, except for tree crops. The small population of about 400 000 (2.5 persons km^{-1}) is concentrated in the coastal plain. In terms of forestry development, the country thus corresponds to the fifth and final situation sketched out in Tropical Forest Action Plan (FAO 1985) that of humid tropical rain forest in thinly populated regions.

Forestry development in Suriname was for a long time influenced by forestry experience in the Dutch East Indies, which was dominated by intensive management of plantation forests on the heavily populated island of Java. This development created conditions inimical to sound policy and regular management of the productive forest estate consisting of rain forest on poor soils, notwithstanding that the first Forest Service in Suriname was established as long ago as 1907.

In the first phase of modern forestry development, after the Second World War, a timber industry was set up, and with it, many small sawmills, a plywood plant, and an infrastructure (mainly gravel roads) for the extraction of logs. Many of the forest areas thus made accessible were creamed and lightly damaged, but not destroyed. Prognoses about the future levels of timber consumption evoked quite justified anxiety about the continuity of supply. It led to several aid programmes geared to forest plantations, designed to compensate for the assumed impossibility of sustained production of timber in natural forest.

However, as time went on, it became obvious that such plantations – while they did produce timber – could not entirely replace the natural forest, as the quality of the timber, as well as the environmental effects, were so different. The plantations, in fact, proved less productive and much more expensive than had been assumed previously. It became increasingly recognized that, apart from the need for timber, the natural forest estate had to be managed permanently, if only to protect it effectively. But the money needed for management had to be generated by the forestry sector.

One response to such a situation is a return to the extraction philosophy; one just takes out more species, as the good trees of preferred species have been removed long ago. A new harvest, of low quality, but also a lower and decreasing yield, would be the result of this elimination of the best and most productive trees from the forest. Suriname has now arrived in this phase of its forestry development.

A choice must inevitably be made between returning to the forest yet another time to take out what is available, or changing strategies to a more sophisticated system of forest management which would make the forest much more productive of a sustained yield, but at the same time would entail costs and inputs in terms of regulations, control and investment.

Such a system of management, elaborated in Suriname, is known as the Celos Management System, named after the agricultural research centre where the experimental research effort was based. The system has two principal components. The first is a strict control of harvesting operations – the Celos Harvesting System or CHS (Boxman *et al.* 1985, Hendrison in press). The second consists of subsequent silvicultural treatment – the Celos Silvicultural System or CSS (de Graaf 1986).

CELOS HARVESTING SYSTEM

The Celos Harvesting System is an improved harvesting technique aimed at reducing logging damage and logging cost (Hendrison 1984, in press). Controlled logging is the key feature of the system and research has shown that not only is controlled logging an economic proposition, but that it serves to enlighten the task of the silviculturist. It also makes the harvesting operation itself more profitable. Important ingredients are as follows. A full inventory of harvestable trees precedes felling by half a year or more. This inventory enables the manager to prepare an exploitation plan for the forest compartment, with planning of skidroads and landings. Trees to be felled are indicated on the maps made from inventory data, and the selection of these trees is done in close co-operation with the responsible silviculturist. The volume to be harvested is restricted to about 20–30 m^3 ha^{-1}.

Harvesting damage, especially skidding damage, is reduced by the planning of skidroads and knowledge about the positions of trees and of logs after felling, which restricts or eliminates the 'wandering' of the machines in search of logs to haul. After felling with powersaws, logs are concentrated by crawler tractors, and are winched to the skidroad whenever possible, thus reducing the number of tracks in the forest. The logs are then skidded along the permanent skidroads by wheeled skidders, to the roadside landing. Logs are registered throughout the operation; they carry a metal number plate attached immediately after felling. Through such means, the total area affected by felling and skidtracks is reduced from 25% in conventional logging to about 15% or less in controlled logging (Hendrison 1984).

CELOS SILVICULTURAL SYSTEM

Exploitation is clearly possible without silvicultural follow-up. In fact, this is the usual situation in tropical lowland forest in South America. The alternative, effective silvicultural management without a firm grip on exploitation, is nearly impossible for natural forest. The opportunities to make the harvested stand again productive and worthwhile for further management are largely determined by biological and ecological circumstances. This is why it is the silvicultural approach that sets the standard for the whole management system.

Field experiments

The Celos Silvicultural System is based mainly on research carried out at two locations (Mapane and Kabo), both about 100 km from Paramaribo. At both sites, the soils are poor, with a high percentage of aluminium at the adsorption complex. Annual rainfall is in the order of 2000–2500 mm. The Evergreen Seasonal Forest is species-rich; 108 tree species belonging to 38 families were recorded in an estimate of trees > 5 cm diameter at breast height (dbh) in a one hectare plot at Kabo (Schmidt 1982). The amount of the living phytomass in the Kabo forest is about 480 t ha^{-1}: 16 t ha^{-1} of leaves, 118 t ha^{-1} of branches, 280 t ha^{-1} of stems and 65 t ha^{-1} of roots (Ohler 1980, Schmidt 1982). Coarse and fine litter accounts for about 34 t ha^{-1}. Analysis of the nutrients in this system indicates that there is an abundant amount of nitrogen, that calcium and potassium are present in fair quantities and that phosphorus and magnesium are scarce. The distribution of these nutrients over various compartments of the ecosystem stresses the importance of the living phytomass (Jonkers and Schmidt 1984). Between 70% and 90% of the amount of phosphorus, potassium, calcium and magnesium present in the ecosystem is incorporated in the living phytomass. Nitrogen is the notable exception (Ohler 1980, Schmidt 1982).

These are among the general characteristics of the natural forest upon which four field experiments were undertaken, with a view to providing insights and data upon which an improved silvicultural system could be developed. The first experiment was set up in 1965, at a time when attention was focused on a monocyclic approach to forest management. Thus, the experiment concentrated on manipulation of seedlings and small trees of commercial timber species to form a homogeneous stand, rather than on increasing short-term output of marketable timber. The results of the experiment over a 14-year period indicated a good potential for increasing the number of saplings and middle-sized trees by eliminating undesirable trees of non-marketable species. By today's standards, the resulting stand cannot be considered to be optimal, in terms of structure and economics, because the monocyclic approach is expensive and leads to undesirably homogeneous forest stands.

In the second experiment, the dynamics of lightly exploited forest with no silvicultural manipulations were studied over a nine-year period. Growth and mortality were found to result in low production of commercial timber volume, too low to be the basis for a management option.

The third experiment studied a variety of treatments and methods of intervention after light exploitation. The range of treatment schedules tested over a 12-year period included an option for polycyclic management. The relatively small size of the tree populations studied per treatment schedule hindered interpretation of results, and girth increment per diameter class was found to be the best parameter to identify the optimal schedule of treatments.

The fourth experiment was a field trial of the optimal treatment schedule in the third experiment. As yet, the period covered has been too short to allow definitive conclusions, but the population of timber trees in the 16 ha plot was sufficient to make an initial prediction of harvestable volume at the end of the first cycle.

Research methodology in these various studies was based on silvicultural manipulation of permanent field plots, and observation and recording over extended periods, rather than construction of theoretical models of tree growth or stand development. Treated field plots have been monitored for 7 to 15 years, mostly with annual recording. Whether the optimal experimental set-up was selected may be questioned, but the range of treatment schedules was well selected. Using conventional methods, an extensive amount of data was collected on a number of variables, including girth at breast height, local species name, form and condition of the tree, and total basal area of all species in a stand. In addition, information about treatment cost and organization has been collected and, in later years, the findings of ecological studies on parameters relevant to silvicultural manipulation have been included. Data were processed by computer, but outputs were kept very simple to allow for recombination of interim results of calculations.

Silvicultural operations

The results of the silvicultural experiments have been presented in detail by de Graaf (1986), with accompanying tables and figures which include forest profile diagrams (Fig. 27.1) and stereo-photographs of the vegetation in experimental plots.

After controlled logging, a few silvicultural treatments must be applied, because, even after well-controlled harvesting, the volume increment of commercial timber and plywood species is very low. In Suriname, it is estimated as about 0.2 m^3 annually, when no silvicultural treatment is given. This would mean that a period of 100 years would be required before it was possible to harvest a second equal yield. The increment of commercial species clearly has to be improved.

In the lightly exploited forest, competition is still high, especially for the smaller trees of commercial species left after harvesting. Refinement (i.e. killing of non-commercial competitive trees) is applied to alleviate competition in the stand. Repeated refinement will increase production to more than 2 m^3 yr^{-1}, a ten-fold increase. The silvicultural philosophy is to do the least possible to reach an economically satisfactory production. Growth is channelled into commercial timber production, but without destroying the ecosystem. Treatments must be easy to apply as they usually have to be organized with mostly unskilled personnel. Ecological risk is minimized by keeping the forest as natural as possible.

A stage of secondary forest is quite undesirable in this scheme of management. Natural forest, once cleared, comes back as impoverished secondary forest. This is optimal neither for quality timber production, nor for continued site productivity. Thus, overshoots in areas cleared should not be permitted, and land-use planning needs to indicate firmly the boundaries to be respected.

The choice of species to be grown in the system is a basic one. Those species already accepted by the timber industry seem a logical starting point. The Celos list of commercial species comprises 20 families, with 49 species. A stand composed of these taxa would form a quite diverse forest, even in the absence of other species. The total number of families in, for example, the Mapane forest is about 41, with some 144 species, not all of which are large trees. Many of these would be conserved in the CSS.

Stem-diameter and volume-diameter distributions are in line with those generally reported for humid lowland tropical forest, with decreasing (normal) distribution for numbers, and a unimodal distribution of volumes, with a peak in the 40–50 cm diameter class. Total timber volume in stems above 25 cm dbh is about 200 m^3 ha^{-1}, of which 20–30% is made up of commercial species using the Celos list, which is admittedly a fairly progressive one compared to that of the timber trade, which is usually more conservative, taking out no more than 10–30 m^3 ha^{-1}.

The sequence of treatments is illustrated by Figure 27.1, in which the development of the stand structure is followed during a 20-year period. Three refinements are applied during this period, reducing the total basal area as shown in Figure 27.2.

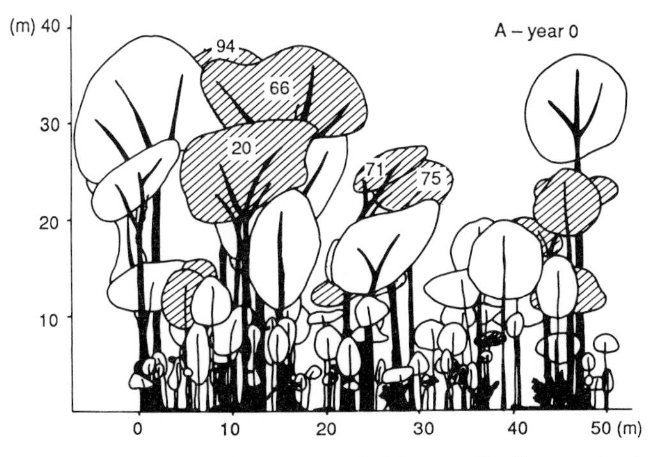

Fig. 27.1 (Continued on following pages) Simplified forest profile diagrams showing changes in hypothetical forest structure at Mapane at stages during a 20-year period of harvesting and refinement (after de Graaf 1986). A – year 0. The hatched crowns are the commercial species. B – year 2. After a light harvesting of two trees (Nos. 66 and 75) and a refinement in which all non-commercial species above 30 cm diameter at breast height (dbh) have been killed to provide growing space for remaining trees. C – year 7. Immediately prior to second refinement. Most dead trees have fallen Commercial species have grown considerably. Non-commercial species have increased, and these need to be reduced in number to improve increment and recruitment of commercial species. D – year 9. After a second refinement in which all non-commercial species above 10 cm dbh have been killed. The method of drawing leads to under-representation of small trees. E – year 15. Immediately before a third, light, refinement. Trees have increased in size and numbers, and much recruitment has passed the diameter limit for representation in this type of profile diagram, in sharp contrast to the forest profile. Note the broken stump of a large commercial tree (No. 71) and the lianas which will be a potential hazard to small trees in the subsequent harvest. F – year 20. After the third light refinement (in year 17, which eliminated mainly lianas and a few small trees of undesirable species) and after a second light exploitation in year 20, during which two trees (Nos. 20 and 94) were removed. The stumps are about 15 m apart and the falling trees caused a fair amount of damage to the stand. The height of the forest is comparable with the original forest, but the more open structure indicates less competition. Commercial species dominate the stand, but in the lower strata, non-commercial species are still represented

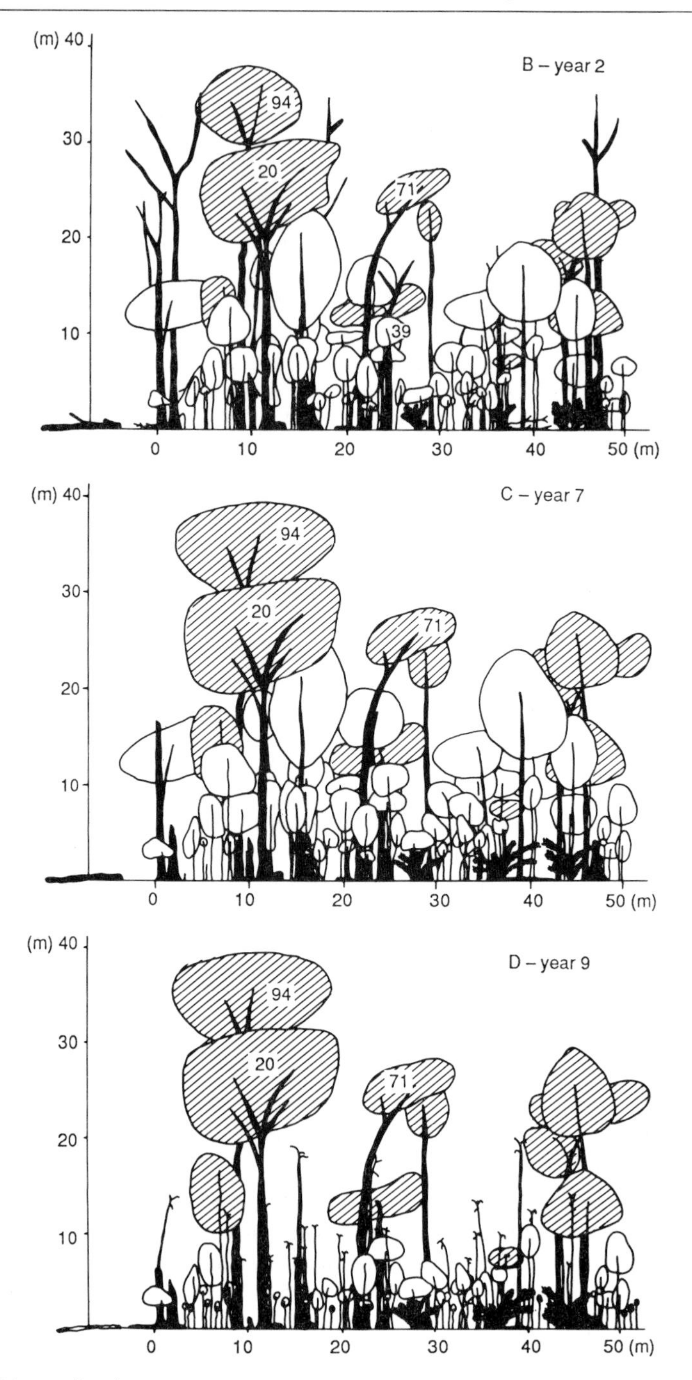

Fig. 27.1 continued

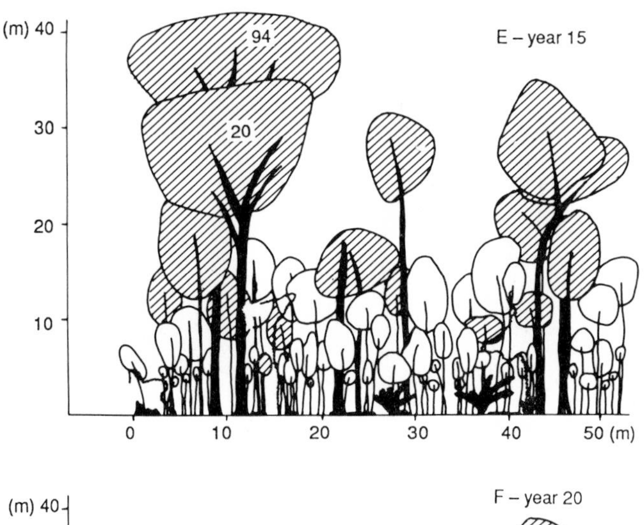

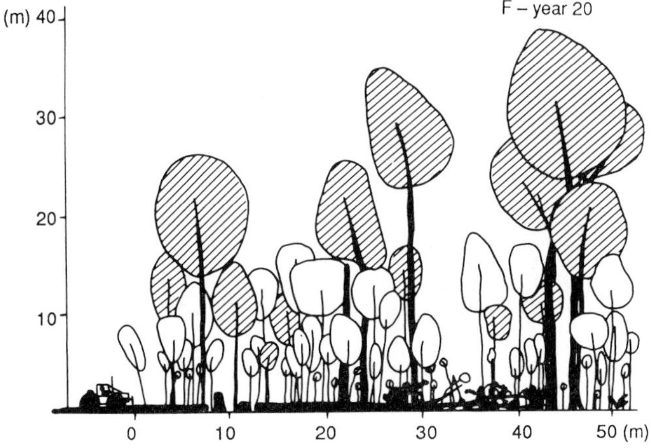

Fig. 27.1 continued

Refinement in itself is applicable in monocyclic as well as polycyclic systems, and the choice between these systems is important to examine. In Suriname, research started in 1958 with a monocyclic systems approach. Reasons for this were: the harvesting damage that accompanies polycyclic systems; the purported economic need for high per area volumes at final felling; the ease of canopy manipulation in monocyclic systems, and the supposed dominance of a single age-class in heavily refined stands. As noted above, a series of field experiments was set up in 1965–67 to test this approach and reasoning.

Results prior to 1965 had indicated that a reasonable amount of advance and small regeneration was usually present in primary or lightly logged

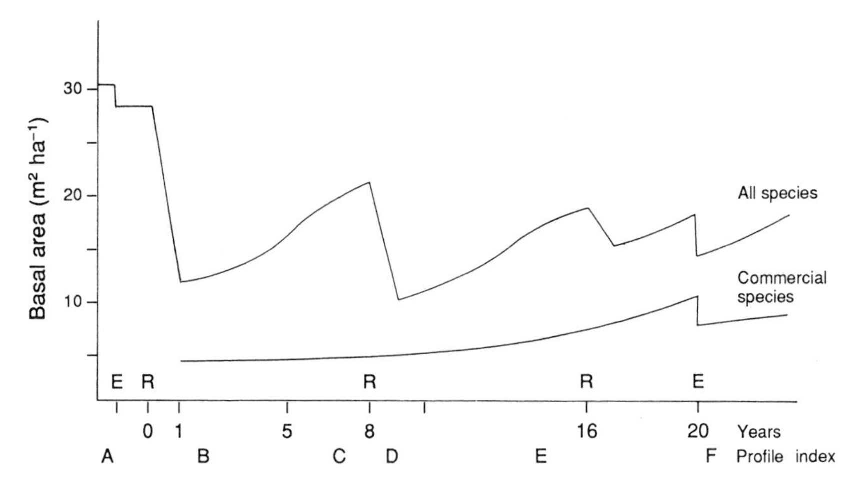

Fig. 27.2 Projected development of basal area of all species and of basal area of commercial species, under the Celos Silvicultural System proposed for the Mapane region. Time of: exploitation = E; refinement = R; forest profile diagrams in Figure 27.1 = A – F

forest. Special treatment to induce more regeneration was assumed to be unnecessary. Heavy refinements had proved quite effective in improving the increment of this regeneration. However, frequent maintenance was necessary to help this commercial regeneration to compete in the rapidly growing vegetation with many secondary species that came to prominence after such drastic refinements. Further change in the frequency of maintenance was needed to reduce costs. The length of the cycle was set at 60–80 years, to produce logs of 60–80 cm dbh.

Arguments to change over to a polycyclic system came forward strongly after 1975. Such arguments had both ecological and economic dimensions. The relatively poor mesophytic forest in Suriname is more fitted to polycyclic than monocyclic working. The diameter distribution always reverts to a 'normal' one, decreasing and meagre. Very few trees reach large dimensions and mortality is high (1.5–2% annually). Further reduction of site potential may result in a stand in which trees will only in exceptional cases grow to good timberlog size.

The most compelling argument in favour of polycyclic working is the harvesting damage at the end of a monocyclic period of 60–80 years. The more successful the management, the more destructive is the final harvesting. A period of quite unattractive, because unproductive, secondary forest will have to be tolerated before a better forest may or may not slowly emerge from the ruins of the ecosystem. The supposed need for high volumes to make harvesting more economic has been less persuasive since modern skidders became available. Also, the high cost of silviculture in a long cycle with intensive initial treatments pleads against a monocyclic approach. Furthermore, the risk of permanent site degradation is quite large when the phytomass is reduced to very low levels periodically. A polycyclic system can maintain plant biomass at fairly high levels at all times, whereas a monocyclic system usually will reduce the total phytomass considerably, particularly at the time of final felling and following regeneration treatments.

Another consideration is the role of the soil component. The traditional view of some soil scientists, that the soil is the most stable component and template of the system which holds the key to land-use planning and development, is perhaps true in some locations, in certain climes. But perhaps the concept needs to be modified in parts of the humid tropics – where the acid soils left after clearing are the dead bones of the destroyed ecosystem, rather than the base of a new system. The forest keeps healthy by way of its highly efficient nutrient cycling, while agricultural systems with low phytomass fail. Rather than inducing rattling bones, the whole living ecosystem might perhaps better be conserved in order to maintain production potential and to prevent site degradation.

CELOS MANAGEMENT SYSTEM: CONSTRAINTS, IMPLICATIONS AND POSSIBILITIES

The Celos Management System is one suited especially for permanent forest estates, as the investment has to be written off over several felling cycles. There is a problem here, in that the timber coming from non-permanent forest areas which are in the process of being converted to non-forestry uses, may compete fiercely in the national or global market place with the timber from permanent forest reserves. In Suriname, such clearing is not yet being done on a large scale, but, in other countries in South America, this is a very real difficulty confronted by those seeking to establish long-term forestry management. Timber harvested from such cleared areas should not compete, but should contribute to the financing of permanent forest reserves.

Regulation of timber harvesting is a political issue, and resistance from existing timber harvesting industries may be expected. Arguments that work in favour of the establishment of managed forest areas are not only the improved efficiency and higher output of timber per hectare, but also the minimization of transport costs. Even if additional timber is available in more distant virgin forest, the cost of infrastructure and its maintenance will eat away at the profits. It soon becomes cheaper to produce timber in a sustained way near to the sawmill, even if this means investment in silvicultural treatments. In the lowland tropical rain forest, transport distances and costs increase rapidly, as per hectare volumes to be harvested are usually low.

Large areas of unbroken high forest, with low actual population pressure, are a basic requirement for the Celos Management System. Once these forests have been put under management, a higher population pressure can be withstood subsequently. But scattered remnants of forest of several hundred or a few thousand hectares, interspersed with agricultural lands, would be far from optimal for such a system. This is primarily because the forest ecosystem would suffer from illegal hunting and felling, the areas being too readily accessible and difficult to guard effectively. The vulnerability of the natural forest ecosystem and derived managed forest systems to such disturbance is much greater than with plantations, as many essential processes are strongly dependent on relatively undisturbed animal life. Also, plantations are more readily seen as someone's property, and thus tend to be respected more than natural forest.

An area of several tens of thousands of hectares would seem to be the minimum for the CMS to work effectively. Such an area would form the territory of a medium-size forest industry, for example supporting a sawmill with plywood complex. For a 50 000 ha net area of productive forest, an estimated 150 jobs in silviculture might be expected. On top of this are jobs in the related timber industry. Each man-day devoted to silviculture results in the growth of an estimated 1–2 m^3 of timber, without input of machines, except for transportation of labour (de Graaf 1986).

In situations where increasing population pressure is expected, incorporating the still existing forest into use within a CMS-type operation would help provide jobs and would provide an alternative to clearing forest for shifting cultivation as a source of cash income. Developing a local tradition of 'naturalistic forestry' would help to stabilize land-use. The role of local labour in this process is significant, and not only as specialized tree-spotters or foremen. Rather the whole philosophy of the local people towards their land would change, once the forest had become a source of income and a stabilizing factor for their economic future. People would pay increasing attention to the well-being of the forest ecosystem, instead of seeing it as the traditional reservoir of potential agricultural land.

Illegal hunting and fishing with poison would become antisocial activities in the opinion of such a population, which would become an active force in restricting negative impacts. An example is the hostile reaction of local forest residents to the ruthless fishing with poison in small rivers in Suriname by roving bands of workers coming from other regions. Deprived of their regular source of fish through such poaching, local residents become the active allies of government in helping to put a stop to it.

Such positive effects of forestry development should help to close the gap between nature conservationists/environmentalists and foresters. Not all development in tropical rain forest lands needs to be viewed in a negative light. Operations such as those proposed within the Celos Management System have an important role to play, in fitting a niche somewhere between nature reserves on the one hand and intensively cultivated areas on the other. As a rough estimate, something like one-third of the total humid rain forest of present-day tropical America might be ultimately destined to be managed under systems of the CMS-type.

REFERENCES

Boxman, O., Graaf, N.R. de, Hendrison, J., Jonkers, W.B.J., Poels, R.L.H., Schmidt,P. and Tjon Lim Sang, R. (1985). Towards sustained timber production from tropical rain forests in Suriname. *Netherlands Journal of Agricultural Science,* **33**, 125–32

FAO. (1985). *Tropical Forestry Action Plan.* Committee on Forest Development in the Tropics. (FAO: Rome)

Graaf, N.R. de. (1986). *A Silvicultural System for Natural Regeneration of Tropical Rain Forest in Suriname.* (Pudoc, Agricultural University: Wageningen)

Hendrison, J. (1984). Harvesting systems. In *Human Interference in the Tropical Rain Forest Ecosystem.* Project LH/UvS 01. Annual Report 1983, pp. 34–7. (Centre for Agricultural Research, University of Suriname: Paramaribo)

Hendrison, J. *Damage-controlled Logging in Managed Tropical Rain Forest in Suriname.* (Agricultural University: Wageningen) (in press).

Jonkers, W.B.J. and Schmidt, P. (1984). Ecology and timber production in tropical rain forest in Suriname. *Interciencia,* **9**, 290–8.

Ohler, F.M.J. (1980). Phytomass and mineral content in untouched forest. In *Celos Rapporten*, No. **132**, pp. 1–43. (Centre for Agricultural Research, University of Suriname: Paramaribo)

Schmidt, P. (1982). Ecology. In *Human Interference in the Tropical Rain Forest Ecosystem.* Project LH/UvS 01. Annual Report 1980, pp. 30–3. (Centre for Agricultural Research, University of Suriname: Paramaribo)

CHAPTER 28

WILDLIFE CONSERVATION AND RAIN FOREST MANAGEMENT – EXAMPLES FROM NORTH EAST QUEENSLAND

F.H.J. Crome

ABSTRACT

Forest management under various regeneration systems can also be excellent habitat for many species of wildlife. Research in northern Queensland has led to guidelines whereby managers can incorporate the needs of wildlife (particularly birds and bats) simultaneously with wood production, under selective logging both with and without silvicultural treatment. Prescriptions include further protection of streamsides, retention of intact canopy areas, maintenance of certain levels of plant species diversity, measures to ensure the continued existence of sufficient individuals of 'pivotal' species (i.e. plants that maintain frugivores through lean periods, over wide areas).

INTRODUCTION

The wildlife values of tropical forest are amongst the most compelling reasons for the conservation of this habitat. Although wildlife conservation is usually included in forest management plans, there is often an unfortunate emphasis placed on reserve systems and an inadequate recognition of the value of forest managed under some sort of natural regeneration system for wildlife conservation. This situation results from several reinforcing influences. First, there is an unfortunately strong 'hard nosed' economic view that tropical forests are not worth managing under a regeneration system but should be replaced with plantations, agriculture, etc. Conservation is relegated to reserves. In such circumstances, wildlife will be greatly reduced, and there is no guarantee that what reserves are left will be adequate. Second, busy forest managers under political and economic pressures can usually pay scant attention to wildlife even where they are sympathetic. They have more immediate concerns, such as producing wood and trying to retain the forest of which they have control. Third, the conservation lobby tends to represent forests that are selectively logged and naturally regenerated as lost to conservation. They are incorrectly lumped with converted forests.

Conversion, preservation and management of natural stands are complementary forms of land-use that can be combined to maximize the land's long-term and multiple values to the population. Reserves are definitely needed to conserve many animal species. This hard fact must be realized. However, forest managed under various regeneration systems can also be excellent habitat, and the two can interact to ensure that a region retains a rich and diverse fauna. Taking this broad approach is best for wildlife conservation in view of the fact that, under foreseeable population and political pressure, most of the world's tropical forests will either be managed for timber and other products, or eliminated. To maximize and publicize the conservation values of managed forests will enhance the value placed on this form of land-use as opposed to more destructive strategies. Many of the chapters and case studies in this volume have shown that naturally managed systems are attractive alternatives to conversion if the multiple values of tropical forest are taken into account. The imperatives to manage natural tropical forest on ecologically sound principles have never been greater.

In this case study, I discuss wildlife conservation in relation to various management practices in the forests of northern Queensland. I refer only to forests managed under our selective logging system and not to clear-fell systems, tree plantations, etc. I wish to demonstrate that research can suggest simple prescriptions so that managers can incorporate the needs of wildlife simultaneously with wood production. I also wish to draw attention to the potential problems that intensive silviculture could generate for wildlife, and the factors to consider if such problems are to be avoided.

THE FORESTS

Although the total area of rain forest in mainland Australia is small, only some 15 000 km^2, it is of singular importance for the Australian vertebrate fauna. Of the 475 species of land birds and 172 mammals on the mainland, 166 and 71 respectively occur in rain forest; of these, 89 and 59 are totally dependent on these forests. Thus, 0.2% of Australia's land area supports 19% of its bird species, and 34% of its mammal species. The wise management of these forests is thus of great importance.

The forests between Townsville and Cooktown are our most extensive and complex examples, support the highest diversity of vertebrates and are managed for timber production. Some 18% of the original 10 000 km^2 of rain forest in this region has been cleared and, of the remainder, 14% is in national park and 61% under the control of the Queensland Forestry Department (Winter *et al.* 1984). Departmental policy is to log selectively all accessible areas of State Forest and to continue recutting on a low yield basis (60 000 m^3 yr^{-1}) for the foreseeable future. Logging is carried out under a polycyclic system with no subsequent stand treatment after logging. Commercial species are harvested to specified lower girth limits and the department has developed environmental guidelines to reduce soil erosion and residual stand damage. There was much early interest and experimentation into post-logging silvicultural treatment for stand improvement, but these are not pursued at present.

MANAGEMENT AND WILDLIFE

This case study is in two parts. The first presents preliminary results of a study to minimize the impact of disturbance by the present selective logging system on wildlife. The second part explores the implications of certain silvicultural treatments for wildlife conservation.

Selective logging without treatment

A study has been carried out by CSIRO into the effects of selective logging on birds, small mammals and bats in upland rain forests on the Windsor Tableland. This is an isolated granite massif, 110 km NW of Cairns, clothed in complex notophyll vineforest and surrounded by various dry, open eucalypt formations. Results for birds and bats are presented here.

Birds

The study site consisted of a 45 ha unlogged control area and an adjoining experimental area of 19 ha that was selectively logged in October 1985. Populations of birds were monitored by mist netting simultaneously on both areas before and after logging. The sampling was designed to test the hypothesis that bird species were differentially distributed according to topography and the availability of moisture, and to assess how their use of the area changed after logging. Three 'units' were sampled within each area – ridges, gullies (i.e. the heads of small streams that only have running water during rain) and the vegetation along permanent streams. These units receive different 'treatments' during logging. Roads tend to be on the ridges and much soil is exposed; the gullies receive no environmental protection and may be completely exposed and/or choked with tree heads and debris; the vegetation along streams receives some protection if the stream is large enough.

Our work has confirmed that birds are differentially distributed according to topography and it appears that the streams and their associated vegetation are particularly significant (Table 28.1). Fourteen of 36 species that were consistently captured were caught more frequently adjacent to streams compared with 12 on ridges. Moreover, three species are bower birds (Ptilonorhynchidae) that mostly use the ridges for their display grounds. Most significantly, the streamside supported about half of the terrestrial species caught, and, surprisingly, a disproportionate number of middle and upper canopy species. Observation, movements of marked birds and time of capture during the day indicated that these and others species commute to the streams to drink and/or to shelter during hot, dry periods. After logging, the use of streamside vegetation increased. For example, capture rates per net metre per day for bridled honeyeaters (*Lichenostomus frenatus*), a mid to upper canopy species, increased from an average (in November) of 0.21 in the 3 years before logging to 0.47 in the 1st year after logging.

It is concluded that no bird species are lost through selective logging, but that there is an increased use of intact canopy areas by the avifauna. Of these intact canopy areas, streamside vegetation is particularly important for both specialist streamside species and for a range of species that commute to the streams daily or in times of drought. Even very small creeks are significant in this respect.

One implication of these findings to forest management is that streamsides should be afforded further protection. At the moment, a shelter belt of vegetation 10 m wide on both sides of a creek is provided on water courses downstream from catchments of at least 60 ha. This shelter belt should also be afforded to the smaller tributaries. Where the stream is simply too small to

Table 28.1 Analysis of bird captures in the Windsor Tableland logging study

	Caught more frequently in			
Habit	*Streams*	*Ridges*	*No difference*	*Not caught*
Terrestrial	4	2	1	2
Low canopy	2	2	3	1
Mid to upper canopy	7	1	4	10
Upper canopy				2
All levels	1	3	5	5
Above canopy				2

(Entries are number of species)

provide a shelter belt, rules to prevent felling into the streamside vegetation should be strictly enforced. These rules are currently in place but not necessarily enforced on small streams. This practice will also benefit the rich and endemic amphibious and aquatic fauna not included in this study.

From the managers' point of view, stream protection and erosion control is an accepted objective. These results indicate the added benefits to wildlife that can accrue from extending this protection, albeit with the loss of some production depending on the application of protection rules in each logging area. This is a matter for skilful balancing by the manager and field staff, if possible assisted by research workers. For example, a percentage of the small streams in each logging area could be protected, depending on the results of a pre-logging survey.

Bats

A transect was established on the Tableland from eucalypt woodland into the rain forest, and bat activity was measured by recording their ultrasonic calls at 21 sites on the transect. Three of these sites were in woodland, one on a river in woodland, one on the ecotone between rain forest and woodland, and the rest in rain forest. Within the rain forest, bats were monitored in paired areas at each site – one in closed canopy and the other in a contiguous gap produced from logging activity. Prior to field work, a set of hypotheses was set up derived from a consideration of the ways in which animals could use gaps as opposed to areas of closed canopy, and from an analysis of bat morphology.

Bats are highly specialized and their wing morphology determines their flight characteristics. Species with long narrow wings (as measured by aspect ratio = wing span^2/wing area) and a high wing loading (= weight/ wing area) are fast flyers with limited manoeuvrability and are poor at avoiding obstacles in highly cluttered environments, whereas those with broad wings and low wing loadings are slow and highly manoeuvrable. I reasoned therefore that: first, there should be strong differentiation between the gaps and canopy areas in their faunas – the fast flyers would be restricted to gaps, and the slow, more manoeuvrable, species to the closed canopy; second, because of conflicting demands on morphology, there would be few 'gap includers' (i.e. species that could use both gaps and intact canopy); third, since the fauna is very small there should be no specialist rain forest gap species (i.e. those that were restricted to gaps in rain forest), and, because the gaps are numerous and highly connected, open habitat invaders from eucalypt woodland should dominate the gaps.

The results confirmed these predictions (Table 28.2). There were 12 species of insectivorous bats and the fauna of gaps was different to that of closed canopy. There was no evidence of any gap specialist and all the species caught in gaps were open habitat invaders recorded from the

Table 28.2 Use of gaps and canopy areas by bats

Species	*Aspect ratio*	*Wing loading*	*Activity in* Gap	Canopy	Eucalypt
Chaerophon jobensis	9.03	0.191	0.7	0	0.3
Chalinolobus nigrogriseus	7.67	0.106	3.1	0	0.9
Hipposideros diadema	6.49	0.144	0.7	0	0.4
Mormopterus beccarii	8.49	0.215	1.7	0	0.5
M. loriae	8.61	0.102	3.8	0	1.0
Nycticeius balstoni	7.22	0.106	0.4	0	0.7
Eptesicus pumilus	6.95	0.087	0.1	0.2	0.3
Unidentified species	–	–	0.1	0.2	0
Nyctophilus bifax	5.85	0.089	0.1	0.5	0
Eptesicus sagittula	7.84	0.09	0	0.8	0
Hipposideros ater	5.52	0.079	0	0.3	0
Rhinolophus megaphyllus	5.62	0.085	0	0.8	0

(Activity measured as average number of passes per half hour per site)

woodland sites. As Table 28.2 shows, gap or canopy use was strongly related to flight morphology. Thus, there are two subfaunas within the insectivorous bats in this rain forest: one in the gaps produced from logging, and the other in the closed canopy areas.

The implications for forest management are that, where a forest is managed for multiple use, conservation requirements would dictate that areas of closed canopy must be left to conserve the endemic rain forest bats. Under our polycyclic system, this practice would involve reducing roading damage and road width, winching logs as far as possible rather than driving tractors to them, and reducing or eliminating very large gaps produced when several trees are felled in a small area. To a certain extent, these measures are already taken to reduce residual stand damage, and a strengthening and extension of these rules would be beneficial. On the debit side, again, is the potential loss of some production in areas where several commercial stems are concentrated and some are left to reduce gap size.

When recutting an area, the fate of the species that need intact canopy will depend upon the time between recuts and the stage of succession at which they can begin to use the regrowth in the gap, the number of intact canopy areas left after the recut, and the amount and extent of openings created by making the timber accessible, in addition to the inevitable reopening of successional areas on the roads and major snig tracks. These are important research topics, particularly the timing of the recut in relation to the age at which canopy species can use the growing successional areas. All that we can say at present is that the rotation cycle should be as long as possible.

Silvicultural treatment

Silvicultural treatment systems have been an important topic in this volume and are an important alternative to converting forests. Their usefulness is well brought out in the Appanah-Salleh case study (Chapter 24). Large-scale silvicultural treatment, following or associated with logging, that aims to alter plant species composition in addition to promoting regeneration, and which encourages the recruitment of woody-fruited plants, may threaten the survival of the specialist frugivores that are such an important component of the forest system.

A characteristic of vertebrate communities in rain forest is the large number of species heavily or totally dependent upon fruit, and which require fruits that supply at least a maintenance level of nutrients through the year. These species are generally divisible into obligate and facultative frugivores. Within these groups, some digest the pericarp only, voiding the seeds intact ('legitimate frugivores' *sensu* Snow (1981)), and some are capable of digesting the seed ('seed predators'). For example, the avifauna of North Queensland rain forests has nine species of obligate and at least 29 species of facultative frugivores, and eight of the nine obligate frugivores are 'legitimate'.

'Legitimate frugivores' require fruits of high nutritive value and are more specialized in their fruit choice than 'seed predators' which can gain much of their nutrition from the seeds and therefore are able to eat fruit with poor quality flesh (Crome 1975*a*, Snow 1981). The range of plant species used by particular frugivores is often narrow and some plants are visited by a limited range of animals. For example, in New Guinea, two trees *Chisocheton weinlandii* and *Gastonia spectabilis* are only visited by birds of paradise (Beehler 1983), and the family Lauraceae is a critical family for fruit pigeons *Ptilinopus* and *Ducula* (Crome 1975*a*, 1975*b*), cassowaries *Casuarius casaurius* (Crome 1976) and quetzals *Pharomachrus* (Wheelwright 1983).

These specialist frugivores are maintained by a succession of appropriate plant species fruiting sequentially. The bulk of studies to date show that fruit is seasonal and sometimes erratic in its availability and frugivores sometimes go through lean periods when they depend upon only a few species of plants called 'pivotal' species. At such times, they are vulnerable to loss of these food sources and the disappearance of these frugivores may, in turn, have profound effects on dispersal and hence demography of plant species. Howe (1984) has recognized this problem in relation to the management of isolated tropical forest reserves. Foster (1982) reported mass starvations of frugivores when crops failed on Barro Colorado Island, Panama, and, in the lowland forests of North Queensland, the wet season is a general time of fruit scarcity. In some years, populations of the canopy-dwelling Wompoo fruit-dove *Ptilinopus magnificus*, for example, may be maintained by as few as two to four plant species between March and April. The shortages are usually erratic in space allowing nomadic species to move away from areas short of fruit, but others such as cassowaries and the arboreal possums are not so mobile and may succumb. The overall distribution of fruit resources in the landscape in general is vital in this respect. When they are widespread, statistical vagaries in the nature and extent of a lean period will ensure that there are some places with food that nomads can reach and that will maintain populations of the more sedentary frugivores, and also provide source areas for reinvasion of patches from which these species have disappeared. Severe alterations in species diversity and composition of essential fruit bearing species by silvicultural treatment have the potential to eliminate frugivores.

There is a long history of silvicultural research in Australia (Volck 1960) that variously involved experiments in underplanting desirable species, poisoning and/or girdling less desirable species, cutting climbers and regrowth, etc. It resulted (in 1954) in a system of logging and treatment rules which have been progressively modified. The basis of the 1954 system was a grouping of compulsory species into four desirability groups A to D. So-called A species were to be preferred over B, B over C, etc.. Unclassified commercial species were regarded as less desirable than D, but were given no official classification. They are included in the group 'U' in Table 28.3. The logging rules attempted to improve the representation of species in the higher classes over broad areas by selectively removing less desirable stems and species. Specific silvicultural experiments were directly aimed to improve markedly the stocking of species of the more desirable groups.

A variety of studies of frugivores (Crome 1975*a*, 1975*b*, 1976, Stocker and Irvine 1983) have provided a good picture of the specific requirements of these species, and Table 28.3 presents the diets of selected species classified according to the timber groups of the 1954 rules. As can be seen, they are all highly dependent upon species in the lowest desirability group and non-commercial species ('U'). This is because the high value groups consist mostly of species with woody fruits and wind-dispersed seeds. Group A, for instance, consists of 11 species of which only one bears fleshy fruit. Conversion of forests to high representation of the more valued groups thus has the potential of adversely affecting frugivores by reducing plant species diversity and increasing the abundance of woody fruited species. It would not be necessary to see the realization of Dawkins' (1955) refining concept or Ryley's (1960) suggestion that all non-commercial species be eliminated before frugivores would be affected. Loss or diminution of populations of trees needed during lean periods is all that would be required. In the Australian situation the large, less mobile species such as cassowaries would be most at risk.

Table 28.3 The percentage of the diets of selected frugivores classified according to timber groups

	Timber Groups 1954 Rules				
	A	*B*	*C*	*D*	*U*
Southern cassowary (1)	0	26	17	1	51
(*Casuarius casuarius*) (2)	0	1	7	13	52
Red-crowned fruit-dove (*Ptilinopus regina*)	0	0	0	43	51
Superb fruit-dove (*Ptilinopus superbus*)	0	0	4	28	48
Wompoo fruit-dove (*Ptilinopus magnificus*)	0	1	34	16	38
Torresian imperial pigeon (3) (*Ducula spillorrhoa*)	0	0	10	14	50
White-headed pigeon (*Columba leucomela*)	0	0	27	38	16
Brown cuckoo-dove (*Macropygia amboinensis*)	0	0	0	0	66

U = non-compulsory or uncommercial species.
Data from:
(1) Crome (1976) based on % of food items;
(2) Stocker and Irvine (1983) based on % of species; and
(3) Crome (1975*b*);
all other data from Crome (1975*a*).

It must be stressed that the current Queensland system of selective logging without silvicultural treatment does not pose a threat to frugivore populations. Clear-felling is not undertaken and the regrowth has a high plant species diversity. It is post-logging management of forests to alter species composition which poses the problem. Only some 5000 ha have been treated in northern Queensland because it has been perceived as being uneconomic and of limited efficacy. However, experimental plots show high levels of conversion to A group species are possible, with one stand being converted to over 70% of A group species (D. Nicholson, *pers. comm.*). Should such manipulation be accepted for forest management in the future, then the following considerations will have to be taken into account.

(1) The size and location of intensively treated areas in relation to existing pristine reserves, less intensively managed areas, the distribution of frugivores and their major movement and migration routes.

(2) The efficiency and efficacy of the silvicultural system in changing species composition. Systems that enrich the forest with timber trees but do not drastically alter species composition probably present no problem. Those that result in drastic alteration would need to be

strategically designed and located and/or be modified to ensure sufficient trees for frugivore survival.

(3) The ensured existence of sufficient 'pivotal species', i.e. those plants that maintain frugivores through lean periods, over wide areas. Food shortages occur unpredictably in space and time and so a wide distribution of species that can play this role, not necessarily the same ones everywhere, would increase the odds for survival of frugivores when the shortage is widespread.

We still lack critical information for the above. This missing information includes a better knowledge of the range of species useful to frugivores, particularly the pivotal species, a better understanding of their movements between the existing forest areas, and an assessment of the range in composition of stands resulting from silvicultural treatment. If treatment is reintroduced, such information would be of great value.

GENERALIZATIONS BEYOND AUSTRALIA

The Australian rain forest fauna is impoverished in comparison to other parts of the world. Groups such as large browsing mammals, primates, and large carnivores are missing or poorly represented. The animals in this study are only a part of the Australian rain forest fauna, and the study cannot purport therefore to provide a comprehensive analysis of forest management and wildlife conservation problems. It does, however, identify areas that are generally important, whilst accepting that each geographic region is going to have an additional unique set of problems, such as controlling illegal hunting, managing elephants, preserving rhinos, etc.

The protection of streamside vegetation should be generally useful beyond Australia. In a completely different and much richer avifauna in Panama, Karr and Freemark (1983) have also recorded differential use of vegetation in wet areas as opposed to dry ridges by birds, and Karr (1982) explains the loss of some species on Barro Colorado Island to be due to the rarity of permanent streams and their associated dense humid vegetation that birds need during the occasional dry period. That a similar phenomenon exists in lowland Panama and montane Queensland strongly suggests that it is of widespread relevance.

The retention of intact canopy areas is undoubtedly useful for canopy species other than Australian bats. It is particularly important in areas where monocyclic systems result in virtual clear-felling. In such situations, if a polycyclic system cannot be introduced, then efforts should be made to maintain some canopy patches in addition to stream vegetation.

The significance of silvicultural treatment is widely relevant and may be more apposite beyond Australia where there has been emphasis on treating forests to alter species representation and where silviculture has been widespread and sometimes intensive, as in Nigeria (Kio 1976, FAO/UNEP 1981) and Sabah (Chai and Udarbe 1977, Lee 1982*a*) or is still being researched and/or implemented as in Malaysia's hill forests (Lee 1982*b*, Jabil 1983).

Where silvicultural treatment is to be practised extensively, it should be assessed critically in relation to the requirements of specialist frugivores, taking particular account of the three points above. This assessment is particularly necessary where there are few rain forest national parks and where the forest is in isolated blocks. Basic, simple information on what frugivores eat, when and where they go during critical periods of fruit shortage, and the nature and extent of forest treatment is needed. If silvicultural treatment returns to favour, such information could be used in modifying treatment rules as necessary to ensure that sufficient plant species diversity is maintained in these treated forests.

Each region has a characteristic flora and fauna, and would have to be assessed individually, so the research itself should be carried out by local ecologists and/or foresters. It could usefully form the basis for doctoral studies with immediate application. Above all, it would be most appropriately done in co-operation with forest managers at all stages.

REFERENCES

Beehler, B. (1983). Frugivory and polygamy in birds of paradise. *Auk*, **100**, 1–12

Chai, N.P. and Udarbe, N.T. (1977). The effectiveness of current silvicultural practice in Sabah. *Malaysian Forester,* **40**, 27–35

Crome, F.H.J. (1975*a*). The ecology of fruit pigeons in tropical north Queensland. *Australian Wildlife Research,* **2**, 155–85

Crome, F.H.J. (1975*b*). Breeding, feeding and status of the Torres Strait Pigeon at Low Isles, north-eastern Queensland. *Emu,* **75**, 189–98

Crome, F.H.J. (1976). Some observations of the biology of the cassowary in northern Queensland. *Emu,* **76**, 49–58

Dawkins, H.C. (1955). The refining of mixed forest. A new objective for tropical silviculture. *Empire Forestry Review,* **35**, 448–54

FAO/UNEP. (1981). *Tropical Forest Resources Assessment Project. Forest Resources of Tropical Africa. II.* Country Briefs. (FAO: Rome)

Foster, R.B. (1982). Famine on Barro Colorado Island. In Leigh E.G. Jr., Rand, A.S. and Windsor, D.S. (eds.) *The Ecology of a Tropical Forest: Seasonal Rhythms and Long-term Changes*, pp. 201–12. (Smithsonian Institution Press: Washington D.C.)

Howe, H.F. (1984). Constraints on the evolution of mutualisms. *American Naturalist,* **123**, 764–77

Jabil, Dato M. (1983). Problems and prospects in tropical rain forest management for sustained yield. *Malaysian Forester,* **46**, 398–408

Karr, J.R. (1982). Avian extinctions on Barro Colorado Island, Panama: a reassessment. *American Naturalist,* **119**, 220–39

Karr, J.R. and Freemark, K.E. (1983). Habitat selection and environmental gradients: dynamics in the stable tropics. *Ecology,* **64**, 1481–94

Kio, P.R.O. (1976). What future for natural regeneration of tropical high forest? An appraisal with examples from Nigeria and Uganda. *Commonwealth Forestry Review,* **55**, 309–18

Lee, H.S. (1982*a*). The development of silvicultural systems in the hill forest of Malaysia. *Malaysian Forester,* **40**, 27–35

Lee, H.S. (1982*b*). Silvicultural management in Sarawak. *Malaysian Forester,* **40**, 485–96

Ryley, T.P. (1960). The timber resources of north Queensland. *Australian Timber Journal*, **26**, 50–60

Snow, D.W. (1981). Tropical frugivorous birds and their food plants: a world survey. *Biotropica,* **13**, 1–14

Stocker, G. C. and Irvine, A.K. (1983). Seed dispersal by cassowaries (*Casuarius casuarius*) in north Queensland's rain forests. *Biotropica,* **15**, 170–6

Volck, H.E. (1960). *Silvicultural Work in the Tropical Rain Forests of North Queensland.* Paper presented to Ninth Commonwealth Forestry Conference.

Wheelwright, N.T. (1983). Fruits and the ecology of resplendent quetzals. *Auk,* **100**, 286–301

Winter, J.W., Bell, F.C., Pahl, L.I. and Atherton, R.G.. (1984). The specific habitats of selected north-eastern Australian Rain Forest Mammals. Report to the World Wildlife Fund, June 1984.

CHAPTER 29

JARI: LESSONS FOR LAND MANAGERS IN THE TROPICS

J.R. Palmer

ABSTRACT

Successes with the 100 000 ha tropical tree plantations at Jari in eastern Amazonia, Brazil, were due to application of conventional management techniques and conventional tropical plantation silviculture. Failures were due to ignorance of the tropical literature, employment of unsuitable personnel and a reluctance to call on external expertise. Successes were sustained by demand-driven continuity of policy and assured finance for flexible management. Rapid feedback from silvicultural research kept plantation management costs low. The large scale of the operation required the use of simple management systems in which small scale ecological effects were submerged. No distinctively 'ecological' studies were undertaken: the long-term studies were those associated with prudent silviculture.

INTRODUCTION

A few introductory words are perhaps necessary on why a case study on the Jari scheme – essentially a project of tropical plantation silviculture – appears in a book on the regeneration and management of mixed rain forest systems.

The Jari operation has been one of the largest-scale attempts at managing humid tropical lands during the last 20 years. As such, there have been many newspaper and magazine articles on the scheme, but relatively few technical papers on its research results, apart from Rollet's (1980) account and several attempts to cover the estate as an enterprise (e.g. Hornick *et al.* 1984, Kalish 1979).

I do not think that the experience of Jari provides a model for development elsewhere. However, there are a number of broadly applicable lessons which can be learned from Jari – lessons which one should learn during an undergraduate forestry course:

(1) You can learn expensively through making mistakes, or you can learn cheaply by using a good library where the elements of tropical land development and plantation forestry are in dozens of texts and thousands of articles.

(2) Any development is to some extent site-specific, and to be successful requires the development, through applied research, of its own unique combination of techniques.

(3) Fast-growing trees require timely silvicultural operations to attain their commercial potential, and that, in turn, means agile and sustained financing, delegated authority and an efficient short-linked command structure.

(4) The skill of a forest manager is in deciding what combination of operations is required for any particular situation, a skill which can be partly taught or book-learned, but is mostly built up by sheer experience; and that this skill is now all too rare in the tropics, but is worth almost any asking salary.

(5) A forestry scheme is long term and spatially extensive, so the land tenure of the estate must be clearly defined and legally secure.

My direct experience of Jari spans 1976–79, before the sale of the forest operations to a consortium of Brazilian companies in 1982. From the beginning of operations in 1967, the scheme has been staffed almost entirely by Brazilians, with only a handful of expatriate senior staff, mostly from the United States. The relative isolation of the Jari estate helped to canalize staff energies into the tasks in hand rather than into non-professional activities. This factor probably accelerated the learning process faster than it was slowed by the limited access to, and knowledge of, the technical literature and the experience of tropical foresters elsewhere. The understandable pride in home-grown solutions led to some manifestations of the NIH (Not Invented Here) syndrome, unhelpful in a somewhat closed community. Any technical error in a scheme where the planting rate reached 13 000 ha per year was bound to be extremely expensive compared with the cost of not making the mistake. As Jari is perhaps better known for its mistakes than for its successes, I shall spend most of this case study on a sample of these mistakes and attempt to explain why they occurred. The contribution is not intended to be a balanced and comprehensive summary or critique of the Jari estate. Readers will notice a number of differences between this paper and the much more elaborate account prepared by Rollet (1980). These differences are due partly to changes in company practices, and partly to the different objectives of the authors.

ORIGINS AND HISTORICAL DEVELOPMENT

The conception of the Jari scheme dates from the 1950s. A shipping magnate with financial interests in many fields commissioned a review of potential multipurpose tropical trees to form the basis for an integrated land development scheme. He was advised of the FAO projections forecasting a world shortage of paper pulp from the 1980s onwards. Since he wished to run his scheme as a wholly-owned operation, it was planned on a sufficiently large scale to sustain the cost of establishing, from scratch, the infrastructural support which is the responsibility of the national government in most countries. A staff member of the New York Botanical Garden proposed *Gmelina arborea* as the preferred species, emphasizing its rapid growth and its ability to grow on a wide range of tropical sites. Trial plantings were made in various of the owner's estates in neotropical countries, or on land purchased or leased for this purpose. These were all abandoned sooner or later because the national governments would not, or could not, give the guarantees required by the owner against invasion by squatters or expropriation.

In 1967, 1.6 million ha were purchased from private individuals on both banks of the Rio Jari in Brazilian Amazonia. The Rio Jari is the last major north bank tributary of the Amazon before its estuary, and is deep enough to admit ocean-going freighters to the 'port' of Munguba, below the first rapids which mark the southern edge of the Guyana Shield. Clear title was obtained to 400 000 ha and the then military government of Brazil agreed that the land reform and colonization agency INCRA would clarify the property titles on the remainder, and transfer them to what later became Cia. Jari Florestal e Agropecuaria Ltda. The shipping magnate's companies were highly experienced in large scale construction projects and Jari was approached essentially in civil engineering terms, with engineers rather than foresters in command for the first 7 years. The engineers were charged with blanketing the forest area with *Gmelina* and the existing forest was simply an obstacle to be removed. For people accustomed to moving mountains to build dams, the solution was obviously to use large tractors and a bush-crusher.

FORESTRY CONSIDERATIONS

Loss of surface soil and compaction of subsoil

One of the penalties for attempting development in a remote place is the lack of a machine culture among the inhabitants. In the interior of Amazonia, there was not much call for ability to drive or maintain a Le Tourneau tree crusher. This machine was designed for clearing heavy hardwood forest, but unskilled drivers overstrained it by attempting to use raw power rather than directed energy when felling and windrowing trees. Crawler-tracked machines with blades were often driven with the blades down, the thin nutrient-bearing soil horizons were scraped off, and the subsoil was often compacted by repeated passes of the tractors. Forest clearing started at the western bank of the Rio Jari and moved west. The serious effect of surface horizon removal and subsoil compaction on tree growth was not noticed until the crews moved from the sandy riverside sediments on to the clayey hills and plateaux. From that point, some 2 years after the start of the plantations, clearing changed to manual methods. From then, at least until 1980, considerable care was taken to avoid topsoil loss and subsoil compaction during plantation management operations. The short- and long-term damage caused by unskilled use of heavy machinery in tropical agricultural schemes was documented long before Jari started, but the civil engineers in charge were not aware of the effects.

Site sensitivity

The same hapless engineers were also unaware of the body of literature on *Gmelina* built up since the turn of the century, especially in India and Burma. Alan Lamb's (1968) monograph was published as Jari was starting, but a company copy did not reach the plantation headquarters in Monte Dourado until 1974. *Gmelina*, like other commercial Verbenaceae, is markedly site-sensitive. Empirical site indices were developed for Jari in the mid-1970s and ranged from 7 to 31 m dominant height (mean total height of the 100 fattest trees ha^{-1}) at an age of 10 years. The plantations on the sandy sediments were at the lower end of the site qualities. The compartments looked more like apple orchards than forest plantations. From 1974, when conventional forest management was initiated, the *Gmelina* was increasingly restricted to the top end of the site classes, as determined by soil sampling during pre-felling forest inventories. Growth of final crop trees of *Gmelina* in twice-thinned stands on '*terra-roxa*' soils approached the best recorded from volcanic clays in Western Samoa. Fast-grown trees have more sustained apical dominance than sluggards, so a higher proportion of the net increment goes into the commercially usable bole rather than into branches.

Stump removal

Once it was clear that the *Gmelina* was not a commercial crop on the sandy soils, trials were undertaken to determine the best method of eliminating it, and replacing it by *Eucalyptus deglupta* or *E. urophylla*. These trials included repeated cutting back and the application of arboricide to the stumps. However, although the Verbenaceae may not grow well on poor sites, they are tenacious survivors. The only satisfactory method of elimination was winching out the stumps, at a cost of US $500 ha^{-1}. Some 64 000 ha of *Gmelina* were planted, but, by 1978, it was clear that only about 25 000 ha of soils in the 'First Forest' of 100 000 ha were suitable for commercial crops of this species. Post-1980, heavy tractors with root rakes have been used to grub out and windrow the stumps. This technique induces the soil compaction which we were keen to avoid in the 1970s.

Crooked versus straight trees

It might be thought that the form of the trees would not matter in a mainly pulpwood scheme, but that biomass would be all important. However, you can load more cylindrical logs onto a trailer than conical and crooked tree boles, whether you handle full tree lengths on Kenworth trucks or 2.5 m bolts on Big-Stick self-loaders. Since harvesting and transport are more expensive operations than silviculture, the advantage of producing cylindrical trees is substantial. From 1975, Jari established silvicultural trials to improve the yield in the current rotation, as well as genetic studies to pave the way for continuous yield improvements in subsequent generations. Four techniques were successful in improving the form of the trees: selection of mother trees of superior phenotype, direct-seeding in the field, intensive mass selection of seedlings in the nursery, and choice of initial spacing in the field. Mass selected plants produced more uniform stands of straighter and more cylindrical trees. Calculations suggested an improvement of 30–35% in net yield of sawn lumber from such trees on soils of site index 27, compared with unselected plants. Spacing trials, from 1.25 to 7.0 m, showed that form declined somewhat between 6 and 9 m^2 initial growing space, and markedly thereafter. Cost accounting for some years showed that the cost of harvesting small diameter trees precluded spacing below 9 m^2, but more recent studies have led to a standard spacing of 3 x 2 m. Improved economic calculations have been possible at Jari only because annually remeasured trials have been summarized soon after dry season data collection and the analyses fed back to the planners with minimum delay.

Site-species matching

The instructions of the owner at the start of the Jari plantations were quite explicit: it was to be a *Gmelina* plantation. *Gmelina* would grow on any soil, he had been advised, so there was no need for soil survey. When the first foresters with tropical experience arrived, it was clear to them that the sands were more suitable for pine. A small number of trial plots were installed well away from the main roads. By 1972, their growth was sufficiently impressive for the owner to be shown the comparison with adjacent *Gmelina*. With this evidence, approval was given for the use of any species which would produce wood pulpable by the kraft process. The mill uses batch digestion, so there is no theoretical problem in switching between species. Short-fibre 'Jaripulp' now contains up to 20% by volume of other species besides *Gmelina*, including more than 120 species of the indigenous Amazonia hardwoods. Nevertheless, the great bulk of the plantations are in the well-known and relatively narrow-crowned exotics which make low demands on the nutrient stores in the soil. By the late 1970s, the soils were being stratified into three classes from the empirical site indices assigned from the soil survey analytical data: SI 7–13 *Pinus caribaea* var. *hondurensis*, SI 14–20 eucalypt species, SI 21 and better to *Gmelina*. A modest range of well-replicated species trials showed no local species as productive as the chosen exotics. However, a number of species were identified as possible replacements should pest or disease attacks require replacement of the first choice species. There was nevertheless some argument within the company as to the value of the species trials. One side argued that a modest investment was a good form of insurance should disaster strike, and that it was the responsibility of national institutions to mount comprehensive trials. Another side argued that the range of species under test was too limited to be more than window-dressing for visiting ecologists, and that the total investment in forestry at Jari justified a much greater insurance effort in view of the weakness of the national institutions in Amazonia, and the increased product options afforded by the opening of the large sawmill in 1979.

Exploitation of native species

During the latter half of the 1970s, conventional forest plantation management was introduced, silviculturally and in terms of staff organization and in-service training. As the expected improvements were realized, the scope of work for the forestry division expanded. For the 1st decade of Jari, little timber was harvested from the natural forest before it was burned. A few species were extracted for the construction of the town and villages of the estate, and for the port and pulpmill complex. At the start of the project, it had been agreed that Jari would not flood the local timber markets with cheap lumber. Also, the owner was keen to avoid having Brazilian government inspectors on his estate, so, by another agreement, Jari did not export sawn lumber outside Brazil. Economic factors, however, caused a change of heart. The many thousands of hectares of *Gmelina* planted on unsuitable soils were naturally giving much less than the initially produced yield, so a diversification of income was highly desirable to maintain cash flow. The forest inventories started in the mid 1970s showed that Jari had mostly forest of mediocre commercial quality, but the large scale of the annual clearing produced a considerable volume of marketable logs. In 1979, a modern sawmill was constructed with an annual log input capacity of 36 000 cubic metres and export shipment began in 1980. The mill also serves to break down the logs harvested from the natural forest for wood fuel into sizes which can be handled by the 60 cm chipper. Steam for the pulp mill and the electricity turbogenerator is raised in three boilers, two of which are fuelled by wood chips, at a vast saving on imported petroleum fuel. A more comprehensive analysis of the estate at the time of purchase, including conventional forest inventories and soil surveys, would have revealed a wider range of options for its development than was initially foreseen or implemented.

Long-term productivity of the estate

In the early years of the estate, no concern was shown for the nutrient status of the soil in the forestry areas. Short-term trials in the early 1970s showed no economic response to fertilization with either nitrogen or phosphorus at the time of planting. However, simultaneously with the initiation of the yield plot programme (1200 plots by 1979), a soil monitoring programme was started, using mainly the analytical methods developed by North Carolina State University. This programme included some permanent plots to trace nutrient status from the undisturbed natural forest through felling and burning, planting, silvicultural operations and harvesting into the next felling cycle. At the same time, Jari began to take an interest in similar studies elsewhere in the tropics and especially in eastern Amazonia, and has facilitated postgraduate thesis studies. When the pulp mill came into operation, a demand was created for 1800 tonnes per day, wet weight, of hog fuel chips. A two stage harvesting operation was then begun from the natural forest, the first to remove the sawlogs, and the second to clear the remaining logs and large branches for boiler fuel. Far less ash than previously is now left after the pre-planting burn to fertilize the new seedlings. Small doses of a commercial compound fertilizer must be applied to secure the same initial growth rate, which is essential to provide rapid canopy closure and weed suppression. Unfortunately, the soil monitoring programme was closed down, together with the pre-planting soil surveys, at the time of financial stringency in the early 1980s. Although soil survey has restarted, I have no information about the monitoring programme. Senior management staff from the mid 1970s realized that some fertilizer application would be necessary in order to maintain long-term productivity, especially for the short felling cycle pulpwood crops. Lop and top were left on site during plantation harvesting, but in-forest debarking was uneconomic compared with adding the same amount of in-bark nutrients from a bag of inorganic fertilizer. *Gmelina* coppice was well able to grow through the lop and top, which provided a sort of mattress to protect the soil from the tyres of the skidders and self-loaders.

A decline in productivity was noticed early on in the main nursery. Heavy watering by sprinklers, combined with tractor-mechanized operations, led to the formation of a plough pan. Continuous cropping of seed-beds led to a decline in organic matter content and an inability to make efficient use of inorganic fertilizer. Ripping the hardpan with a subsoiler, fallowing the beds one year in three with a ploughed-in green manure crop, mainly of sorghum, and a more considered application of irrigation restored the productivity of the nursery.

Pests and diseases

One serious insect pest has affected the plantations from the beginning and will probably continue to do so. One serious fungal pest has reduced the silvicultural options. Other pests have been noted from time to time, the most serious being a 1976 defoliation of 200 ha of *Gmelina* by a geometrid. This outbreak was controlled by natural agents and has recurred annually, but not in epidemic proportions. The normal range of pests and pathogens attack the nursery and are either controlled chemically or the losses are too small to justify treatment. Good nursery hygiene and timely operations to reduce stress on the plants contribute to a minimal use of chemicals.

The serious pest is the leaf-cutter ant. Species of both *Atta* and *Acromyrmex* infest the plantations. Elimination is impossible, control is essential: three defoliations in the first 2 years can kill even a vigorous *Gmelina*, while the pine and eucalypts are even more susceptible. All plantation trees tested at Jari in the species trials are liable to attack. The best defence is rapid stress-free growth to build a large leaf area capable of recovery from partial defoliation. The most economical control is by Mirex baited with fresh citrus waste and sealed in 10–20 g polythene sachets. The ants cut open the sachets and remove the bait to their underground nests.

The common Latin American practice of dribbling Aldrin or Mirex along an ant trail is wasteful and inefficient. The bait rapidly loses its attractive odour in the open air and is rendered unpalatable by rain. Manual patrolling for ant nests and their repeated baiting to destruction was costing the company about US $1 million annually in the late 1970s. Mirex is a very persistent insecticide and the sachet method is thought to be not only the most efficacious, but also the most selective control.

The fungal pathogen is *Ceratocystis fimbriata*. Like the leaf-cutters it is indigenous and is the cause of the widespread 'machete' disease, which makes the cultivation of *Theobroma cacao* so problematical in Amazonia. Scrupulous chemical sterilization of pruning/thinning tools and pruning/thinning wounds is essential to prevent ingress of *C. fimbriata*. There is some evidence that the fungus is also carried by a bark beetle which attacks shaded parts of tree stems.

At one time, the pruning of selected *Gmelina* plantations or individual stems within the high SI compartments was contemplated with a view to producing peeler logs for plywood. *Gmelina* is a preferred species for plywood, being easy to peel and having exceptionally good properties as a base for surface treatments (painting, printing, paper and film overlays, etc.). Surveys showed an almost unlimited market in the United States for flush door blanks manufactured from *Gmelina*. However, cylindrical logs are required for economic veneer peeling and early pruning could have produced both the necessary small knotty core and a better distribution of wood increment along the tree bole. The major experiment to test pruning intensities and frequencies was destroyed in only a few months by *C. fimbriata* entering the pruning wounds. Control of bole form must thus depend on selection of good genetic material, planting method and adjustment of tree growing space by choice of initial espacement, and by later thinning. Jari knew of *C. fimbriata* before the pruning experiments were started, but the virulence of the fungus was not appreciated. Some work has been done on the selection of *Gmelina* trees which show resistance to the fungus.

Fire control

The dry season is rather variable in length from year to year, but averages 4 months having less than 100 mm. The chief cause of fires is the human with a box of matches. Most of the 7000 workers employed annually by the seven contractors operating at Jari are young men from the north-east of Brazil. Fires are set in the plantations to drive out game animals, as an expression of antipathy towards the employers, and for the sheer delight in watching a forest burn. Fire precautions include inter-visible watch towers, radio links, fire-fighting teams and standby tractors ready loaded on transporters during the fire season.

Plantation management blocks are divided into compartments of about 50 ha (500 x 1000 m where the topography permits a rectangular layout). The compartments are separated by wide gravel-surfaced main roads and well-graded earth minor roads. However, the main defence against large fires is the substantial area of almost fireproof natural forest left between the plantation blocks. Forest is not cut beside water courses nor on steep or rocky slopes, where fire would run fast.

Conservation

Brazilian law prohibits the removal of arboreal cover from more than 50% of a property. Detailed planning of forestry at Jari was confined to the 100 000 ha First Forest with preliminary mapping and inventory for a second 100 000 ha, and outline plans for two further Forests. The other Jari companies, such as the fully mechanized double-cropped padi rice farm, the kaolin mine, and the farms to supply a high proportion of the estate food requirements, occupied relatively tiny areas. Certainly, in the 1970s, no plans existed to convert anything like half the 1.6 million ha estate.

An area of particularly fine natural forest near the northern boundary of the estate was closed to access by vehicles and hunters. It was notably rich in vertebrate wildlife. Nevertheless, the company was not disposed towards conservation except in so far as Brazilian law required it and good land management dictated it. One might add that, if these simple precepts were followed throughout Brazil, there would be a lot more forest than stands now.

CONCLUDING REMARKS

What has this brief account of a few forestry aspects of the Jari enterprise to do with rain forest regeneration and ecology? Regeneration studies reviewed in many of the contributions of this volume seem to be worlds away from the problems tackled by the Jari forest managers. The successful silviculture of plantation species at Jari has involved no new techniques, and no particular knowledge of ecology as distinct from conventional silviculture. Forestry failures at Jari have been caused by the employment of the wrong kind of personnel, by a reluctance to read before doing, and by a certain hermetic attitude fostered by the estate owner. Technically, by the late 1970s, the new plantations must have rated as among the most successful anywhere in the tropics, and certainly the most successful in private ownership.

The Jari enterprise failed for a variety of reasons. A wholly-owned foreign company in a relatively remote part of Amazonia; a company headquartered at the unfortunately named and non-auriferous Gold Hill (in a country fascinated by gold); rigorous controls by guards on access to the estate; these features were bound to arouse suspicion and dislike in an increasingly confident and nationalistic Brazil. The prolonged reluctance of the owner to engage in dialogue with even the more responsible members of the Press was apparently due to his belief in the power of his personal contacts with an increasingly tottery military regime. When he finally did permit the entry of journalists (and even ecologists), the Press reports were as complimentary as he could have wished. By then it was too late. The issue of the land tenure over the bulk of the estate was never resolved, thus greatly restricting his collateral for borrowing. World prices for paper pulp when Jari started large scale production were much lower than had been predicted. The cost of construction of the pulp mill itself was greatly increased by the strengthening of the Japanese yen. A combination of rapidly rising social costs to support a population of 20 000 or more, and cash flow difficulties stemming in part from his unwise early decision to plant only *Gmelina*, left his dream unfulfilled.

ACKNOWLEDGEMENTS

A number of corrections have been made to the original paper on the basis of information kindly supplied by a former Chief of the Forest Management Division at Jari, Dr. C.B. Briscoe.

REFERENCES

Hornick, R., Zerbe, J. and Whitmore, L. (1984). Jari's successes. *Journal of Forestry,* **82 (11)**, 663–7

Kalish, J. (1979). Jari. *World Wood,* **20 (2,3,4)**, 15–21, 36–8, 20–4

Lamb, A.F.A. (1968). *Gmelina odorata.* Fast Growing Timber Trees of the Lowland Tropics, No.1. (Commonwealth Forestry Institute: Oxford)

Rollet, B. (1980). Jari: succès ou échec? Un example de développement agro-sylvo-pastoral et industriel en Amazonie brésilienne. *Bois et Forêts des Tropiques,* **192**, juillet–août 1980, 3–34

CHAPTER 30

THE PRESENT STATUS OF RESEARCH INTO MANAGEMENT OF THE RAIN FORESTS OF AMAZONIAN BRAZIL

J.C.L. Dubois

ABSTRACT

Sustained yield forest management based on natural regeneration, with or without enrichment planting, is expected to become more common in the Brazilian Amazon, in part as a result of a new law and its regulation which were approved in 1986. Research linked to the management of Amazonian forests has been hampered by insufficient financial support, understaffed research teams, and untimely delays affecting the flow of financial resources required for field work. Nonetheless, much more research is being carried out than is generally appreciated outside the country, because most research has been reported in Portuguese in publications of rather limited distribution. A short account is provided of continuing studies on natural regeneration at several sites and research stations in the Brazilian Amazon, including research at Curua-Una Forestry Research Station (where silvicultural research applied to natural forests started in 1958), Palhão Forest Reserve, Tapajós National Forest, Belterra, Jari, and Manaus.

BRAZILIAN AMAZONIA, ITS FOREST RESOURCES AND THEIR MANAGEMENT

The vast Amazonian watershed, covering approximately 750 x 10^6 ha, is the largest single extension of tropical forest in the world. The Brazilian portion amounts to an estimated 350 x 10^6 ha, with a standing timber volume in excess of 70 x 10^9 m^3. The area is served by an exceptionally extensive network of navigable rivers, which, particularly in the Brazilian portion of the basin, is one of greatest assets of the region as a sustained source of wood products, and of increasing potential value in terms of national and world timber trade. Notwithstanding these advantages, and even when liberal allowance is made for the unrecorded commercial operations, the annual amount of timber exported from the Brazilian Amazon is less, indeed much less, than timber exportations from certain smaller tropical countries like Malaysia, Indonesia, Côte d'Ivoire or western Nigeria.

The first sawmills established in the region (Belém, Santarém, Manaus) have worked on the basis of highly selective commercial fellings, involving a very restricted number of mainly cabinet timber species. The distances over which these species have to be transported from stump to mill has rapidly increased. Larger and more modern timber industries, established in the more recent past, use a much larger number of species, mainly as a consequence of a growing national demand for Amazonian timber.

The present forest destruction for the establishment of agricultural projects in the Brazilian Amazon, and the substantial volume of current timber exploitation, as well as the increasing rate of genetic erosion affecting economic timber species, require much more effort to be made in order to ensure the prompt renewal of timber resources.

The Tropical Forestry Action Plan, prepared in 1985 by FAO's Committee on Forest Development in the Tropics, proposed an investment of US\$325 million, for a 5-year period (1987–1991) to meet two objectives for forestry development in the Brazilian Amazon: first to define a forest resources management policy for the region, and second to bring under effective control and management 5×10^6 ha of high value forest in areas close to industrial centres such as Manaus.

So far, *c.* 180 000 ha of plantation forest have been established in the Brazilian Amazon (mainly, *Eucalyptus* spp., *Gmelina arborea*, *Pinus caribaea* var. *hondurensis*, *Virola surinamensis*, *Schefflera* (*Didymopanax*) *morototoni*, *Swietenia macrophylla*), but no large-scale forest management based on natural regeneration has been started. The first phase of implementation of a forest management plan based on natural regeneration is being carried out in the State of Maranhao (Forest Reserve Burticupu, 6000 ha), by Companhia Vale Rio Doce (CVRD). The objective of the management plan is to obtain fuelwood on a sustained yield basis (expected yield 250 steres ha^{-1} yr^{-1} on a ten years rotation) and sawlogs (*c.* 15 m^3 ha^{-1} sawlogs at the first commercial felling in the original forest). The preliminary forest management working plan is based on the assumption that the maintenance and simple silvicultural treatment of the 30–50 desirable trees per ha of 15–35 cm diameter at breast height (dbh) which remain after commercial fellings, will ensure the next timber crop.

As regards the two Timber Production Forest Reserves existing in the State of Pará (the Caxiuana and the Tapajós National Forests, of respectively 200 000 and 600 000 ha) commercial timber felling contracts might be signed in the near future and forest operations started within 1 or 2 years. To this effect, pre-felling commercial inventories have been carried out, so far in two forest blocks (4000 ha at Tapajós and 1000 ha at Caxiuana). Forest management, including silvicultural treatments, would be implemented under the authority of the Brazilian Forestry Development Institute (IBDF, now IBAMA – Brazilian Institute for the Environment and Renewable Natural Resources).

A new Law (Lei no. 7511) and its regulation (IBDF/Portaria no.486/86-P) were approved in 1986. They charge any commercial timber felling in privately owned natural forests with the condition that a forest management plan be approved by IBDF and implemented by the owner of the forest. As a consequence of this law, it is expected that sustained yield forest management based on natural regeneration, with or without enrichment plantings, might become more common.

CURRENT RESEARCH ON RAIN FOREST MANAGEMENT IN BRAZIL

Notwithstanding several shortcomings (especially insufficient financial support, understaffed research teams and untimely delays affecting the flow of financial resources required for field-work), research linked to the management of natural Amazonian forests has been the object of comparatively much more work and has provided many more useful results than is generally realized outside the country, because most research papers and reports have been published in Portuguese in publications of rather limited distribution. Silvicultural research applied to natural forests started in 1958 at the Curua Una Forestry Research Station through the joint efforts of the Brazilian Government and FAO. FAO experts remained in charge until 1967. Other institutions (mainly EMBRAPA, the Brazilian Agency for Agronomic Research, and INPA, the National Institute for Amazonian Research) joined later, providing conditions for a much larger geographical spread of research. The main contributions and the present state of research programmes are as follows.

Curua-Una

Natural regeneration research at the Curua-Una Forestry Research Station (71 250 ha, State of Pará, 2°45'S, 53°48'W) is under the charge of SUDAM (Superintendency for the Development of the Amazon), with the technical assistance of the Forestry Department of the Agrarian Sciences

Faculty of Pará University (from 1958–1967, with technical assistance of FAO). All the natural regeneration experimental plots are in dryland high forest on the Curua planalto, and among the main lines of research are phenological studies of some economic species, and survey and analysis of pre-existing natural regeneration. Comparative experimental work on distinct natural regeneration systems includes pre- and post-exploitation Tropical Shelterwood System (total area 15 ha), stripwise natural regeneration after clear-cutting and burning (total area *c.* 5 ha), and Stratified Uniform System (total area 300 ha; no pre-felling treatments; two commercial felling intensities; after exploitation, enrichment plantings where needed; some refining and liberation; continuous inventory with, so far, two periodic measurements made in the yield plots).

The effect of felling and logging damage on the residual stand and pre-existing natural regeneration are being assessed, and there is incipient research on biological inhibitors acting on natural regeneration inducement and seedling growth. A preliminary assessment is being carried out on the ecological tolerance of upper storey species (gregariousness, reciprocal exclusion). Assessments are also underway on silvicultural correlations, between canopy opening (caused by commercial felling) and induced natural regeneration; classes of soil disturbance due to logging and classes of emerging pioneers (timber species, weeds, climbers, etc.); botanical composition of upper canopy and composition of induced natural regeneration.

Palhão

Natural regeneration research is underway at the Palhão Forest Reserve (1200 ha, State of Pará, 2° 33'S, 54° 33'W), between the Curua Una and Tapajós rivers. Experimental plots have been set up in planalto dryland high forest. SUDAM is again the responsible institution, and research includes work on the natural regeneration of local commercial species, mainly *Euxylophora paraensis* (Rutaceae) (50 ha, no pre-felling treatment, commercial felling carried out in 1971 followed by enrichment plantings where needed; some refining and some liberation; continuous inventory yield plots).

Tapajós

Research on natural regeneration is also being carried out at Tapajós National Forest (about 600 000 ha, State of Pará, northern limit *c.* 65 km south of Santarém; between the Santarém-Cuiaba Road and the Tapajós

River). All experimental plots are located in planalto dryland high forest and the institution in charge is EMBRAPA, through its regional centre (CPATU), with headquarters in Belém.

The principal research lines include phenological studies (on a much larger scale than in Curua-Una) and experimental forest management based on natural regeneration. Several experimental plots have been established:

> Km 67 Experimental Plot (64 ha without surrounds): pre-felling climber cutting and selective removal of undesirable and proscripts in the upper understorey by simple ring-barking (76% effectiveness); commercial exploitation in 1979 with two degrees of felling intensity; assessment of felling and logging damage to the residual stand; study of natural regeneration (abundance, frequency, distribution models); autecological analysis of forest inventories; continuous inventory by yield plots:
>
> Km 114 Experimental Plot (144 ha): commercial felling in 1982 involving four treatments (T1: commercial fellings removing trees ≥45 cm dbh; T2: removing commercial trees ≥55 cm dbh + additional (30%) reduction of basal area; T3: removing commercial trees ≥55 cm dbh + additional (50%) reduction of basal area; T4: removing commercial trees ≥55 cm dbh + additional (70%) reduction of basal area); assessment of felling and logging damage to residual stand; study of phytosociological and ecological parameters of the stand and existing natural regeneration; continuous inventory:
>
> Km 83 Experimental Plot (1000 ha): to assess technical and economical feasibility of management under a polycyclic system. Logging initiated in 1987 at average to light felling intensity (35–40 m^3 ha^{-1}) corresponding to an average of 6 trees ha^{-1}. Poison-girdling to reduce basal area to about one third will be carried out after exploitation. Refinements to be performed at 8–10 year intervals through the felling cycle (expected to be of 20–25 years).

Belterra

EMBRAPA/CPATU is in charge of natural regeneration research at Belterra, located between Santarém and the northern limit of the Tapajós National Forest. Underway is study and analysis of silvicultural parameters of a secondary forest, with dominance of *Jacaranda copaia*, *Vochysia maxima* and *Schefflera* (*Didymopanax*) *morototoni*. So far, no silvicultural treatment has been carried out. The experimental plot has an extension of 132 ha.

Jari

Experimental areas for natural regeneration research have been set up within the forest estate of the Jari Company, State of Pará, 2°57'S, 52° 23'W (see also contribution by Palmer in this volume). The institution in charge is again EMBRAPA/CPATU, through agreement with the Jari Company. One of the principal research lines concerns forest management as a source of alternative energy and/or pulpwood, with natural regeneration after complete clear-cutting (three blocks of 400 ha each, testing separately forest management models for energy, for pulpwood and for energy plus pulpwood). Basic forest inventories have been carried out, together with commercial felling followed by complete clear-cutting of residual stand. A study of early secondary growth dynamics is in progress. The results will be used to define silvicultural treatments (up to eight treatments in each block). Once the first treatments have been implemented, yield plots will be established.

Silvicultural research for polycyclic management (500 ha) is also underway, to find the best technical/economic combination of felling intensity and basal area reduction. Logging was carried out in 1985 under two intensities: *c.* 20 m^3 ha^{-1} and 60 m^3 ha^{-1}, to be followed by four levels of basal area reduction (down to 0%, 30%, 50% and 70%). Poison-girdling in 1987 will be followed by refinements, to be performed according to the results of the continuous forest inventory.

Manaus

Natural regeneration research is being undertaken by INPA in experimental areas near to city of Manaus (State of Amazonas; 2°51'S; 60°39'W). Phenological research was started in 1961, as were studies on natural regeneration after clear-cutting and burning on *c.* 1 ha. Experimental forest management is based on natural regeneration comparing two models:

(1) Sawlog production (commercial felling of primary forest, started in 1987), and

(2) Biomass/energy production (original stand already cut).

Special studies include postgraduate thesis work, on natural regeneration and phytosociological status of *Aniba duckei* (Lauraceae) (a rosewood species) and natural regeneration of *Copaifera* sp. (Leguminosae).

CHAPTER 31

SCIENTIFIC RESEARCH AND MANAGEMENT OF THE CARONI RIVER BASIN

L. Castro Morales and S. Gorzula

ABSTRACT

Rain forest management forms but one component of the management plan for the Caroní River Basin, the river with the largest hydroelectric potential in Venezuela. Water quality, hydrological regime and sedimentation rates are three of the principal issues that need to be taken account in considering environmental dimensions of dam construction. River basin management is an integral part of the operation of a large-scale hydroelectric project. In the case of the Caroní, current and potential uses of the river basin include gold and diamond mining, agriculture, cattle ranching, lumber extraction, subsistence hunting and tourism – all activities that either are, or could be, in conflict with the protection of the catchment area.

INTRODUCTION

The Caroní is the river with the greatest hydroelectric potential in Venezuela. Guri Final Stage, together with three dams that are to be constructed downstream (Tocoma, Caruachi and Macagua II), will produce 17 980 MW of electricity in the lower Caroní alone. Construction of the first stage of Guri was begun in 1963. At the end of 1986, the lake was brought from a previous operating level of 216 m up to its final elevation of 270 m above sea level, covering 4300 km^2 and storing 140 km^3 of water (Castro Morales and Gorzula 1986).

Hydroelectric schemes are long-term projects that are designed on the basis of values of certain environmental parameters which become "historical" at the moment that the final project design is accepted.

Changes in their values, whether natural or man-induced, have an effect on the project's functioning, its efficiency or on its useful life. The Caroní River Basin is large (95 000 km^2) and has a low population density (15 000 people in the catchment area). However, population expansion over the next few decades and a consequent increase in human activities that have direct impacts on the environment are probable. C.V.G. Electrificación del Caroní (EDELCA) is evaluating the present ecological scenario together with the impacts of man's activities, in order to predict future impacts, and to implement an effective long-term management plan for the Caroní River Basin.

ENVIRONMENTAL IMPACT ON LARGE DAMS

There are three main areas where the environment may affect a hydroelectric project: water quality, hydrology and sediments.

Water quality

Direct corrosion of both cement and steel structures of hydroelectric dams may be caused by water. In Guri, water quality has been monitored, and, in general, the lake has retained the original oligotrophic characteristics of the Caroní River. Flooding by stages apparently reduced the possibility of excessive eutrophication affecting the impoundment. Colonization of the Guri Lake by aquatic plants has been restricted to the southern part of the lake, and wave action and the seasonal changes in water level of the lake have also had a controlling effect on the aquatic plants (Alvarez *et al.* 1986).

Hydrology

Large-scale hydroelectric projects require a constant water supply and a drop in height. In the tropics, where there are annual dry and rainy seasons, the flow rate of a river fluctuates seasonally. The mean monthly flow rate of the Caroní at San Pedro de las Bocas of 8490 m^3 sec^{-1} in the rainy season (July) drops to 1380 m^3 sec^{-1} during the dry season (March). To make efficient use of the hydro-power of the Caroní, the Guri Dam (Represa Raúl Leoni) was constructed in order to store excess rainy season water, and use it during the dry season.

Some 65 000 km^2 of the catchment area of the Caroní are covered by forests. Forests play an important role in the maintenance of the water balance of a region, by introducing into the atmosphere a considerable quantity of water vapour, which in turn is recycled to the region as rainfall

(Leopoldo *et al.* 1984). Possible general ecological effects of deforestation have been summarized by Goodland and Irwin (1975) who postulate that massive deforestation of a tropical forested area would result in a net reduction of rainfall, due to increased convection and decreased evapotranspiration and water retention.

This scenario is believed by hydrologists to be overstated. Any continuous vegetation cover is likely to have higher evapotranspiration compared to bare ground. But massive deforestation is often associated with erosion, which creates bare ground as well as silting up watercourses. Regional and global effects of deforestation and land clearance remain under active discussion. They have been examined in detail by Dickinson (1981), Henderson-Sellers (1981) and Henderson-Sellers and Gornitz (1984).

EDELCA operates a network of 67 weather and hydrology stations. Computer model studies are being carried out in order to relate hydro-metereological characteristics of the catchment area to the major terrestrial ecosystems and to predict the potential effects of habitat destruction.

Sediments

The useful life of a hydroelectric project depends upon the rate at which the empoundment fills up with sediments. In Guri, where the upstream sediment load is trapped in the lake, a point will eventually be reached when sediment begins to pass through the turbines. This will result in scouring and damage. Eventually, the production of electricity will no longer cover the cost of maintenance and repair and the generating losses caused by shutdowns.

One of the principal activities that could result in increased sedimentation of the Guri dam is the gold and diamond mining operations that occur both within the lake and upstream. Possibly a third of the population of the catchment area depend partially or totally on these activities for their subsistence. EDELCA is currently quantifying both suspended and bottom sediments in the rivers of the catchment area, and identifying those sub-basins which have major problems with erosion.

RIVER BASIN MANAGEMENT

River basin management is an integral part of the operation of a large-scale hydroelectric project. In the case of the Caroní, EDELCA has to evaluate the impacts of other current and potential uses of the river basin. Gold and diamond mining, agriculture, cattle ranching, lumber extraction, subsistence hunting and tourism are all activities that either are, or could be, in conflict with the protection of the catchment area. The wildfire problem in the Gran Sabana is considered to be the most serious environmental problem in the catchment area at the present time. It is estimated that nearly 70% of these fires are caused by hunting activities of the indigenous people (EDELCA 1984).

Fires, which lead to forest being replaced by savannas, which then themselves become degrading (Fölster 1986) are large-scale problems in the south-east of the basin which could have serious long-term consequences by altering hydrological cycles, and exposing soils to erosion. A rough estimate of the area of secondary savannas in the upper Caroní is 8 000 to 10 000 km^2 which represents 12% of the basin. Progressive elimination of forest and degradation of the resulting savannas could increase the amplitude and frequencies of variations in flow rates of the rivers, increase their rate of sediment transport and also alter the biochemical characteristics of the water.

In 1981, EDELCA began a forest fire control programme involving detection, combat and prevention. Between 1981 and 1985, 401 out of 963 forest fires were extinguished. The programme for prevention includes environmental education and the formation of volunteer fire brigades (by 1986 there were 32). Research was first aimed at determining priority areas for fire prevention and control within the Gran Sabana (EDELCA 1984). In 1986 research was begun for the preparation of combustible biomass maps, and a modification of the Canadian Fire Weather Index was tested.

REFERENCES

Alvarez, E., Balbás, L., Massa, I. and Pacheco, J. (1986). Aspectos ecológicos del Embalse Guri. *Interciencia,* **11** (**6**), 325–33

Castro Morales, L. and Gorzula, S. (1986). The interrelations of the Caroní River Basin ecosystems and hydroelectric power projects. *Interciencia,* **11** (**6**), 272–7

Dickinson, R.E. (1981). Effects of tropical deforestation on climate. In Sutlive, V.H., Altshuler, N. and Zamora, M.D. (eds.) *Blowing in the Wind: Deforestation and Long-range Implications*, pp. 411–41 Studies in Third World Societies Publication 14. (College of William and Mary, Williamsburg: Virginia)

EDELCA (1984). *La Protección de la Cuenca del Rio Caroní.* C.V.G. Electrificación del Caroní C.A. and Editorial Arte, Caracas.

Fölster, H.. (1986). Forest-savanna dynamics and desertification processes in the Gran Sabana. *Interciencia,* **11**(**6**), 311–16

Goodland, R. and Irwin, H. (1975). *Amazon Jungle: Green Hell to Red Desert?* (Elsevier: Amsterdam)

Henderson-Sellers, A. (1981). The effects of land clearance and agricultural practices on climate. In Sutlive, V.H., Altshuler, N. and Zamora, M.D. (eds.) *Blowing in the Wind: Deforestation and Long-range Implications*, pp. 443–85. Studies in Third World Societies Publication 14. (College of William and Mary, Williamsburg: Virginia)

Henderson-Sellers, A. and Gornitz, V. (1984). Possible climatic impacts of land cover transformations, with particular emphasis on tropical deforestation. *Climatic Change,* **6**, 231–57

Leopoldo, P.R., Franken, W. and Matsui, E. (1984). Hydrological aspects of the tropical rain forest in central Amazon. *Interciencia,* **9**(**3**), 125–31

Section 4

In guise of a conclusion

CHAPTER 32

RAIN FOREST REGENERATION AND MANAGEMENT: STRENGTHENING THE RESEARCH–MANAGEMENT CONNECTION

M. Hadley

That only a very small proportion of existing tropical rain forests is currently managed in any real sense of the term, is a conclusion of Schmidt's report earlier in this book on the management of humid tropical forests. Even when management is attempted, any shortcoming – whether silvicultural, socio-economic, political or institutional – rather effectively prevents success. So intractable might seem these factors that there has been a tendency among many involved in resource management and land-use planning to discuss mixed tropical forest management as unrealistic, unworkable or impractical.

Examples are, however, available to help counter such scepticism. Though few in number, they do suggest that management of mixed tropical forests can be made to work, given the right circumstances and mix of ingredients. Examples of approaches for timber production described in this volume include the Malayan Uniform System and its modifications and the Selective Management System, experimental management plots in three forest areas in the Côte d'Ivoire, and the Celos Silvicultural System in Suriname. At the same time, management for the sustained production of timber should not be isolated from the search for workable multiple-use management systems, which seek to take advantage of the inappropriately named "minor forest products" such as rattans, resins, latex, drugs, fruits, etc.

Policy formulation is shaped by such factors as the stage of national development and the weight of human population pressure on tropical lands. The social context of densely populated countries is clearly different to that in low density areas, and such differences in time and place will, in part, determine the objectives of a management agency and its needs for information. Even so, it may be difficult for managers to articulate their information needs, sometimes because they may find it embarrassing to explain or defend the current practices of their governments in respect to forest management, in other cases because there is a lack of clear policy in respect to management, or because time might just not be available for putting existing tools and approaches into use.

One such approach is through decision charts and checklists, which are useful for training forest managers and have been produced by Colyer Dawkins for Ghana and Uganda, by Frank Wadsworth for Puerto Rico and extended to the rest of the neotropics, and by John Palmer for the Unesco/UNEP/FAO State-of-knowledge report on tropical forest ecosystems (Unesco 1978). The charts are useful also to managers for explaining their needs and time scales to politicians and to Ministries of Finance. One highly simplified chart on what a manager of tropical rain forest needs to know, in approximately chronological sequence, is shown as Figure 32.1 (Palmer 1988).

For simplicity the chart is confined to a primary management objective, the production of valuable timber on a large scale (tens of thousands of hectares) to feed a capital-intensive forest industry. The chart is limited further to the permanent forest estate, that is, the forest which is reserved legally for the supply of forest products to fulfil the national domestic (and perhaps export) requirements, in accordance with the national forest policy.

The principal steps identified in this chart would clearly need to be modified if the management goals were different – for example, if the concern was with forestry for the recovery of water catchments damaged through improper land-use, or with the management of forests scheduled for conversion to other forms of land-use on the basis of land capability surveys and subsequent zoning, or with management of small communal or privately owned woodlots. Such goals would require separate and somewhat different treatment, but a common feature would be that ecological knowledge implicitly underpins the forest management, though in an explicit form it is only one of a number of factors influencing that management.

Ecologists might have more influence if they interpreted their research in terms of potential impact on management, while managers should articulate their research needs more clearly and phase them into grant-sized projects as understood by ecologists. The land manager should be able to indicate his research needs for any one land management system in any one area of ecological research, as suggested by the matrix shown in Figure 32.2. Such approaches would facilitate collaboration between forester and scientist in tackling particular management problems.

From the research viewpoint, it is recognized that many of the management problems are not ecological problems as such, or at least are not amenable to an ecological solution. Increased effort should be made to define areas where ecological inputs can be put to good use in

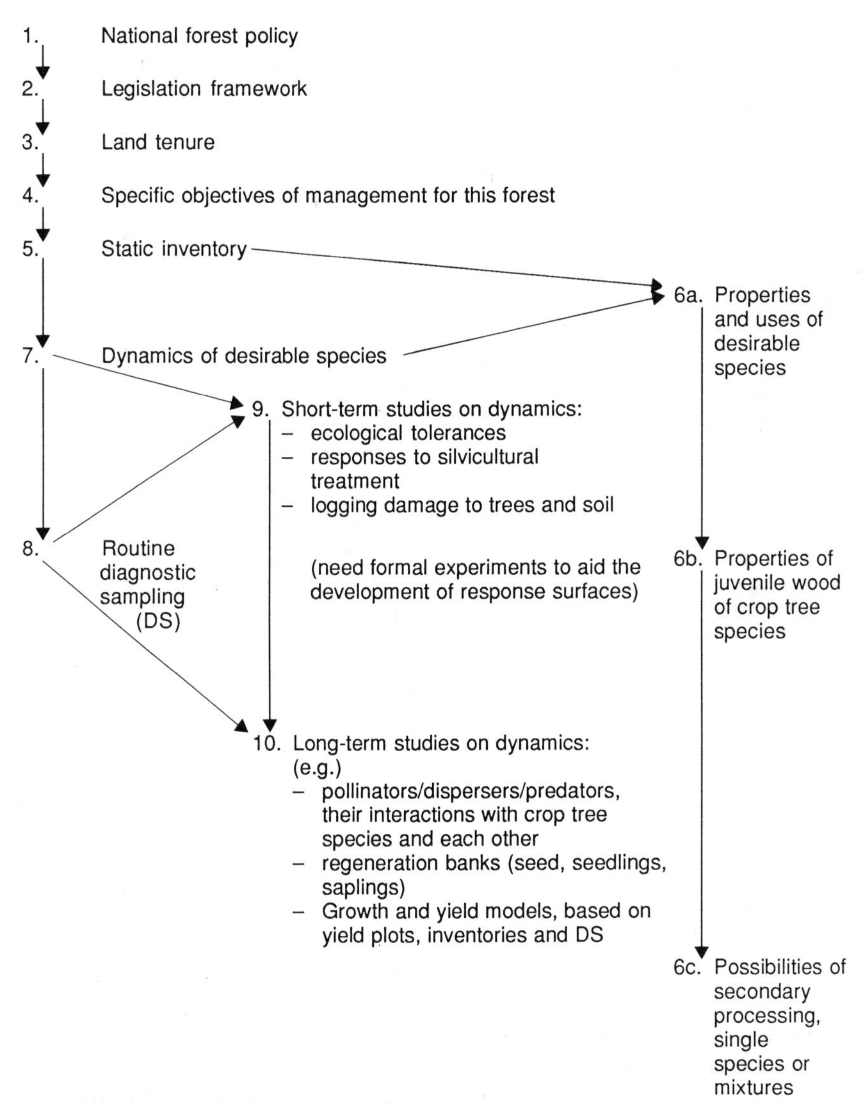

Fig. 32.1 What a manager of tropical rain forest needs to know (from Palmer 1988)

management, to translate existing ecological information on the behaviour of rain forest into forms in which it can be incorporated into the management process, and to tailor future research to provide predictive knowledge of processes inherent in future management options.

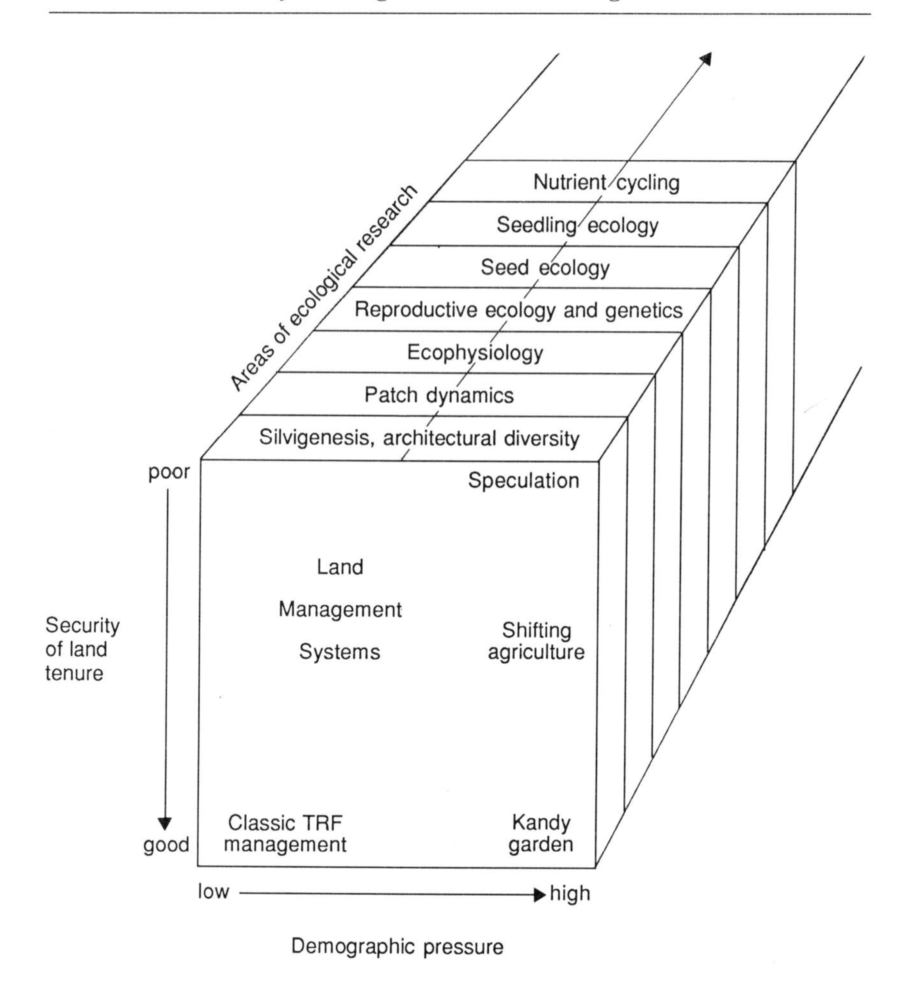

Fig. 32.2 Relating ecological research and different management needs in the humid tropics (from Palmer 1988)

Much of the earlier ecological research on rain forest ecosystems was descriptive, fragmentary and focused on "pristine" situations. In future, greater emphasis might usefully be given to prescriptive, co-operative efforts, with a focus on sustained management and multiple use of mixed tropical forest ecosystems as well as on the redevelopment of human impacted and degraded systems. Efforts should also be made to extend and reinforce the network of biosphere reserves in the humid tropics as areas for research, education, training and demonstration of sound management practices, including the conservation of genetic materials.

Differences in the scale of interest of researcher and manager remain a major stumbling block to the applicability and application of research findings. The focus of much research is at a more detailed scale than those of concern to the planner and manager. Many researchers working on rain forests are interested in processes and phenomena acting at a scale of a few square metres to a few hectares, while management and decision-making take place within a context of several tens, hundreds or thousands of hectares or square kilometres. The sheer size of areas under the control of a tropical silviculturist or a forest manager forces acceptance of a high level of heterogeneity in their forests, augmented by the effects of logging operations. This difference in scale surely accounts for much of the difference in approach to problems between the tropical rain forest manager and the ecologist. In designing new research activities, attention might thus be given to ways of increasing the time and space scales that are considered, with a view to linking up with the concerns of the manager and planner. A related aspect is that of assessing the interrelations and conflicts between ecological, social and economic processes and of seeking ways of "internalizing" dimensions that are presently considered as "externalites". Such measures might help in building bridges between scientist and manager, as well as in exploring how "fixed variables" might be transformed into variables that are no longer fixed.

At the same time as relating research to larger space and time frameworks, there is also need for new research initiatives at a more fundamental, finer scale, designed to gain a better understanding of the mechanisms and processes that underpin rain forest regeneration. This would be in line with what could be called a hierarchical view of life – the notion that the natural world can be profitably viewed as a multilayered system, hierarchical in space and time, and that one needs to look up a step in the hierarchy (e.g. in terms of space and time) to understand the constraints under which a phenomenon occurs and descend a step to determine causality (Allen and Starr 1982, O'Neill *et al.* 1986, Salthe 1985). An example of switching scale in rain forest regeneration is the detailed, fine-scale study by Smits (1985) of the role of mycorrhizae in dipterocarp regeneration, and the subsequent testing of the insights gained within large-scale reafforestation schemes in East Kalimantan.

Improved communication between researchers and managers is needed for putting the results of research into practice. The primary channel for the communication of the results of much scientific research remains the scholarly or professional journal, with contributions subject to rigorous peer review. At the same time, there is need to encourage scientists to make available their findings in a form comprehensible to and usable by those responsible for land-use planning and resource management. For example, a detailed scientific paper in an international journal might be complemented by a shorter communication geared to the local forest-planner – what Salleh (1986) has called a policy of "double publishing".

Field workshops on tropical forest management that bring together researchers and managers – as equal partners present in approximately equal numbers – also have a role to play at the national and regional level. Such workshops can serve to acquaint researchers with the practical problems and constraints of tropical forest management, and familiarize forest managers with recent ecological findings and scientific methodology. Personal interaction may be fostered by focusing on problems as perceived within local, national and regional contexts. In seeking out a dialogue and interface between researcher and manager, discussion should generally focus on topics and terms that impinge on the daily concerns of the forest manager and planner. For example, an examination of the effect of logging damage on trees and soil is likely to be of greater interest to the manager than a discussion on gap phase dynamics.

Achieving sound management will not be easy, and will depend on the concerted efforts of national governments, management agencies, research institutions and non-governmental groups in the tropical countries themselves. The broader international community also has a role to play in providing expertise, co-operation and financial assistance, working through frameworks such as the Tropical Forestry Action Plan, co-ordinated by FAO, and Unesco's Action Plan for Biosphere Reserves.

There is increasing support for natural forest management in scientific and conservation communities. There are also indications of heightened awareness by government management agencies and private commercial enterprises of the desirability of promoting – and to be seen to be promoting – approaches leading to a more sustained use of the resources of tropical forest ecosystems. Advantage should be taken of these opportunities for generating support for management and research projects, with a priority given to the demonstration of the economic and ecological viability for sustained management of mixed tropical forests.

Borrowing from Schmink (1987), the challenge of seeking sustainable development in the humid tropics depends on the best creative efforts of people in many different fields, willing to work together on the practical task of designing feasible policies and land management systems. Given a willingness to acknowledge the complexity of the problems involved – and to search for broad principles as well as small, incremental solutions – there is still room for optimism.

REFERENCES

Allen, T.F.H. and Starr, T.B. (1982). *Hierarchy, Perspectives for Ecological Complexity.* (University of Chicago Press: Chicago)

O'Neill, R.V., DeAngelis, D.L., Waide, J.B. and Allen, T.F.H. (1986). *A Hierarchical Concept of Ecosystems.* (Princeton University Press: Princeton)

Palmer, J. (1988). What the manager needs to know. In Hadley, M. (ed.) *Rain Forest Regeneration and Management.* Report of a workshop. Guri, Venezuela, 24–28 November 1986, pp. 5–9. Biology International – Special Issue 18. (IUBS: Paris)

Salleh, M.N. (1986). The need for a common interface. In *Proceedings of Tropenbos Consultation on a United Approach for Research in Humid Tropical Forests.* Tiel, 27–31 October 1986, pp. 82–90. Tropenbos, Ede.

Salthe, S.N. (1985). *Evolving Hierarchical Systems: Their Structure and Representation.* (Columbia University Press: New York)

Schmink, M. (1987). The rationality of tropical forest destruction. In Figueroa, J.C., Wadsworth, F.H., and Branham, S. (eds.) *Management of the Forests of Tropical America: Prospects and Technologies*, pp. 22–7. (US Department of Agriculture Forest Service, Institute of Tropical Forestry: Rio Piedras, Puerto Rico)

Smits, W. (1985). Dipterocarp mycorrhiza. In Davidson, J., Tho Yow Pong and Bijleveld, M. (eds.) *The Future of Tropical Rain Forests in South East Asia*, pp. 51–4. Commission on Ecology Papers No.10. (IUCN: Gland)

Unesco. (1978). *Tropical Forest Ecosystems.* A state-of-knowledge report prepared by Unesco, UNEP and FAO. Natural Resources Research Series **14**. (Unesco: Paris)

INDEX

MAN AND THE BIOSPHERE SERIES

VOLUME 1
THE CONTROL OF EUTROPHICATION OF LAKES AND RESERVOIRS
Edited by **S.-O. Ryding** and **W. Rast**

VOLUME 2
AN AMAZONIAN RAIN FOREST
By **C.F. Jordan**

VOLUME 3
EXPLOITING THE TROPICAL RAIN FOREST
By **D. Lamb**

VOLUME 4
THE ECOLOGY AND MANAGEMENT OF AQUATIC-TERRESTRIAL ECOTONES
Edited by **R.J. Naiman** and **H. Décamps**

VOLUME 5
SUSTAINABLE DEVELOPMENT AND ENVIRONMENTAL MANAGEMENT OF SMALL ISLANDS
Edited by **W. Beller, P. d'Ayala** and **P. Hein**

VOLUME 6
RAIN FOREST REGENERATION AND MANAGEMENT
Edited by **A. Gómez-Pompa, T.C. Whitmore** and **M. Hadley**

VOLUME 7
REPRODUCTIVE ECOLOGY OF TROPICAL FOREST PLANTS
Edited by **K. Bawa** and **M. Hadley**

VOLUME 8
REDEVELOPMENT OF DEGRADED ECOSYSTEMS
Edited by **H. Regier, L. Nilsson** and **F. Toth**

VOLUME 9
PASTORALISM IN TRANSITION
By **J.N.R. Jeffers**